Overview of Cellular Processes

Regardless of from where an animal acquires energy, whether it's from another animal or a plant, an animal must convert that acquired energy into a form that its cells can use because that energy is needed to fuel all the other processes that occur in a cell. Metabolism is the sum of all those processes that break down fuel to release energy, as well as use energy to create other products. Here's a list of some of the most important things going on in cells (note that these are in no particular order — all of these processes are occurring simultaneously in the body):

- The digestive system takes in food and drinks that contain proteins, carbohydrates, and fats. The proteins are broken down into amino acids, the carbohydrates are broken down into monosaccharides (such as glucose), and the fats are broken down into fatty acids and glycerol.
- The respiratory system takes in oxygen.
- The circulatory system transports oxygen, amino acids, glucose, fatty acids, and glycerol to every cell in the body via capillary exchange.
- Inside the cell, glucose is converted to pyruvic acid via glycolysis.
- Pyruvate is shuttled into the cell's mitochondria, where it is converted to acetyl coenzyme A (acetyl CoA).
- Acetyl CoA combines with water and oxaloacetate to begin Krebs cycle.
- Krebs cycle converts citric acid, which contains six carbon atoms, to oxaloacetate, which contains four carbon atoms. The two "lost" carbon atoms aren't really lost; they end up in the waste product carbon dioxide. The Krebs cycle also results in the production of the energy-rich molecule adenosine triphosphate (ATP), as well as the cofactors NADH and $FADH_2$.
- The process of oxidative phosphorylation converts the cofactors NADH and FADH2 given off during the Krebs cycle into more ATP. After two turns of Krebs cycle and one pass through the electron transport chain of oxidative phosphorylation, the result is 36 molecules of ATP, as well as carbon dioxide and water, which are waste products (see Chapter 6). The waste products are put into the bloodstream via capillary exchange and transported to the appropriate spots: carbon dioxide is deposited in the lungs so that it can be exhaled; water is deposited in the excretory system so that it can be excreted through urination or reabsorbed back into the body if necessary.
- The cell uses the ATP it produces to supply energy for other functions, such as mitosis, meiosis, DNA replication, transcription, translation, not to mention the energy needed to shuttle products into and out of the cell.
- Mitosis is crucial for cell division, which is crucial for cell replacement. Mitosis goes through the stages of prophase, metaphase, anaphase, and telophase to replicate chromosomes in the nucleus, one full set for each daughter nuclei (see Chapter 11).
- Meiosis is the process of reducing the chromosome number by half, which is crucial for the production of gametes (sex cells). Each gamete has half the number of chromosomes, which contain one of each "type" of chromosome (such as one of the mother's alleles for eye color, one of the mother's alleles for height, and so on). See Chapter 11.
- DNA replication, transcription, and translation are covered in Chapter 14.

Biology For Dummies®

The Circle and Cycles of Life

The cycles and pathways provided in this cheat sheet are some of the most crucial biological processes. These cycles and pathways are what keep life moving on. If you understand these processes, you'll have a pretty good idea of how energy is passed around and how energy is converted to allow life forms to function. And the conversion and transfer of energy from the original source of energy (the sun) to a living thing to another living thing and so on, and the release of energy from living things back to the earth is the circle of life. It is what defines life. It is what sustains life. It is what has been done for billions of years and what will continue to be done until the end of the universe. The cycles that use the transferred energy are the just the detailed ways that different living things use the bits of energy stored in and around the entire world.

Photosynthesis

This all-important process harnesses energy from the sun so that it can be passed on to other living things. Water and minerals travel up the stem of a plant and diffuse into the cells of the leaves. When light from the sun strikes the leaves, the chemical reaction of photosynthesis takes place. The plants also take in carbon dioxide from the atmosphere, which is used in the reaction. In the end, photosynthesis allows plants to convert energy from the sun into food for the plant in the form of carbohydrates, and it causes plants to release oxygen into the atmosphere (which animals need). The carbohydrates that a plant stores in the form of cellulose are transferred to any animal that consumes the plant. Thus, the energy from the sun was converted to fuel by the plant, and it is then transferred to another living thing or is released back into the soil if the plant dies before it's consumed by an animal. Here's the reaction of photosynthesis (note that it is the reverse of what happens when an animal breaks down carbohydrate during aerobic respiration):

$$\text{Energy (light)} + 6\,H_2O + 6\,CO_2 \longrightarrow C_6H_{12}O_6 + 6\,O_2$$

The Nitrogen Cycle

The nitrogen cycle (see Chapter 17) passes the important element nitrogen from its gaseous form in air to a soluble form in water to its elemental form in soil. Biological nitrogen fixation occurs when plants take in nitrogen from the water and soil so that it can be transferred to animals; decaying animals and plants transfer nitrogen back into the earth. Nitrogen also is released into the atmosphere by natural sources, such as volcanoes and lightning, as well as unnatural sources, such as the burning of fossil fuels. Rain and snow force nitrogen in the atmosphere down into the soil and water.

Item 5326-7.
For more information about Wiley Publishing, call 1-800-762-2974.

TM

References for the Rest of Us!®

BESTSELLING BOOK SERIES

Do you find that traditional reference books are overloaded with technical details and advice you'll never use? Do you postpone important life decisions because you just don't want to deal with them? Then our *For Dummies*® business and general reference book series is for you.

For Dummies business and general reference books are written for those frustrated and hard-working souls who know they aren't dumb, but find that the myriad of personal and business issues and the accompanying horror stories make them feel helpless. *For Dummies* books use a lighthearted approach, a down-to-earth style, and even cartoons and humorous icons to dispel fears and build confidence. Lighthearted but not lightweight, these books are perfect survival guides to solve your everyday personal and business problems.

"More than a publishing phenomenon, 'Dummies' is a sign of the times."

— The New York Times

"A world of detailed and authoritative information is packed into them..."

— U.S. News and World Report

"...you won't go wrong buying them."

— Walter Mossberg, Wall Street Journal, on For Dummies books

Already, millions of satisfied readers agree. They have made For Dummies the #1 introductory level computer book series and a best-selling business book series. They have written asking for more. So, if you're looking for the best and easiest way to learn about business and other general reference topics, look to For Dummies to give you a helping hand.

5/09

by Donna Rae Siegfried

Wiley Publishing, Inc.

Biology For Dummies®

Published by
Wiley Publishing, Inc.
111 River St.
Hoboken, NJ 07030
www.wiley.com

Library of Congress Cataloging-in-Publication Data:

Library of Congress Control Number: 2001092749

ISBN: 978-0-7645-5326-4

Manufactured in the United States of America

15 14 13 12 11

About the Author

Donna Rae Siegfried has been writing and editing medical information for 14 years. She began her college years at Moravian College in Bethlehem, Pennsylvania, with the goal of attending medical school and then discovered her knack for writing about biology and medicine. The field of science writing was in its infancy then, but while working at nearby Lehigh University, she found "science writing" in one of Lehigh's course listings. Her career path became clear at that moment. She added journalism to her biology major, took the science writing and medical ethics courses at Lehigh, and began to write.

Donna's journalism internship was done at Rodale Press, in Emmaus, Pennsylvania, where she worked with the editor of gardening books. That led to a full-time job with *Organic Gardening* magazine before her college graduation. However, as much as Donna loved publishing and writing about biology topics, gardening (prior to home ownership) just wasn't her forte. Donna became an information analyst at Rodale, where she got to read about 500 medical journals every month (more her "thing"), selected and abstracted important articles for Rodale's files, created biweekly newsletters, and wrote special reports (such as how supplying vitamin A to third-world countries can dramatically decrease the incidence of blindness in those countries). She began to write small bits for *Runner's World* magazine and then left Rodale Press for an opportunity at a medical publishing company, the pursuit of a master's degree in science and technical communication, and the creation of a marriage.

Donna attended Drexel University in Philadelphia, and worked for Williams & Wilkins, Inc. in Media, Pennsylvania, which is now Lippincott/Williams & Wilkins (Philadelphia). Donna worked on staff at Williams & Wilkins as a development editor, working directly with authors of textbooks to add to, change, or correct their manuscripts and making them fit the format of the National Medical Series of review books. She traveled to medical conferences and conducted focus groups of medical students where she gathered more information on how to improve the NMS books and kept up with changes in the United States Medical Licensing Examination (USMLE).

After 5 years on staff at Williams & Wilkins, Donna, her husband, and their then 18-month-old son moved to the tiny mountain village of Germania (Potter County), Pennsylvania, to try the work-at-home lifestyle. There, Donna launched her freelance career and company, Synergy Publishing Services. She develops and edits books for several different medical publishing companies and has written articles on the drugs Avonex and Copaxone for those with multiple sclerosis, the heart surgery technique called transmyocardial laser revascularization, and some alternative medicine treatments.

She has edited dozens of basic science and clinical medicine books and articles (and thousands of USMLE-type questions!). While doing so, she has been fortunate enough to work with some of the leading MDs and PhDs in the country.

Donna also held a position as an instructor of anatomy and physiology at the Pennsylvania College of Technology in Wellsboro, Pennsylvania. She discovered that she absolutely loved teaching science as much as writing and editing the information, and was just about to start a master's in education program when she, her husband, three children, and two dogs relocated to the Atlanta suburb of Alpharetta, Georgia. There, Donna is a member of two tennis teams and is the flute player in a woodwind group. She still plans to pursue that master's in education and work to promote science education and wellness and to make science fun for kids so that they may choose an area of science for their careers.

Dedication

This book is dedicated to all the young budding scientists I know, including my daughter Abby, my son Ryan, and my son Steven and his friends at Medlock Bridge Elementary School. I dedicate it to the students I had when I taught anatomy and physiology at the Pennsylvania College of Technology; I learned as much from teaching them as they did by being biology students. I dedicate it to my husband, Keith, who certainly is the love of my life, my best friend, and my intellectual and physical challenger. (I *will* have the winning record in Scrabble and tennis someday!) I also dedicate this work to my mother, Gail Bonstein, and my late grandfather, Henry Bonstein, both of whom not only gave me my musical talent but also my ability to comprehend science well. Above and beyond their genetic contributions, they both encouraged me to use my talents and share them with others, to continually learn, to educate, and to communicate. And, isn't that what life is *really* about? I hope I can inspire others to learn with as much enthusiasm as they motivated me to do.

Author's Acknowledgments

I would not have learned about the opportunity to do this project if it were not for Sue Mellen of YourWriters.com. Her confidence in my abilities, her unwavering support, and her pitching in to help me meet some deadlines will never be forgotten. I also would like to thank the staff at Wiley Publishing, Inc., especially former acquisitions editor Susan Decker and editor Kelly Ewing, for giving me the chance to write about my favorite subjects and giving me some great suggestions. Donald Sittman, Ph.D. cannot go without a big written hug, either. I edited his books, he reviewed mine, and I am much appreciative of his comments and suggestions. I would also like to thank the biology professors I had at Moravian College — Drs. Frank Kuserk, Karen Kurvink, Donald Hosier, and John Bevington — all of you gave me such a great foundation in biology. Thanks also to Dr. Joel Wingard of Moravian College for pointing out my talent in writing about science, encouraging me to add a journalism major, and helping me to find my niche in the world. Moravian might not be a big school, but all of you had a big impact on me. Thanks also go to Kathryn Born for her excellent artwork and patience with my changes. And last but certainly not least, I must thank my family. This book truly was a family project in the sense that my husband, Keith, pitched in unasked — and my children usually were understanding, patient, and quiet — so that I could have more time to work on writing. My family is my strength. Without them, I would not have made it through the writing schedule. Thank you, everyone.

Publisher's Acknowledgments

We're proud of this book; please send us your comments through our Online Registration Form located at `www.dummies.com/register`.

Some of the people who helped bring this book to market include the following:

Acquisitions, Editorial, and Media Development

Project Editor: Kelly Ewing

Acquisitions Editor: Roxane Cerda

General Reviewer: Donald B. Sittman, PhD, Professor, Department of Biochemistry, The University of Mississippi Medical Center

Illustrator: Kathryn Born

Editorial Manager: Jennifer Ehrlich

Editorial Coordinator: Michelle Hacker

Cover Photo: © Andrew Brookes/Corbis Stock Market

Composition

Project Coordinator: Ryan Steffen

Layout and Graphics: Joyce Haughey, Jill Piscitelli, Jacque Schneider, Betty Schulte, Erin Zeltner

Proofreader: Aptara

Indexer: Aptara

Publishing and Editorial for Consumer Dummies

Diane Graves Steele, Vice President and Publisher, Consumer Dummies

Joyce Pepple, Acquisitions Director, Consumer Dummies

Kristin A. Cocks, Product Development Director, Consumer Dummies

Michael Spring, Vice President and Publisher, Travel

Brice Gosnell, Associate Publisher, Travel

Suzanne Jannetta, Editorial Director, Travel

Publishing for Technology Dummies

Andy Cummings, Vice President and Publisher, Dummies Technology/General User

Composition Services

Gerry Fahey, Vice President of Production Services

Debbie Stailey, Director of Composition Services

Contents at a Glance

Introduction 1

Part I: Biology Basics: Organizing Life 7

Chapter 1: Just How Life Is Studied 9
Chapter 2: The Fundamental Units of Life: Cells 25
Chapter 3: A (Very) Quick Review of Basic Chemistry 37
Chapter 4: Macromolecules (The Big Ones) 51

Part II: Living Things Need Energy 69

Chapter 5: Acquiring Energy to Fill the Tank 71
Chapter 6: Using Energy to Keep the Motor Running 103

Part III: Living Things Need to Metabolize 127

Chapter 7: Taxi! Transport of Nutrients 129
Chapter 8: Take a Deep Breath: Gas Exchange 151
Chapter 9: Throwing Out the Trash: Eliminating Waste
to Maintain Homeostasis 159
Chapter 10: Better Living Through Biology 171

Part IV: Let's Talk About Sex, Baby 199

Chapter 11: Dividing to Conquer: Cell Division 201
Chapter 12: Making More Plants 217
Chapter 13: Making More Animals 225
Chapter 14: Making Mendel Proud: Understanding Genetics 243

**Part V: Ch-, Ch-, Ch-, Changes:
Species Development and Evolution 263**

Chapter 15: Differentiating Differentiation and Development 265
Chapter 16: Changing World, Evolving Species 277

Part VI: Ecology and Ecosystems 293

Chapter 17: Sharing the Globe: How Organisms Get Along 295
Chapter 18: Living with Little Buggers: Bacteria, Viruses, and Insects 305

Part VII: The Part of Tens 321
Chapter 19: Ten Great Biology Discoveries 323
Chapter 20: Ten Great Biology Web Sites 329
Chapter 21: Ten Interesting Biology Facts 333

Appendix A: Classification of Living Things 337
Appendix B: Units of Measure 345

Index 347

Cartoons at a Glance
By Rich Tennant
The 5th Wave
By Rich Tennant
BIOCAFETERIAOLOGIST
"I'll have the cheese sandwich with the interesting mold on the bread, and the manicotti with the fungal growth, and that really, really old dish of vanilla bread pudding."
page 7
The 5th Wave
By Rich Tennant
"For the last time— pregnant vegetarians do NOT give birth to Cabbage Patch Dolls!"
page 263
The 5th Wave
By Rich Tennant
IRONICALLY, THE LAST THING PROF. CARUTHERS REMEMBERED WAS EXPLAINING HOW ENERGY IS PRODUCED WHEN MOLECULES COLLIDE.
MOLECULAR INC.
page 69
The 5th Wave
By Rich Tennant
LOUIS PHILBIN DISCOVERS THE JERK GENE
Hey, Crawford—there's something interesting here in this analysis of your DNA...
Get outta da way and lemme look! You ain't found nuttin'! Your data's all screwy! Dat's why you're called Screwy Louie...
page 127
The 5th Wave
By Rich Tennant
"OOO-KAY, LET'S SEE. IF WE CAN ALL REMAIN CALM AND STOP ACTING CRAZY, I'M SURE I'LL EVENTUALLY REMEMBER WHAT NAME I FILED THE ANTIDOTE UNDER."
page 199
The 5th Wave
By Rich Tennant
Oh, those silly things? They've been coming up every June since anyone can remember. We just stay out of the backyard for about a month and keep the dog on a leash.
page 321
The 5th Wave
By Rich Tennant
He's a mix.
page 293
Cartoon Information:
Fax: 978-546-7747
E-Mail: richtennant@the5thwave.com
World Wide Web: www.the5thwave.com

Table of Contents

Introduction 1

About This Book 1
Conventions Used in This Book 1
What You're Not to Read 2
Foolish Assumptions 2
How This Book Is Organized 3
    Part I: Biology Basics: Organizing Life 3
    Part II: Living Things Need Energy 4
    Part III: Living Things Need to Metabolize 4
    Part IV: Let's Talk About Sex, Baby 4
    Part V: Ch-, Ch-, Ch-, Changes: Species Development and Evolution 4
    Part VI: Ecology and Ecosystems 5
    Part VII: The Part of Tens 5
    Part VIII: The Appendixes 5
Icons Used in This Book 5
Where to Go from Here 6

Part I: Biology Basics: Organizing Life 7

Chapter 1: Just How Life Is Studied 9

So Just What Is Biology? 9
Science at Work: New Information, Conflicting Reports 10
Relating the Areas of Biology: Bio Lingo 10
A Day in the Life of an " . . . ologist" 11
    Corporate scientists 11
    University scientists 12
    Specialists 12
Gathering the Information 13
    Designing experiments using the scientific method: The value of variables 13
    Plain and simple: Observing nature 18
Equipment Used in Basic Experiments 18
    Under the looking glass: Microscopes 18
    What do I do with this stuff? Slides, test tubes, and petri dishes ... 19
    Clarifying with color: Dyes and other indicators 20
    Careful, it's sharp: Dissecting forceps, probes, and scalpels 20
    In the mix: Beakers, flasks, and Bunsen burners 21

Finding Science Information 21
Journals present research, not record dreams 21
The text of textbooks 22
Popular press: not so popular with scientists 22

Chapter 2: The Fundamental Units of Life: Cells 25

Windows on the World's Organisms 25
Examining Eukaryotes 26
Cells and the Organelles: Not a Motown Doo-Wop Group 27
Holding it all together: The plasma membrane 27
Controlling the show: The nucleus 31
The factory of the cell: The endoplasmic reticulum 32
Preparing for distribution: The Golgi apparatus 33
Lysosomes really clean up 33
Peroxisomes break down hydrogen peroxide 34
The powerhouses of the cell: The mitochondria 34

Chapter 3: A (Very) Quick Review of Basic Chemistry 37

Why Matter Matters 37
What Matter Is 38
Matter has mass 38
Matter takes up space 38
Matter comes in three forms 38
Matter has two categories 39
What's the Difference? Elements, Atoms, and Isotopes 39
Elements of elements 40
"Bohr"ing you with atoms 41
I so dig isotopes 43
Molecules, compounds, and bonds (oh, my!) 44
Electrolytes 44
Eyes on ions 45
Molecules and compounds 45
Acids and bases (no, this is not a heavy metal band) 46
"Ph"iguring out the pH scale 47

Chapter 4: Macromolecules (The Big Ones) 51

Organizing Organic Chemistry Basics 51
Carbon Is Key 52
Long carbon chains = low reactivity 52
Forming functional groups based on properties 53
Loading You Up on Carbohydrates 53
Making sense of monosaccharides 54
Dissecting disaccharides 55
Pulling apart polysaccharides 56
Storage Forms of Glucose 56
You store it: Glycogen 57
Starch it, please: Storing glucose in plants 57

Selling You on Cellulose 58
Building Organisms: Proteins 59
Amino acid chains link information to build proteins 59
Speeding up reactions: It's enzymatic, not automatic 60
What I learned about collagen in college 61
Around the hemoglobin (in less than 80 days) 62
Links of Life: Nucleic Acids 63
Deoxyribonucleic acid (DNA) 64
Ribonucleic acid (RNA) 65
Fats Aren't Forbidden: The Truth About Lipids 65
Combinations of Three Macromolecules: Blood Group Antigens 68

Part II: Living Things Need Energy 69

Chapter 5: Acquiring Energy to Fill the Tank 71

How Animals Get Nutrients and Oxygen 71
The hunt for food 71
Heterotrophs on the hunt 72
You are part of the food chain (luckily, you're near the top) 72
Breaking Down the Breaking Down of Food 74
Ruminating over ruminants 74
The daily grind of a gizzard 75
The truth about teeth 75
What Is There to Eat? Nutrition and Wise Choices 75
Counting and cutting calories 77
Mining for information on minerals and vitamins 81
Craving (knowledge about) carbohydrates 82
Proteins: You break down their chains; they build yours 83
Fats: Yes, you need some (but don't overdo it) 87
Aspiring to Be Respiring: How Animals Breathe 90
Integumentary exchange 91
Going over gills 91
Tracheal exchange systems 93
Longing for lung info? 93
How Plants Acquire Their Energy 96
Presenting basic plant structure 96
Making energy from the ultimate energy source 98
Flowin' through the xylem and phloem 99
Transporting water from cell to cell 100
The inspiration for transpiration 101
Mighty important minerals (to a plant, anyway) 101

Chapter 6: Using Energy to Keep the Motor Running 103

Down the Hatch: Types of Digestion 103
Intracellular digestion 104
Extracellular digestion 104

The Ins and Outs of Digestive Systems 105
Incomplete versus complete digestive tracts 105
Continuous versus discontinuous feeders 106
An Inside Look at the Human Digestive System 107
Mmmm. . . . Your mouth is a busy place 107
Breaking the chain 108
The long and winding road 109
Absorbing the Good Stuff, Passing the Junk 110
Worthy nutrients remain in your system 110
Callin' on the colon to dump the junk 111
Back to the liver 112
Plants Digest, Too 112
How Plants Absorb Nutrients and Create Fuel 113
Photosynthesis and transpiration 113
Photophosphorylation elation 115
Splitting the light: Photolysis 117
Making sugar in the dark: The Calvin-Benson cycle 118
Respiring Plants Break Down Glucose 121
Where it happens 121
How it happens 122

Part III: Living Things Need to Metabolize 127

Chapter 7: Taxi! Transport of Nutrients 129

Circulating Circulation Information 130
Open circulatory systems 130
Closed circulatory systems 130
Traveling the Highways and Biways: How Blood Circulates 131
Ya gotta have heart 131
The cardiac cycle: Path of blood through the heart 133
Path of blood through the body 134
What makes a ticker tick: Heartbeat generation 139
Hearts Can Be Broken: Heart Disease 139
What's Blood Got to Do with It? 140
The elements of formed elements 141
Plasma puts the "stream" in bloodstream 143
Draining, fighting, and absorbing: The lymphatic system 143
Clotting to prevent leaking 144
Transporting Materials Through Plants 145
Stemming from the seeds 145
Moving fluids and minerals through plants 147

Chapter 8: Take a Deep Breath: Gas Exchange 151

Life's a Gas 151
Oxygen in air versus oxygen in water 152
Why body size and shape matter 153
What does the metabolic rate have to do with it? 153

Cycling Through the Cycles....154
Aerobic respiration....155
Where it all happens....155
The chemiosmotic theory....157

Chapter 9: Throwing Out the Trash: Eliminating Waste to Maintain Homeostasis....159

What's in the Trash?....159
Homeostasis....160
Chew on This: Workings of the Digestive System....161
The scoop on poop....162
Back to the bloodstream....163
Nitrogenous Wastes....164
Kidney structure and function....164
How animals other than humans excrete their wastes....166
Figuring out how plants excrete wastes....167

Chapter 10: Better Living Through Biology....171

Revving Things Up with Enzymes....171
Getting things going with catalysts and activation energy....172
Cofactors and coenzymes: Coexisting with enzymes....173
Controlling the enzymes that control you: Allosteric control and feedback inhibition....174
All Hormones Are Not Raging....175
General functions of hormones....176
How hormones work....177
Ouch! Nerve Impulses....178
Just passing through: Neuron structure....179
Getting off here? Last stops for impulses....180
Creating and carrying impulses....181
What a Sensation! The Brain and the Five Senses....186
The brain....186
The five senses....188
Shake Your Groove Thing: Moving Those Muscles....193
Muscle tissue and physiology....194
Looking at a muscle contraction....195
Movement in other organisms....197

Part IV: Let's Talk About Sex, Baby....199

Chapter 11: Dividing to Conquer: Cell Division....201

Keep On Keepin' On....201
Reproduction and Life....202
What Is Cell Division?....203
Getting organized: Interphase....204
Mitosis: One for you, and one for you....206

Cytokinesis 208
Meiosis: It's all about sex, baby 208
Viva la Difference! 213
Mutations 213
Crossing-over 213
Segregation 213
Independent assortment 214
Fertilization 214
Nondisjunction 214
Pink and blue chromosomes 216

Chapter 12: Making More Plants 217

Asexual Reproduction 217
Sexual Reproduction 218
Life cycles of plants 219
Flowering plants 219
Pollination and fertilization 222
Developing the zygote into an embryo 222
Seed production 223

Chapter 13: Making More Animals 225

Asexual Reproduction: This Budding's for You 225
Sexual Reproduction — The Ins and Outs 226
Getting to know gametes 226
Mating rituals and other preparations for the big event 228
The act of mating: The big event 233
How Other Animals Do It 235
How the birds do it: It's a yolk 236
How the bees do it: Parthenogenesis 237
Developing New Beings 238
From single cells to blastocyst 238
Go, go, embryo 239
Fetuses 241

Chapter 14: Making Mendel Proud: Understanding Genetics 243

Jumping Into the Gene Pool: Some Definitions to Springboard You 243
"Monk"ing Around with Peas: Mendel's Law 245
Bearing Genetic Crosses 247
Copying Your DNA: Can't Wait to Replicate 248
Mistakes Can Happen 250
Producing Proteins Promptly 252
Rewriting the DNA's message: Transcription 253
Processing the RNA 255
Putting the code into the right language: Translation 256

Exploring the Unknown: Genetic Pioneering and Genetic Engineering....258
Genetic pioneering....258
Mapping ourselves: The Human Genome Project....258

Part V: Ch-, Ch-, Ch-, Changes: Species Development and Evolution....263

Chapter 15: Differentiating Differentiation and Development....265

Defining Differentiation and Development....265
Genetic equivalence of the nucleus and totipotency....268
Factors That Affect Differentiation and Development....268
Embryonic induction....269
Cytoplasmic factors....269
Homeotic genes....270
Hormonal control....270
Developing Beyond Birth and Throughout Life....274
Aging and ailing: Changes in the immune system....275
Growing wild: Cancer and aging....275

Chapter 16: Changing World, Evolving Species....277

What People Used to Believe, Believe it or Not....277
How Charles Darwin Challenged the Thinking of the Mainstream....278
Darwin's theory of organic evolution....280
Darwin's evidence....280
Darwin's Other Theory: Survival of the Fittest....281
Mitochondrial DNA Provides a Link to the Past....283
So, Who Were Your Ancestors?....285
Keeping Up with the Joneses....289
Okay, So How Did the World Form in the First Place?....290

Part VI: Ecology and Ecosystems....293

Chapter 17: Sharing the Globe: How Organisms Get Along....295

Populations Are Popular in Ecology....296
Population ecology....296
Humans are increasing exponentially....298
Ecosystems, Energy, and Efficiency....299

Decomposers and the Biogeochemical Cycles 300
The hydrologic (water) cycle 300
The carbon cycle....................................301
The phosphorus cycle....................................301
The nitrogen cycle....................................301
How Humans Affect the Circles of Life 303

Chapter 18: Living with Little Buggers: Bacteria, Viruses, and Insects 305

Bacteria and Viruses: They Really Make Me Sick!....................................305
A Bacteria by Any Other Name Would Still Be a Prokaryote....................................306
The good guys 307
Bacteria in black....................................309
Are humans losing the fight? 310
An old weapon revisited....................................311
Viruses: Those Pinheaded Little Buggers 312
How a virus attacks a cell....................................313
The AIDS virus: Why it's been a bugger to beat 315
Insects: The Bugs You Can See....................................316
My, how you've changed! Reproduction and metamorphosis 318
Learning to Love (or at Least Live with) Those Little Buggers....................................320

Part VII: The Part of Tens....................................321

Chapter 19: Ten Great Biology Discoveries 323

Seeing the Unseen....................................323
Beautiful Loser 323
Eradicating Smallpox: Jenner Had Help 324
DNA's Many Discoverers 324
And Speaking of the Human Genome Project 325
Mendel and Genes: Like Peas in a Pod 325
Darwin's Evolutionary Ideas....................................326
Schwann's Cell Theory 326
Krebs' Cycle (No, It Wasn't a Harley) 327
Gram's Reputation Is Stained Forever....................................327

Chapter 20: Ten Great Biology Web Sites 329

Cellbio.com 329
Tripod.com 329
Euronet.nl....................................330
Hoflink.com....................................330
Arizona.edu....................................330

Biochemlinks.com....331
Astrobiology.com....331
Madsci.org....331
Meer.org....332
Discover.com....332

Chapter 21: Ten Interesting Biology Facts 333

The Kangaroo Rat Is a Mammal That Doesn't Need to Drink....333
The Lungfish Breathes Air and May Be an Evolutionary Link....334
Fish Can Drown....334
Plants Can Eat Animals....334
A Woman's Uterus Expands More Than Six Times Its Normal Size During Pregnancy....334
Humans and Chimpanzees Have 99 Percent of Their Genetic Material in Common....335
For as Advanced as Humans Are, Human Offspring Are Helpless....335
Earthworms Can Be Strangled to Death by Parasites That Live in the Soil....335
Plants Look Green Because They Reflect the Green Light Rays from the Sun....336
Chopping a Starfish into Pieces Won't Kill It....336

Appendix A: Classification of Living Things 337

Kingdom Monera....337
Kingdom Protista....337
Kingdom Fungi....338
Kingdom Plantae....339
Kingdom Animalia....340

Appendix B: Units of Measure 345

Length....345
Mass....345
Volume....346
Concentration....346
Useful Constants, Conversions, and Definitions....346

Index....347

Introduction

Congratulations on your decision to study life! Doing so will only make the journey through your own life more fascinating. Living beings are complex, but learning about them does not have to be. Science *can* be fun.

Living organisms are works of art and machines rolled into one. A living being can be looked at superficially, or it can be studied in great detail. It can be taken apart and put back together figuratively and literally. It can be altered and fixed. Although finding out about biology sounds similar to learning car maintenance and repair, biology has an extra facet to it. In the study of biology, the goal is to discover what the parts are, how the parts work together, and how living organisms function — not only how they function on their own, but how they share the world with other organisms.

A car can function completely on its own; it does not rely on another car on the road or in the parking lot to run. (Forget about jumpstarting right now.) Living beings — whether tiny cells or old, grown men — rely on other cells or living organisms to survive. And that is the beauty of life. Good luck on your passage through your life and study of biology. I hope you find out enough about it to appreciate the wonder of it all.

About This Book

One of the most amazing things about biology is that although humans and other animals seem to be so intricate, they can be looked at system by system. Breaking down information in each system makes all animals seem pretty similar. For example, all living things need to do the same basic things to survive: (1) they need to acquire energy, (2) they need to use energy to function, (3) they need to eliminate waste, (4) they need to reproduce to continue their species (whether that species is dandelion, slug, or human), and (5) they need to live in a world with many other living things. This book is broken down by those basic needs.

Conventions Used in This Book

Instead of focusing on plants in one part of the book and animals in another, I organized this book by the basic things that all living organisms need to do to survive (for example, eat, breathe, and so on). Then, under each aspect related to survival, the information that applies to plants and animals (and, yes, humans are animals) is given. In addition to organizing the text this way, which I hope makes the functioning of organisms a little easier to understand, I have supplemented the text with lots of tables and illustrations to make the information more visual. The icons provide you with extra information or extra explanations to also help you discover and strengthen your knowledge. Although this book focuses mainly on plants and animals, information about microorganisms (bacteria, viruses, fungi) is provided where appropriate. The information in the appendixes also is intended to supplement your grasp of biology. Very few genus and species names are given in the text. Instead, for those of you who are interested in taxonomy, Appendix A lists the genus and species names of common plants and animals.

What You're Not to Read

Do not feel obligated to read every word I have written. The sidebars, for example, are meant to supplement the information in the text. They are not required reading and are not essential to your understanding the rest of the material in the chapter.

Foolish Assumptions

I am assuming that you are one of three people:

- A high school student preparing for an advanced placement test or college entrance examination
- A college student trying to make sense of or review the tons of material you learned throughout your courses
- An adult learner either taking a biology course, considering taking a biology course, or trying to get through a course

Of course, you could just be a living organism with a passion for knowledge. You may be the proud owner of a functioning body who just wants to know how you relate to other organisms in this world. Maybe you were one of those kids who picked up bugs, put them in a jar, and stared at them until you felt you explored every part of its body. Maybe you still do! Perhaps you just love to explore nature, and you marvel at how humans and other animals work. Maybe you just want to know more. Whatever your reason for picking up this book, I have done my best to explain the topics of biology simply and effectively. I hope it works for you.

How This Book Is Organized

Here's a brief "map" of what subjects are where in the book. The book is arranged by the basic needs of living things: acquiring energy, converting energy to useable fuel, eliminating waste, reproducing and evolving, and living among other organisms. Whole organisms and individual cells have these basic needs. Many of the chapters present information at the cellular level, but I have a reason for that: Every living organism is made up of many cells, but if you learn what goes on in just one cell, you'll have a pretty good understanding of what goes on in the whole organism. For example, like the entire organism, each cell must acquire energy, use it to maintain health, get rid of the waste produced by using the energy, and reproduce. Each cell also must survive in a "world" (to the cell, the body it lives in is its world) filled with a variety of other cells and organisms.

Part I: Biology Basics: Organizing Life

Biology is the study of life, but as I'm sure you know, life is complex. To simplify it, I've broken the all-encompassing biology into smaller, more palatable chunks. To start, the way of studying biology is explained. The scientific method holds not only for biology but also for chemistry, psychology, physics, geology, etc. Knowing the way that research is done and redone, challenged, checked, and rechecked makes it easier to accept the facts as facts rather than theories.

Then, I spotlight the basic unit of life: the cell. Every organism, whether it is a human, a dog, a flower, a strep throat bacterium, or an amoeba, has at least one cell; most have millions. The cell is a tiny factory unto itself; it absorbs supplies (nutrients), creates energy, eliminates waste, and produces materials (such as hormones) that are used elsewhere. The body of the organism, then, can be considered the "industrialized nation" filled with "cities" (organs) containing "factories" (cells) making different products for distribution throughout the country.

Once you have a grasp of how cells are the powerhouses of bodies, I give you a review of the types of molecules that are important to their functioning. Foods are composed of carbohydrates, proteins, and fats: Each of these is described. You find out how each of these molecules are digested, used, and stored, as well as what they do for the body.

Also included in this first part is the often-dreaded, but oh-so-necessary review of basic chemistry. Man has broken science into subject areas, such as biology, chemistry, and physics, but Mother Nature says they all work together, and sometimes the borders overlap. To learn biology, you must understand some basic principles of how chemicals function. After all, your body and the bodies of every other living organism, are big sacs of chemicals. Every process that occurs in your body is generated by chemical reactions. So, chemistry is pretty darn essential to understanding these biological processes.

Part II: Living Things Need Energy

This part begins the four-part group of chapters that cover the basics of survival. Part II explains that all living things require energy. Humans acquire energy by eating. Plants acquire energy by absorbing nutrients from the soil and using light from the sun. After that, how plants and animals use energy is fairly similar. Nutrients are broken down into smaller and smaller parts until they can be transported through an animal's circulatory system or through a plant's stem and leaf veins to individual cells. Each cell in the organism absorbs the nutrients and uses them for the processes necessary for proper functioning, maintenance of health, and the production of any substances made by the cells.

Part III: Living Things Need to Metabolize

This part covers the metabolic processes that occur in plants and animals. You delve into gas exchange, waste elimination, and things like enzymes, hormones, transmission of nerve impulses, and how muscles (and thus body parts) move — all things that require ions and chemicals.

Part IV: Let's Talk About Sex, Baby

Sex. They say that the word sex really gets people's attention. Well, I hope it did. Sexual and asexual reproduction methods are explained in these chapters, and the basics of genetics are explained. After all, in real life, genetics immediately follows the first step in reproduction — fertilization. And, if reproduction had not occurred when life began with that first cell billions of years ago, the world would be an entirely different place today.

Part V: Ch-, Ch-, Ch-, Changes: Species Development and Evolution

Once you understand how organisms are structurally organized, how they function, and how they reproduce, you can start to learn how they adapt to their environments. Before reproduction they might mutate, or during development they might change slightly. Over time, these little changes add up to evolution of a species.

Part VI: Ecology and Ecosystems

Parts I through IV basically focus on individual organisms within a species. Part V starts to explain how the organisms of a species can change individually to contribute to the evolution of the entire species. In Part VI, the focus is away from individual organisms and on the effect of species to ecological systems. Perhaps you will come to understand how the effects of humans on different environmental communities can upset the ecology of the entire world.

Part VII: The Part of Tens

This part of *For Dummies* books contains short chapters containing lists of ten or so items. In this book, I give you ten interesting biology discoveries and ten great Web sites.

Icons Used in This Book

Some of the familiar *For Dummies* icons are used here, but some new ones that apply to biology have been added. They should help you really grasp the material in the book or give you new insights.

This bull's-eye symbol lets you know what you need to do to get to the heart of the matter at hand. At these icons, you might find information to help you remember the facts being discussed or suggest a way to help you commit it to memory. Most biological information can stand alone as it is being learned, but some topics do not become clear until the building blocks of information are stacked. In those instances, information that may have been explained in an earlier chapter and is crucial to understanding the topic in the current chapter is given at these icons.

The jargon used in science *can* be deciphered and explained in normal words. When you see this symbol, technical language is translated for you. Next to these icons lies information that gives you extra information but is not necessary to understanding the material in the chapter. If you want to take your biology learning to a higher level, incorporate these paragraphs into your reading. If you want just the basics and do not want to be confused by the details, skip them.

Instead of bogging you down with chemical reactions and equations in the text, any chemistry that is pertinent to understanding the information you are reading is given next to these little icons.

This little icon serves to jog your memory. Sometimes the information spotlighted here is just pointing out information I think you should permanently store in your biology file. Other times, the icon makes a connection between what you are reading and related information that is discussed elsewhere in the book. If you want a quick review of biology, scan through the book reading the remember icons. No need for a chunky yellow highlighter.

Where to Go from Here

Where you want to begin your journey through studying life is up to you. If you just want to know about ecology and how organisms live together in this big world, start at Part VI. If you need to get a better picture of what exactly encompasses biology before you dive into the information, read Chapter 1 first. If you have a good idea of what biology is but need to learn the basics, start with Part I. If you understand clearly that cells are the smallest units of life and you know how they function, go to Parts II, III, IV, and V to find out how whole organisms function and how they fulfill the four basic premises of survival: getting/making energy, using energy, eliminating waste, and reproducing. If you need a quick review of chemistry before you read about biology, go to Part I first. And, don't forget about those useful tools called the index and the table of contents! If you are interested in a certain topic, use the index to find where it is covered throughout the book, then read all the discussions that mention that topic. If you want a really brief overview of the most important biological processes, check out the Cheat Sheet inside the front cover.

Wherever you decide to jump in, I hope you will stay in the pool long enough to meet the goals you had when you bought this book.

Part I

Biology Basics: Organizing Life

"I'll have the cheese sandwich with the interesting mold on the bread, and the manicotti with the fungal growth, and that really, really old dish of vanilla bread pudding."

In this part . . .

Before you delve into all the good and fascinating stuff biology has to offer you, you've got to get a little bit of grounding. Because biology deals with living things, and living things were born out of a glop of chemicals so many billions of years ago, you need to find out a little about chemistry. Just a little, I promise. You need to understand what a cell is and how it functions, because all the other parts of the book deal with what occurs in the cells, and therefore in entire organisms. Then, I give you some basic information about the common molecules that exist in and are used by living things. These things are the core basics of biology.

Then, if you come to love biology as much as I do, maybe you'll want to get in on the action. So I give you some basics on what scientists do and how they do it. Who knows? Maybe you'll become the next Darwin, Pasteur, or Mendel? Go for it.

Chapter 1

Just How Life Is Studied

In This Chapter

- Discovering prefixes and suffixes of common fields of biology
- Understanding the scientific method
- Looking at basic equipment used in experiments
- Finding out where to find science information

Like all chapters in this book, it really doesn't matter when you read through this one. In this chapter, I cover information that gives you a broader understanding of what people do in different areas of biology and then help you understand how they focus their studies. As a result, you can read this chapter whenever it suits you.

So Just What Is Biology?

Biology literally means the study of life, and life is pretty gosh-darn complex. There are so many different types of living organisms, many different types of environments, millions of different combinations of genetic material. All the information related to studying living things falls under biology, but a *biologist* does not and cannot study every facet of all living things. It would just take too much time. So, biologists specialize in certain areas of biology and focus their research. With each specialist studying details of certain biological areas, the information can be pooled (usually at big conferences) and shared to make the knowledge base a bit wider. And that's what *science* is: a continually growing knowledge base focused on things in nature, whether those natural things are banana trees, kangaroos, swordfish, dinosaurs, rocks, gases, or chemicals and cells that make up all of those things.

Biology is the branch of science that deals with living organisms. *Chemistry* focuses on the chemicals that comprise matter. *Physics* focuses on the laws that Mother Nature set for all matter: living and nonliving. Within those three major branches of science, the mysteries of life can be found and deciphered. This book focuses on the study of living things.

Science at Work: New Information, Conflicting Reports

It may be aggravating when the media reports conflicting findings. After all, one day margarine is better for your cholesterol level, and then the next day, margarine produces harmful fatty acids that contribute to heart disease. However, when you hear those news reports, you are witnessing science at work. For example, years ago, when scientists figured out that high cholesterol levels contributed to heart disease, they correctly determined that a product created from vegetable oil rather than animal fat — margarine — was a healthier choice if you were trying to lower your cholesterol level.

But scientists don't just leave things alone. They keep wondering, questioning, and pondering. They are curious fellows. So they keep researching margarine. And, recently, they discovered that when margarine breaks down, it releases *transfatty acids* (see Chapter 4), which were found to be harmful to the heart and blood vessels. Yes, this makes your decision in the grocery store a bit tougher, but just be thankful the knowledge base is wider. Scientific information is continually evolving, just like the scientists who are gathering it.

Relating the Areas of Biology: Bio Lingo

Who exactly are those curious fellows exploring the mysteries of life? Although they all are scientists, they are usually referred to by their specialist name. Figure 1-1 shows you several different prefixes and suffixes used in biology.

These prefixes and suffixes can help you figure out many biology terms and also can help you understand what the people in these subspecialties do. For example, *hemato-* is the prefix meaning "blood." Therefore, a *hematologist* is a scientist who studies blood (and *hematocrit* is a measurement of blood cells).

Understanding what the focus of each subspecialty is will help you relate the different areas under the big biology umbrella and give you the bigger picture. For example, if you have an infection in your bloodstream, a hematologist may be called to help you. However, the hematologist may need to work with an immunologist ("immuno" refers to the immune system) and a microbiologist ("micro" and "bio" refer to small living organisms, such as bacteria or viruses) or cardiologist ("cardio" refers to the heart) to completely cure you. All these specialties, because they deal with living things, fall under the heading of biology.

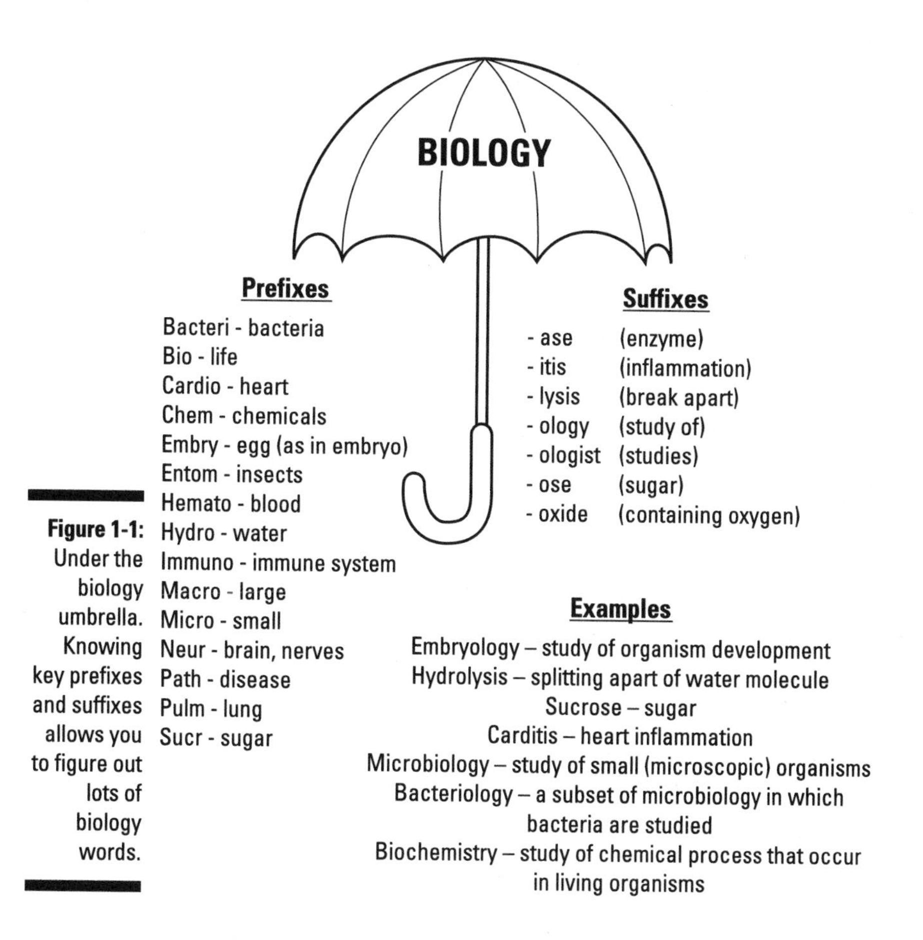

Figure 1-1: Under the biology umbrella. Knowing key prefixes and suffixes allows you to figure out lots of biology words.

A Day in the Life of an ". . . ologist"

Scientists do not just mix chemicals in different types of glassware and perform experiments on animals. Some scientists, such as graduate students, those in postdoctoral programs, and technicians, spend most of their time doing experiments, but they also have many other tasks, depending on what type of scientist they are.

Corporate scientists

If a scientist works for a research company, he or she has meetings to attend just like in any other corporation. Scientists rarely work alone. A corporate scientist must keep the goals of the corporation in mind and work with others to gather information relating to those goals. After the information is gathered (through experiments and reading other studies in the field), the facts must be presented. Scientists read a lot and write often. They attend conferences to talk with other scientists in their field, and they try to develop products or services, such as a test, that their company can sell. They must keep track of financial information (science is a business, after all), and sometimes they must deal with personnel issues and manage people on their research team. They must write proposals and try to obtain research grants or funds from other sources such as venture capitalists. And, scientists also must take care of their equipment, performing routine cleaning and maintenance, as well as sometimes making repairs.

University scientists

If the scientist is employed at a university, he or she may perform experiments that are personally interesting, but many universities have research goals, just as corporations do. University scientists perform many of the functions that corporate scientists perform, but with less of an inclination toward generating profits and more of an inclination toward generating knowledge (which may then be used by or sold to big business). In addition, university scientists must teach classes and publish papers in research journals. And they must attend meetings and conferences. Sometimes, they write or review textbooks or are hired by corporations to do research.

Specialists

Sometimes, the "ologist" is involved in the care of living things rather than just the study of living things. The nature of their work is more clinical — that is, they apply the information that is gathered rather than just focus on gathering it. Often, these "ologists" work together. For example, an *ecologist,* who studies the way that organisms live in their environments, may work together with a microbiologist to improve the quality of a river and the organisms who call it home. An *embryologist,* who studies the development of organisms from conception, may work with a *molecular biologist*, who studies organisms at the cellular level and focuses on genetics, to try to determine the cause of a birth defect. Or, an *entomologist,* who studies insects, may work with a *pathologist,* who studies abnormal cells and tissues, to create a pesticide that does not pose a cancer risk.

One successful corporate venture

In 1980, two research scientists formed a corporation called Genentech in the very infantile genetic engineering industry. As corporate scientists, they raised capital to perform their work of developing recombinant DNA products. The first commercial product that Genentech offered was genetically engineered insulin for people with diabetes. This recombinant DNA product meant that people with diabetes no longer had to take insulin that was produced in the pancreas of a cow or pig, which potentially could cause an allergic reaction. On the day that Genentech went public with its stock, the stock value doubled.

Gathering the Information

How do the scientists know what they know? How do they figure it out? The *scientific method* usually is employed in the task of gathering scientific information. The scientific method is basically a plan that is followed in performing a scientific experiment and writing up the results. It is not a set of instructions for just one experiment, nor was it designed by just one person. The scientific method has evolved over time after many scientists performed experiments and wanted to communicate their results to other scientists. The scientific method allows experiments to be duplicated and results to be communicated uniformly.

Scientists can use the method of observation, or they can design an experiment that follows the scientific method. Really, many people solve problems and answer questions every day in the same way that experiments are designed. The format of the scientific method is very logical. Observe.

Designing experiments using the scientific method: The value of variables

When preparing to do research, a scientist must form a *hypothesis,* which is basically an educated guess, about a particular problem or idea, and then work to *support* it, and prove that it is correct, or *refute* it, and prove that it is wrong.

Whether the scientist is right or wrong is not as important as whether he or she sets up an experiment that can be repeated by other scientists, who expect to reach the same conclusion.

Not all science is driven by hypotheses

Whereas some scientists have a hunch about something and work to prove or disprove their hunch, one project has changed the paradigm of hypothesis-driven research. The Human Genome Project had scientists working diligently in laboratories all over the world to acquire data on the human genome. The human genome is the collection of all the genes found in humans, and genes are what provide information about inherited traits. The Human Genome Project set out to map where a specific trait is found on each human chromosome. The traits range from little things, such as whether you can curl your tongue or not, to truly important things, such as whether you could develop breast cancer or cystic fibrosis. By finding out where genes are located, scientists can now turn their attention to using the newfound information to develop hypotheses about cures and gene therapies. Therefore, the Human Genome Project has been called hypothesis-generating research instead of hypothesis-driven research.

Experiments must have the ability to be duplicated because the "answers" the scientist comes up with (whether it supports or refutes the original hypothesis) cannot become part of the knowledge base unless other scientists can perform the exact same experiment(s) and achieve the same result; otherwise, the experiment is useless, making any data from the experiment essentially worthless.

"Why is it useless," you ask? Well, there are things called *variables.* As you may expect, variables vary: They change, they differ, and they are not the same. A well-designed experiment needs to have an *independent variable* and a *dependent variable*. The independent variable is what the scientist manipulates in the experiment. The dependent variable changes based on how the independent variable is manipulated. Therefore, the dependent variable provides the data for the experiment.

Experiments must contain the following four steps in order to be considered "good science."

1. **A scientist must keep track of the information by recording the data.**

   The data should be presented visually, if possible, such as through a graph or table.

2. **A control must be used.**

   That way, results can be compared to something.

3. **Conclusions must be drawn from the results.**

4. **Errors must be reported.**

Here's an example. Suppose that you wonder whether you can run a marathon faster when you eat pasta the night before or when you drink coffee the morning of the race. Your hunch is that you think loading up on pasta will give you the energy to run faster the next day. A proper hypothesis would be something like, "The time it takes to run a marathon is improved by consuming large quantities of carbohydrates prerace." The independent variable is the consumption of pasta, and the dependent variable is how fast you run the race.

Think of it this way: How fast you run depends on the pasta, so how fast you run is the dependent variable. Now, if you eat several plates of spaghetti at 7 p.m. the night before you race, but then get up the next morning and drink two cups of coffee before you head to the start line, your experiment is useless.

"Why is it useless," you ask again? Well, by drinking the coffee, you introduce a second independent variable, so you will not know whether the faster race time is due to the belly full of pasta or the belly full of coffee. Experiments can have only one independent variable. If you want to know the effect of caffeine (or extra sleep or improved training) on your race time, you would have to design a second (or third or fourth) experiment. The second experiment would have the hypothesis of "Consuming caffeine the morning of a 26-mile marathon improves running time." If you want to know the effect of extra sleep the night before a marathon, you would have to design a third experiment with the hypothesis of, "Sleeping an extra three hours the night before a marathon improves running time." If you want to know the effect of improved training for six months before a marathon, you would have to design a fourth experiment with the hypothesis of, "A six-month period of improved intensive training improves marathon running time." Get the picture?

And, of course these experiments would have to be performed many times by many different runners to demonstrate any valid statistical significance. *Statistical significance* is a mathematical measure of the validity of an experiment. If an experiment is performed repeatedly and the results are within a narrow margin, the results are said to be significant when measured using the branch of mathematics called statistics. If results are all over the board, so to speak, they are not all that significant because one definite conclusion cannot be drawn from the data.

Once an experiment is designed properly, you can begin keeping track of the information you gather through the experiment. In an experiment testing whether eating pasta the night before a marathon improves the running time, suppose that you eat a plate of noodles the night before and then drink only water the morning of the race. You could record your times at each mile along the 26-mile route to keep track of information. Then, for the next marathon you run (boy, you must be in great shape), you eat only meat the night before the race, and you down three espressos on race morning. Again, you would record your times at each mile along the route.

So, what do you do with the information you gather during experiments? Well, you can graph it for a visual comparison of results from two or more experiments. The independent variable from each experiment is plotted on the *x*-axis (the one that runs horizontally), and the dependent variable is plotted on the *y*-axis (the one that runs vertically). In experiments comparing the time it took to run a marathon after eating pasta the night before, getting extra sleep, drinking coffee, or whatever other independent variable you may want to try, miles 1 to 26 would be labeled up the *y*-axis. The factor that does not change in all the experiments is that a marathon is 26 miles long. The time it took to reach each mile would be plotted along the *x*-axis. This data might vary based on what the runner changed before the race, such as diet, sleep, or training. You can plot several independent variables on the same graph by using different colors or different styles of lines. Your graph might look something like the one in Figure 1-2.

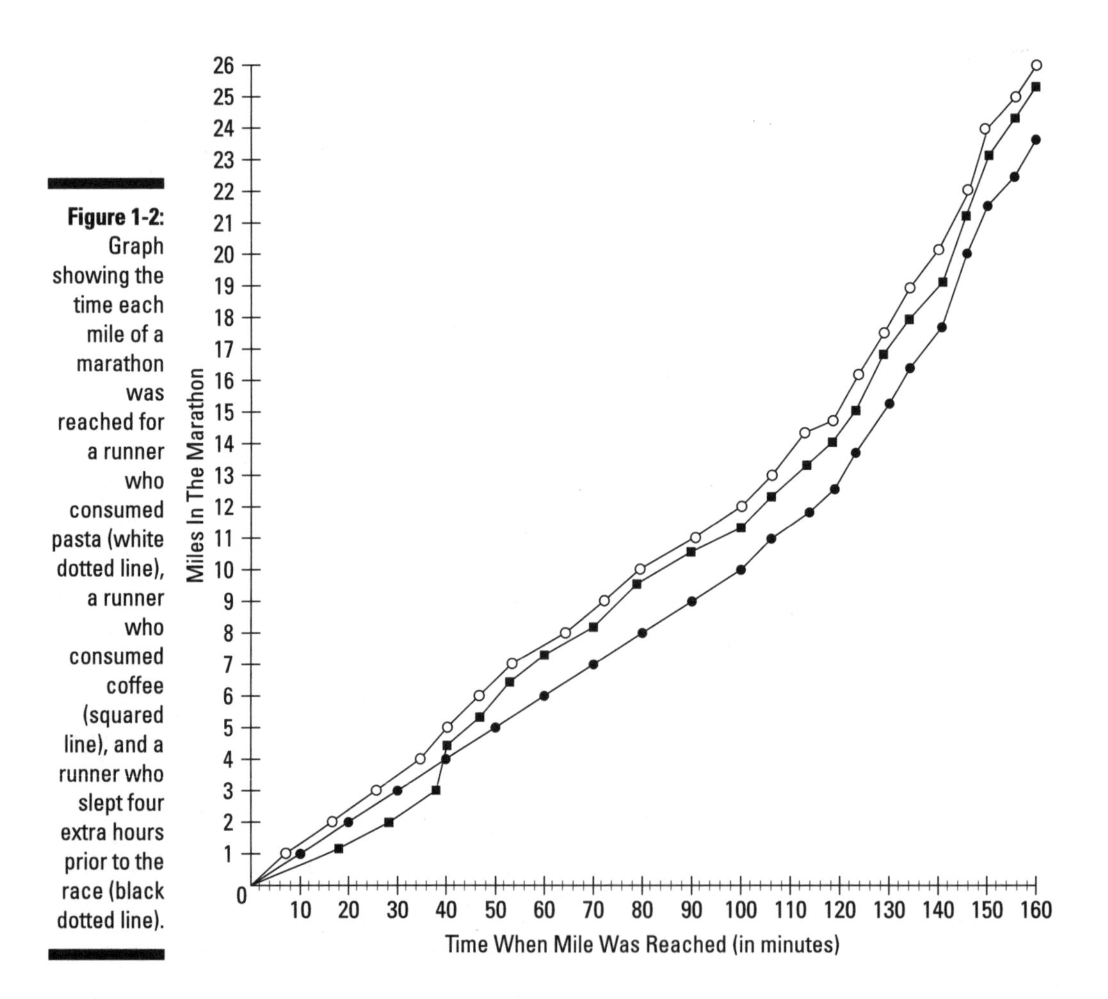

Figure 1-2: Graph showing the time each mile of a marathon was reached for a runner who consumed pasta (white dotted line), a runner who consumed coffee (squared line), and a runner who slept four extra hours prior to the race (black dotted line).

How would you know if your race times were improved either by eating pasta or drinking coffee? You would have to run a marathon without eating pasta the night before or drinking coffee the morning of the race. (Exhausted yet?) This marathon would be your *control*. A control is a set of base values against which you compare the data from your experiments. Otherwise, you would have no idea if your results were better, worse, or the same.

Okay. So, maybe it took you less time to reach each mile along the marathon route after the night of pasta eating, but your race times after drinking the coffee matched those of the control. That would support your initial hypothesis, but it would refute your second hypothesis. There's nothing wrong with being wrong, as long as the information is useful. Knowing what doesn't work is just as important as knowing what does.

Your *conclusion* to these two experiments would be something like: "Consuming pasta the night before a 26-mile marathon improves race time, but consuming caffeine has no effect."

However, in scientific experiments you have to confess your mistakes. This confession lets other scientists know what could be affecting your results. Then, if they choose to repeat the experiment, they can correct for those mistakes and provide additional beneficial information to the knowledge base. In the pasta-caffeine-race experiment, if you had consumed the pasta the night before and then the caffeine the morning of the race, your major *error* would be that of including more than one independent variable.

One man's error is another man's starting point

In the early 1900s, a Russian researcher named A.I. Ignatowski fed rabbits a diet of milk and eggs. He found that the rabbit's aortas developed the same kind of plaques that form in people with atherosclerosis. Ignatowski was not ignorant, but he assumed that the atherosclerosis was caused by the proteins in the milk and eggs. He was wrong. However, a younger researcher who was working in the same pathology department at the time, a Russian named Nikolai Anichkov, knew of Ignatowski's work. Anichkov and some of his colleagues repeated Ignatowski's study, except that they split the rabbits into three different groups. The first group received a supplement of muscle fluid, the second group was fed only egg whites, and the third group was fed only egg yolks. Only the yolk-eating rabbits in the third group developed plaques in their aortas. The two young researchers repeated the experiment, this time analyzing the atherosclerotic plaques to look for any concentrated chemical substances. In 1913, Anichkov and his colleagues discovered that cholesterol in the egg yolk was responsible for creating plaques in the aorta. And, although the connection between cholesterol and atherosclerosis has now been known for nearly 90 years, people are still eating high-cholesterol foods. Eggs contain ten times as much cholesterol as any beef, fish, or chicken of the same quantity. And, people don't just eat plain old eggs; they also consume many foods made with eggs — sauces, soups, bread, baked goods, pasta, ice cream, and so on. Perhaps one man's error will be someone else's starting point for a better diet!

Another error would be having too small of a sample. A more accurate determination could be made by recording the race times at each mile for many runners under the same conditions (i.e., having them eat the same amount of pasta the night before a race or consuming the same amount of caffeine the morning of a race). Of course, their individual control times without those variables would have to be taken into account. Science. It's all in the details.

Plain and simple: Observing nature

In comparison to designing an experiment, performing an experiment, recording the data, and realizing any mistakes you may have made, just observing a living thing sounds so easy. However, depending on what is being studied, observation provides useful information as well. For example, observation is useful in learning about animal behavior or life cycles. If a researcher does not want to disturb a habitat, he or she may use observation to learn about the way a certain animal lives in its natural environment. However, observation requires patience and time. The scientist must make detailed notes in journals or log books about the routine of the animals for a long period of time (usually years) to be sure that the observations are accurate.

Equipment Used in Basic Experiments

You may not have an image of biologists as being handy, but they do use tools. Their daily routine doesn't involve a hammer, chop saw, or level, but it does include some of the following equipment, as well as some really high-tech pieces and computers. The equipment I tell you about in this section is just the bare-bone basics that you'd find in any laboratory. This equipment is necessary for doing the basic studies of biology: visualizing cells and organelles, as well as preparing samples of cells or body fluids for testing or visualization, dissecting specimens, or mixing chemicals.

Under the looking glass: Microscopes

Microscopes are extremely important to the study of living things. Biologists use microscopes of differing powers to see organisms and samples more closely. And, I'm not talking about those simple *light microscopes* your high school supplied for you, although those allow you to view cells. I'm talking about high-powered, extremely expensive, and sensitive pieces of equipment that can make even the smallest parts of a single cell seem clear.

Instead of using beams of light to illuminate the specimen being viewed, as the light microscopes do, *electron microscopes* use beams of electrons, which are negatively charged particles. (I discuss electrons in more detail in

Chapter 3.) The beams of electrons bring the finest details of the cell into focus and can allow even large molecules to be seen. (I discuss molecules in Chapter 3, too.)

The smallest size you can see with your naked eye is 0.2 mm, which is equal to 200 micrometers. This size is equivalent to one ridge on your fingerprint. Light microscopes magnify cells up to 1,000 times. Although the glass lenses in those microscopes may be able to magnify at a higher power, light rays have a limited spectrum, or range of light, that is visible. Using the shortest ray of light, which allows the highest resolution, light microscopes can view things as small as 0.2 micrometers in width — that is, 0.0002 mm.

For objects smaller than 0.2 micrometers, an electron microscope must be used. Electron microscopes allow you to see objects that are as small as 0.2 nanometers (nm), which is equal to 0.000000002 mm. In comparison to a light microscope being able to magnify 1,000 times, electron microscopes can magnify objects 200,000 times (to approximately 0.2 nm, which translates to 0.000000005 mm, or five one-billionths of a mm).

Electron microscopy has two types: transmission electron microscopy and scanning electron microscopy.

- *Transmission electron microscopy* relies on an electron beam that passes through the sample being studied. The sample is treated with heavy metals, such as uranium, and electromagnets are used to pull the electron beam through the sample. However, electrons do not pass through the heavy metal, so cell structures treated with a heavy metal appear dark on the screen.
- *Scanning electron microscopy* relies on an electron beam that just hits the surface of the object. The surface of the specimen is coated with a thin layer of metal. The electron beam displaces the electrons on the thin layer of metal, which creates an image on the screen of the topography of the surface being studied.

What do I do with this stuff? Slides, test tubes, and petri dishes

To examine a specimen, biologists must place a sample — whether the sample is blood, mucus, saliva, skin cells, or urine — in or on something. You can't just hold this stuff in your hand.

If the sample is going to be viewed under a microscope, some of the cells are gently smeared onto a glass slide, treated with a fixative so that the cellular components don't move, and covered with a glass cover slip. If the sample needs to be *centrifuged* — spun very rapidly to separate fluid and particles — or needs to have solutions added to it, then the sample most likely is placed in a test tube. If a sample must be grown before it can be identified (for example, if you have a bacterial infection, but your physician isn't quite sure which bacteria to treat and therefore is not sure what antibiotic to prescribe), the sample must be *cultured.*

To culture a sample, a petri dish containing a culture medium — for example food for the sample in a nourishing, solidified gel, such as nutrient agar — is *inoculated,* or smeared and pressed, onto the medium. Then, the scientist must keep the petri dish at normal body temperature for the species being studied (humans: 98.6ºF, 37ºC) for approximately 24 to 72 hours and wait for the specimen to grow. A series of tests can then be done on the cultured specimen to determine what organism it is.

Clarifying with color: Dyes and other indicators

Dyes are agents that color structures of the cell, which allow the structures to be more easily viewed when using a microscope. In some cases, stains make usually invisible structures visible. Some common stains include iodine and methylene blue. If iodine is placed on a sample that contains starch, such as a piece of potato, it will turn the sample dark blue.

Indicators are premade solutions or papers that are used to determine chemical characteristics, such as acidity and composition). Litmus paper is a common example. When dipped into a solution, litmus paper will turn red if the solution is acidic and blue if the solution is basic. Strips of pH paper have a range of colors that can be matched up to estimate the approximate pH of a solution.

Careful, it's sharp: Dissecting forceps, probes, and scalpels

Yes, sometimes animals are *dissected,* or cut apart in an orderly fashion, to find out more about structure or to teach the person doing the dissecting. And, yes, scientists already know volumes of information on the structure of animals, but dissection not only teaches you structure, it teaches you technique.

To perform a dissection, the organism (which has been euthanized and preserved with formaldehyde) is pinned to a dissecting tray. A *scalpel* is an extremely sharp bladed instrument that can neatly split open skin and cut through muscle and organs. *Forceps* are used to hold tissue out of the way or to pick up a structure. A *probe* can be used to remove connective tissue or to lift a structure before it is dissected.

In the mix: Beakers, flasks, and Bunsen burners

As you can see in Chapter 3, chemistry and biology sometimes overlap. The equipment that is common in a chemistry laboratory often is seen in a biology laboratory, too. Biologists also mix solutions and chemicals.

Beakers are used when the solution mixed in it is going to be poured into something else. (They have a lip on them for pouring.) *Flasks* have a narrow neck and are used when the solution may splash out of a beaker or when the container of solution needs to be plugged at some point in the experiment. *Bunsen burners* are heat sources. They basically are cylinders attached to a gas line. When the gas line is opened, a spark ignites a flame in the Bunsen burner, which is then used to heat solutions. Why do solutions need to be heated? Sometimes solutions need to be boiled to release gases or to dissolve a solid into the solution.

Finding Science Information

Scientists must publish their research. They must put out their work, flaws and all, for all the other scientists to see. Otherwise, nobody would know the work was done. Other scientists may be working on the same thing and could benefit from seeing how someone else approached the problem. Scientists need to see each other's work, but they don't all work in the same laboratory. Scientists need ways of communicating to other scientists around the world.

Journals present research, not record dreams

Hundreds of scientific journals cover every topic and niche imaginable in the fields of biology, chemistry, physics, engineering, and so on. Some professional organizations publish journals, some universities or medical centers publish newsletters or journals, scientific corporations may publish newsletters, and medical and scientific publishing companies publish journals. There is no information shortage.

Journals are considered the primary source of scientific information. Anyone researching a topic, whether it is for a term paper in college or to develop a new experiment in the field, consults the journals first. The journals contain the original research papers, so the latest information in a specific field is always found in a journal. The research papers are written following the scientific style of an abstract (summary) of the research; statement of the hypothesis; description of materials used; how the experiment was designed and performed; results of the experiment, including raw data, graphs, tables; conclusions; and errors.

Some examples of major journals are *Nature*, *Science*, the *Journal of the American Medical Association*, the *British Medical Journal*, *The Lancet*, and the *New England Journal of Medicine*. The major journals are peer reviewed, which means that before a research paper is accepted for publication, other scientists in that field review the research to make sure the science behind it is thorough and that the research adds to the knowledge base. If stringent criteria are not meant, a scientist is unable to publish his or her research, which means the research needs to be performed again (which costs money and time).

The text of textbooks

Textbooks are considered secondary sources of information. Although they do not contain research papers, textbooks often are written by experts in the field. Textbooks present the knowledge base of a specific topic or field at a certain point in time, so they are a good source to turn to for history of a topic, basic facts about a certain subject, and summaries of important research that has furthered the field.

Popular press: not so popular with scientists

If scientists hate one thing, it's being misquoted and misunderstood (Okay. Two things.) Scientists are ever-so-careful in planning their research and taking their time to do it right. Some scientists working in the same field may compete in a way to be the first to publish results of similar studies. But, generally, scientists do not rush research. Usually, research cannot be rushed, even if the scientist did want to.

The Internet was not intended for advertisements

If you look at Web pages on the Internet today, you see borders of advertisements, you receive spam e-mail, and you come across plenty of junk. But, originally, the Internet was started as a way to link universities all around the world so that researchers could share their information with one another more quickly than a publishing schedule allows. Once the possibilities for such instant worldwide access were imagined, however, the Internet developed into its current state. University researchers still use the Internet to share information, but now a lot more traffic (and billboard pollution) is on the information superhighway.

So, when a journalist doesn't take the time to double-check facts and work to make sure that information is not misconstrued, scientists get frustrated. If their body of research was meant to contribute to the knowledge base of a certain field, but a journalist labels their research as "groundbreaking" or an "amazing breakthrough," the scientist(s) get angry. And, with good reason. Amazing breakthroughs and groundbreaking research happens very, very rarely. Most often, research just adds knowledge that can be used as a basis for more research. Or, the research contributes to the development of a product. The results might be "amazing" to the journalist, but the scientist does not want to be scorned by his colleagues, whom he relies on for more information.

Popular press items, such as from a newsstand magazine, newspaper, television, or radio program, are considered tertiary (meaning third-level) sources. These sources provide information, of course, but the validity of the information is not as certain as it is from the original research. There is always the chance that something may be misconstrued by the journalist trying to interpret the information presented in the research, which would mean that the presentation of the journalist may have errors. It's like that old childhood game where the information given to the first person is usually changed by the time it gets relayed to the last person. Best to stick with the original source.

Chapter 2

The Fundamental Units of Life: Cells

In This Chapter

- Understanding why cells are the basic units of life
- Differentiating prokaryotic and eukaryotic cells
- Finding out the structures of the cell
- Discovering how cells function

In this chapter, you explore the tiniest bits of yourself and other living things. You take a look at cells — the smallest level of a living organism. The cellular level of an organism is where the metabolic processes occur that keep the organism alive. And that, my friend, is why the cell is called the fundamental unit of life.

Windows on the World's Organisms

Every living thing has cells. The smallest animals have only one, yet they are as alive as you and I. What exactly are cells? *Cells* are sacs of fluid surrounded by membranes. Inside the fluid float chemicals and *organelles,* which are structures inside the cell, that are used during metabolic processes. Yes, an organism contains parts that are smaller than a cell, but the cell is the smallest part of the organism that retains characteristics of the entire organism. For example, a cell can take in fuel, convert it to energy, and eliminate wastes, just like the organism as a whole can. But, the structures inside the cell cannot perform these functions on their own, so the cell is considered the lowest level.

Each cell is capable of converting fuel to useable energy. Therefore, cells not only make up living things; they *are* living things. Cells are found in all plants, animals, and bacteria. Many of the basic structures found inside all types of cells, as well as the way those structures work, fundamentally are the very similar, so the cell is said to be the fundamental unit of life.

However, the most important characteristic of a cell is that it can reproduce by dividing. (Chapter 11 covers cell division.) If cells did not reproduce, you or any other living thing would not continue to live. *Cell division* is the process by which cells duplicate and replace themselves. If you did not replace your red blood cells, for example, you would have a life span only as long as that of red blood cells — a mere 120 days.

Viruses (like those that cause the flu, a cold, or AIDS) are similar to bacteria, but they are not truly living organisms because they lack one crucial characteristic: They cannot grow and divide by themselves. Viruses are more like parasites in that they need to take over the cells of an animal (a host) to reproduce.

Increasingly more complex organisms are made up of increasingly more groups of cells (for example, in humans, groups of cells make up each organ and muscle tissue), and the organisms survive based on products that the cells make. For example, cells in the pancreas make insulin, which is necessary to ensure that the blood glucose level doesn't skyrocket. Without insulin, the blood glucose can reach a level that is lethal. So, without that cellular product, you would die.

As scientists developed ways of studying cells more in depth, they began to unravel mysteries of life. Certainly, there is more to be learned. But, what is known already is simply fascinating.

Examining Eukaryotes

Cells fall into two major categories: eukaryotes and prokaryotes. *Prokaryotes* are cellular organisms that do not have a "true" nucleus. A *nucleus* is the control center of a cell. A nucleus contains the genetic material packed into chromosomes, and it is associated with other organelles that function in the production of amino acids and proteins based on what the genetic material dictates. Prokaryotes have some genetic material, but it is not as well organized as it is in eukaryotes. Still, prokaryotes are able to reproduce. Examples of these organisms include bacteria and blue-green algae. *Eukaryotes* are organisms that contain chromosomes, including plants and animals, as well as fungi (like mushrooms), protozoa, and most algae. Eukaryotes have the following characteristics:

- They have a nucleus that stores their genetic information.
- Animal cells have an organelle called a *mitochondria* that effectively combines oxygen and food to convert energy to a useable form.
- Plant cells have *chloroplasts,* which use energy from sunlight to create food for the plant.

- Eukaryotic cells have internal membranes, which create compartments inside the cells that have different functions.
- Plants cells have a cell membrane and a cell wall, which is rigid; animal cells have only a cell membrane, which is soft.
- The cytoskeleton, which reinforces the cytoplasm of the cell, controls cellular movements.

Although bacteria (prokaryotic cells) will be looked at later in the book, the discussions on cell structure and function in this chapter focus on eukaryotic cells.

Cells and the Organelles: Not a Motown Doo-Wop Group

You have organs and are made up of cells. Cells have organelles. You have organ systems that perform certain functions in you, the entire organism. Cells have organelles that perform certain functions in the cell. And, although it takes millions and millions of cells to create you, each cell functions on its own. An organelle in one cell does not do the work for another cell. Each cell metabolizes individually. This section gives you information on all the organelles that help the cell metabolize.

Holding it all together: The plasma membrane

The fluid inside a cell *(intracellular fluid)* is called *plasma* or *cytoplasm* (*cyto-* means cell). The membrane holding the fluid in the cell is called a *plasma membrane* (see Figure 2-1), also called the cell membrane. The cells themselves are floating in a type of fluid, called a *matrix*. The matrix is insoluble, meaning that substances do not dissolve in its fluid. The matrix just supports the cells. The fluid that squeezes in between each and every cell is called *extracellular fluid* because it is outside of the cell. The job of the plasma membrane is to separate the chemical reactions occurring inside the cell from the chemicals that are floating in the extracellular fluid.

If the plasma membrane didn't separate the inside and the outside of the cell, waste products excreted from the inside of the cell to the outside could flow back inside.

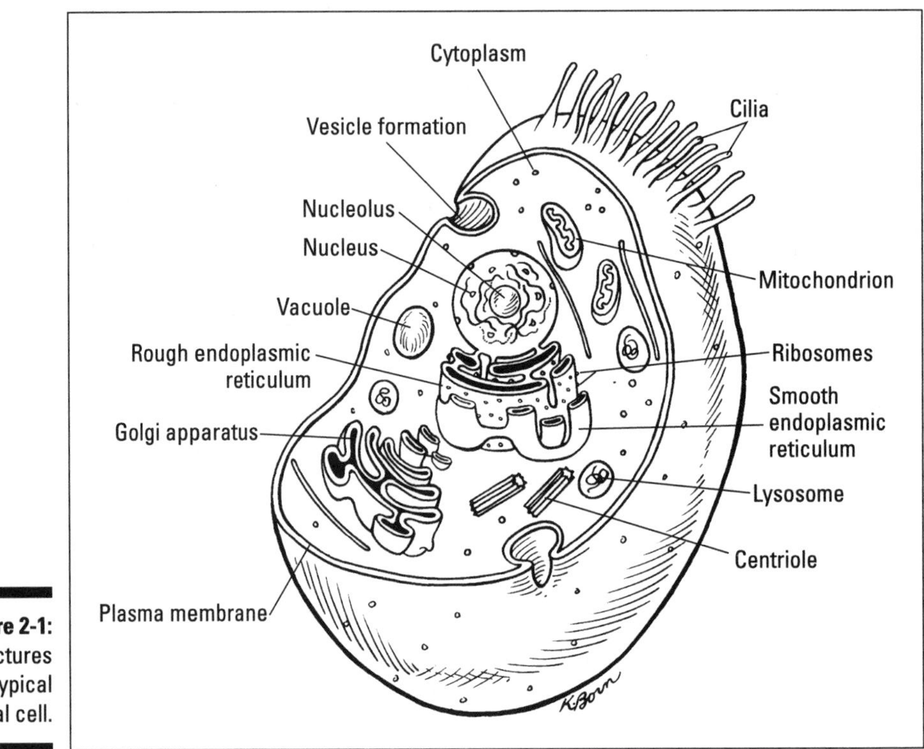

Figure 2-1: Structures in a typical animal cell.

The fluid-mosaic model of the plasma membrane

The plasma membrane has two layers (a *bilayer*) of *phospholipids* (fats with phosphorous attached), which at body temperature are like vegetable oil (fluid). And the structure of the plasma membrane supports the old saying, "Oil and water don't mix."

Each phospholipid molecule has a head that is attracted to water (*hydrophilic: hydro-* = water; *philic-* = loving) and a tail that repels water (*hydrophobic: hydro-* = water; *phobic-* = fearing). Both layers of the plasma membrane have the hydrophilic heads pointing toward the outside; the hydrophobic tails form the inside of the bilayer (see Figure 2-2). Because cells reside in a watery solution (extracellular fluid), and they contain a watery solution inside of them (cytoplasm), the plasma membrane forms a circle around each cell so that the water-loving heads are in contact with the fluid, and the water-fearing tails are protected on the inside.

Proteins and substances such as cholesterol become embedded in the bilayer, giving the membrane the look of a mosaic. Because the plasma membrane has the consistency of vegetable oil at body temperature, the proteins and other substances are able to move across it. That's why the plasma membrane is described using the *fluid*-mosaic model.

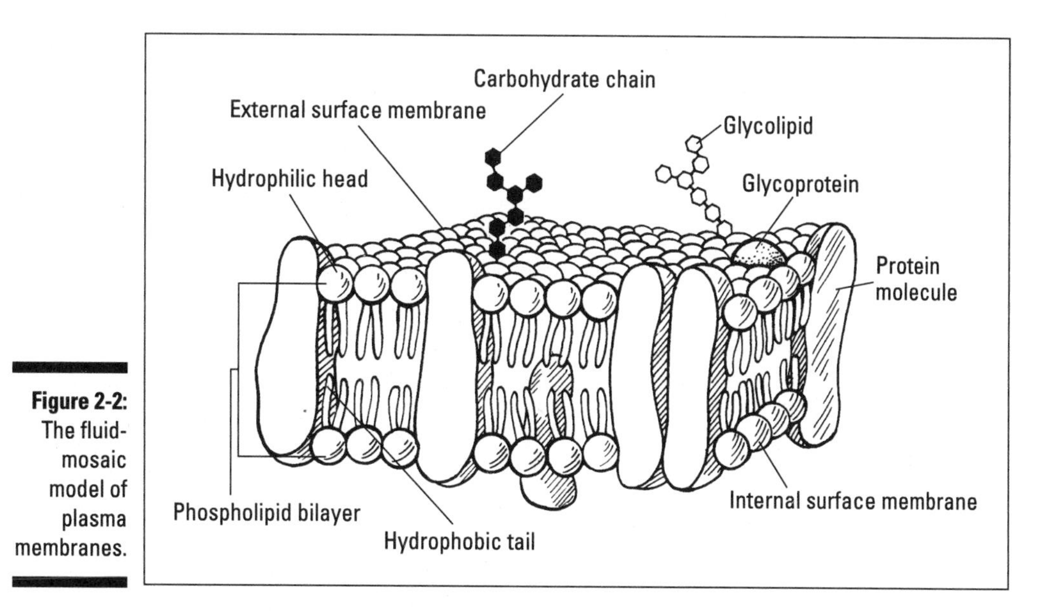

Figure 2-2: The fluid-mosaic model of plasma membranes.

The molecules that are embedded in the plasma membrane also serve a purpose. For example, the cholesterol that is stuck in there makes the membrane more stable and prevents it from solidifying when your body temperature is low. (It keeps you from literally freezing when you're "freezing.") Carbohydrate chains attach to the outer surface of the plasma membrane on each cell. These carbohydrates are specific to every person, and they supply characteristics such as your blood type.

Getting through: Transport through the plasma membrane

Some substances need to move from the extracellular fluid to the inside of the cell, and some substances need to move from the inside of the cell to the extracellular fluid. The plasma membrane is where these exchanges take place.

Some of the proteins that are stuck in the plasma membrane help to form openings *(channels)* in the membrane. Through these channels, some substances such as hormones or ions are allowed to pass through. No, the lucky substances don't utter the magic word; they either are "recognized" by a *receptor* (a protein molecule) within the cell membrane, or they attach to a carrier molecule, which is allowed through the channels. Because the plasma membrane is choosy about what substances can pass through it, it is said to be *selectively permeable*.

Permeability describes the ease with which substances can pass through a border, such as a cell membrane. *Permeable* means that most substances can easily pass through the membrane. *Impermeable* means that substances cannot pass through the membrane. *Selectively permeable* or *semipermeable* means that only certain substances are able to pass through the membrane.

Transporting substances across the plasma membrane can require that the cell use some of its energy to help heft the substance across the border. If energy is used, the transport is called *active*. If molecules can pass through the plasma membrane without using energy, the molecules are using *passive* transport.

Helping the molecules across: Active transport

Sometimes, the molecules are just too big to easily flow across the plasma membranes or dissolve in the water so that they can be filtered through the membrane. In these cases, the cells must put out a little energy to help get molecules in or out of the cell. Remember that embedded in the plasma membrane are protein molecules, some of which form channels through which other molecules can pass. Well, some proteins act as *carriers* — that is, they are "paid" in energy to let a molecule attach to itself and then transport that molecule inside the cell. It's like having to pay to take the Staten Island Ferry. The ferry is the carrier protein, and you are the big molecule that needs help getting from the bloodstream (New York Bay) to the inside of the cell (New York City). The fee that you'd pay is equivalent to the energy molecules expended by the cell.

Quietly going: Passive transport

A membrane can allow molecules to be passively transported through it in three ways: diffusion, osmosis, and filtration.

- **Diffusion:** Sometimes organisms need to move molecules from an area where they are highly concentrated to an area where the molecules are less concentrated. This transport is much more easily done than moving molecules from a low concentration to a high concentration. To go from a high concentration to a low concentration, in essence the molecules need to only "spread" themselves, or *diffuse,* across the membrane separating the areas of concentration.

  In the human body, this action occurs in the lungs. You breathe in air, and oxygen gets into the tiniest air sacs of the lungs, the *alveoli.* Surrounding the tiniest air sacs of the lungs are the tiniest blood vessels — *capillaries.* The capillaries in the lungs, called *pulmonary capillaries,* contain the lowest concentration of oxygen in the body, because by the time the blood gets to the tiniest vessels, most of the oxygen has been used up by other organs and tissues. So, the tiniest air sacs of the lungs have a higher concentration of oxygen than do the capillaries. That means that the oxygen from the alveoli of the lungs can spread across the membrane between the air sac and the capillary, getting into the bloodstream.

- **Osmosis:** This term is used when talking about water molecules diffusing across a membrane. Basically, the diffusion of water (osmosis) works as described in the preceding bullet. However, with osmosis, the concentration of substances in the water is taken into consideration. If a solution is *isotonic*, that means the concentrations of the substances *(solutes)* and water *(solvent)* on both sides of the membrane are equal. If one solution is *hypotonic*, there is a lower concentration of substances (and more water) in it when compared to another solution. If a solution is *hypertonic*, there is a higher concentration of substances in it (and less water) when compared to another solution.

For example, the blood in your body contains a certain amount of salt. The normal concentration is isotonic. If suddenly there is too high a concentration of salt, the blood becomes hypertonic (too many salt molecules). This excess of salt forces water out of the blood cells in an attempt to even things out. But the effect this action has is actually that of shrinking the blood cells. (Think of when you pour salt on a slug — it shrivels up.) This shrinking of cells is called *crenation* (not cremation). If too much fluid is in the bloodstream, the blood cells have too few molecules of salt in comparison, making them hypotonic. Then, the blood cells take in water in an attempt to normalize the blood and make it isotonic. However, if the blood cells need to take in too much water to bring everything back into balance, they can swell until they burst. This bursting of cells is called *hemolysis* (*hemo-* = blood; *lysis* = break apart).

- **Filtration:** The last form of passive transport is used most often in the capillaries. Capillaries are so thin (their membranes are only one cell thick) that diffusion easily takes place through them. But remember that animals have a blood pressure. The pressure at which the blood flows through the capillaries is enough force to push water and small solutes that have dissolved in the water right through the capillary membrane. So, in essence, the capillary membrane acts as filter paper, allowing fluid to surround the body's cells and keeping large molecules from getting into the tissue fluid.

Controlling the show: The nucleus

Every cell of every living thing has a nucleus, and every nucleus in every living thing contains genetic material. The genetic material directs the production of proteins that make the entire organism function; the nucleus makes the entire cell function.

In the nucleus of cells that are not currently dividing, clumps of thread-like genetic material called *chromatin* appear. However, right before a cell divides, the chromatin bunches up into *chromosomes*, which contain DNA (deoxyribonucleic acid).

The DNA has two strands, each of which has sequences of nitrogenous bases that form the *genetic code*. The genetic code, which is derived from the nucleotide bases in the genes on strands of DNA, is "interpreted" by a ribonucleic acid (RNA) molecule called *messenger RNA* (mRNA). The mRNA uses the information from the genetic code to create amino acids — the building blocks of protein — in the cell. The amino acids are then taken by transfer RNA (tRNA) to an organelle called a ribosome, where the final proteins are made. (See Chapter 14 for more information on genetics and protein synthesis.)

Every cell of every eukaryote has a nucleus, and every nucleus in every eukaryote contains genetic material. In the nucleus of cells that are not currently dividing, there are clumps of thread-like genetic material called *chromatin.* Chromatin refers to all of the DNA in the cell and its accompanying proteins. Chromatin cannot easily be seen prior to cell division, at which point the chromatin bunches up into *chromosomes.*

The DNA of eukaryotes is double stranded. (The DNA of prokaryotes is also double stranded. Only a few viruses have single-stranded "genomes," and these aren't always single stranded.) Each strand of DNA has sequences of nitrogenous (for example, containing nitrogen) bases that form the *genetic code*. The code is "interpreted," and then a ribonucleic acid (RNA) molecule called messenger RNA (mRNA) is produced from the DNA template. The mRNA uses the information from the genetic code for certain amino acids in the cell, which are then taken by transfer RNA (tRNA) to a ribosome, where the final proteins are made. How this code is read and converted to a messenger that is carried to the cytoplasm, where it is translated to produce a protein, is discussed further in Chapter 14.

Proteins either contribute to the structure of the cell, or they contribute to the function of the cell, meaning that they are used as enzymes in metabolic processes. Either way, it is the genetic material housed in the nucleus that ultimately controls the structure and function of each and every cell in all eukaryotic organisms.

Each nucleus has a round mass inside it called a *nucleolus.* The nucleolus produces the third type of RNA molecule — that is, ribosomal RNA (rRNA). This type of RNA helps to make *ribosomes,* which get transferred from the nucleus to the cytoplasm to help in making proteins.

Surrounding each nucleus is a double layer formed from proteins and lipids that separates the nucleus from the cytoplasm. This two-layered structure is called the *nuclear envelope* or *nuclear membrane*.

The factory of the cell: The endoplasmic reticulum

Refer to Figure 2-1. See the organelle that looks like it is folded-up sheets resembling a piece of coral? That's the *endoplasmic reticulum* (ER). The ER actually is a series of canals that connects the nucleus to the cytoplasm of the cell. The part of the ER that is dotted with ribosomes is called *rough ER*; the part of the ER that has no ribosomes is called *smooth ER*. Ribosomes on the rough ER serve as the place for the synthesis of proteins that are directed by the genes to be put together in the ER. (Other proteins are put together on ribosomes attached to other organelles or floating free in the cytoplasm.)

The smooth ER contains transport vesicles that shuttle cellular products from cytoplasm to organelle, from organelle to organelle, or from organelle to plasma membrane. In addition to protein synthesis, the ER is involved in the metabolism of lipids (fats).

The main function of ER is to make and transport proteins. The ER is essentially the "womb" for new protein chains. Protein synthesis, or production, begins in the nucleus, with the mRNA molecule carrying the genetic information as to what amino acids (proteins) should be produced. The tRNA molecules bring the amino acids from the cytoplasm to the ribosomes, which are produced by rRNA. At the ribosomes, the amino acids are joined together to form a protein, and the protein is stored in the ER until it can be moved to the Golgi apparatus. (For more information on protein synthesis, see Chapter 14.)

Preparing for distribution: The Golgi apparatus

In biology, as well as other sciences, structures usually are named for the person who found them. In this case, the Italian scientist Camillo Golgi finds fame. The *Golgi apparatus* is very close to the ER; in Figure 2-1, it looks like a maze with water droplets splashing off of it. The "water droplets" are transport vesicles bringing material from the ER to the Golgi apparatus.

Inside the Golgi apparatus, products produced by the cell, such as hormones or enzymes, are packaged for export to other organelles or to the outside of the cell. The Golgi apparatus surrounds the product to be secreted with a sac called a *vesicle*. The vesicle finds its way to the plasma membrane, where certain proteins allow a channel to be produced so that the products inside the vesicle can be secreted to the outside of the cell. Once outside the cell, the products such as hormones or enzymes can enter the bloodstream and be transported through the body to where they are needed.

Lysosomes really clean up

Lysosomes are special vesicles formed by the Golgi apparatus to "clean up" the cell. They are the garbage men (or sanitation engineers) of the cell. Lysosomes contain digestive enzymes, which are used to break down products that may be harmful to the cell and "spit" them back out into the extracellular fluid. Lysosomes also remove dead organelles by surrounding the dead organelle, breaking down the proteins of the dead organelle, and releasing them to reconstruct a new organelle. Because the lysosome acts upon its own cell, the process is called *autodigestion*.

Peroxisomes break down hydrogen peroxide

Peroxisomes are little sacs of enzymes produced by smooth ER to help protect the cell from toxic products. You know how hydrogen peroxide is helpful when you use it to clean out a wound because it kills bacteria? Well, too much hydrogen peroxide inside you could kill you. Hydrogen peroxide is normally produced in some metabolic reactions, so it is inside you. However, hydrogen peroxide becomes harmful to the cells of the body if too much accumulates, so the key is to keep breaking it down to keep it from accumulating. The peroxisomes break down excess hydrogen peroxide.

The chemical formula for hydrogen peroxide is H_2O_2 — pretty close to water (H_2O), eh? Well, that's what peroxisomes turn hydrogen peroxide into: plain old water plus an extra oxygen molecule, both of which are always needed by the body and can be used in any cell.

The powerhouses of the cell: The mitochondria

The ER supplies the products, the Golgi apparatus distributes the products, and the mitochondria supply the energy for all of those processes to take place.

When you get a bill for electricity, the amount of electricity your household used in the past month is measured in kilowatt hours. Inside an organism, the amount of energy a cell uses is measured in molecules of adenosine triphosphate (ATP). The mitochondria produce the ATP, and to do it, mitochondria use products of glucose metabolism as fuel.

Food as fuel

I'm sure you've heard before that food is the fuel for the body. Well, the mitochondria are the converters; they convert the fuel into useable energy. When food is digested, or broken down into its smallest molecules and nutrients, and air is taken in, or *inspired,* the smallest molecules and nutrients cross into the bloodstream (see Chapter 9). These molecules and nutrients include things such as glucose (a sugar molecule derived from carbohydrates) and oxygen. How organisms acquire food and oxygen is covered in Chapter 5.

Use food only as a fuel. Otherwise, you will consume more fuel than is necessary to make the machine called your body function. And you know what happens to the excess fuel? It gets stored for later (the machine stays prepared for times of low fuel intake) as fat.

Defining aerobic respiration

Just as fire burns oxygen and gives off carbon dioxide and water, mitochondria act like furnaces when they convert glucose into ATP: They "burn" (use) oxygen and give off carbon dioxide and water. Because the process uses oxygen, it is said to be *aerobic* (as in aerobic exercise). This chemical process of respiration occurs in every cell, so it is called *aerobic cellular respiration*. The steps that occur in this process are described by the Krebs cycle (also called the tricarboxylic acid [TCA] cycle) and are covered in detail in Chapter 8. The Krebs cycle is an extremely important biological cycle; it is a cornerstone to understanding how cells function.

Aerobic cellular respiration can be diagrammed like this, with each step breaking down the products in the step preceding it:

1. **Food (ingested) + Air (inhaled)**
2. **Carbohydrate + Oxygen and Nitrogen**
3. **Glucose + Oxygen (final products of digestion and inhalation)**
4. **ATP (energy) + Carbon Dioxide (exhaled) + Water (exhaled and excreted)**

Do not confuse respiration with breathing. Breathing is just a part of respiration. Breathing actually is the act of inspiring and expiring (see Chapter 5); respiration is the exchange of oxygen and carbon dioxide between cells and the atmosphere. So, people respire, but it happens at the cellular level.

Chapter 3

A (Very) Quick Review of Basic Chemistry

In This Chapter

- Understanding why chemistry is important in biology
- Figuring out the difference between atoms, molecules, compounds, isotopes, ions, and electrolytes
- Finding out about ionic and covalent bonds
- Understanding acids, bases, and pH

This chapter is intended to give you an overview of chemistry. If you have never studied chemistry before, hopefully this chapter can help you grasp a few major concepts. If you have studied it before, hopefully this chapter can give you a good review of the basics. The chemistry I discuss in this chapter is largely the inorganic variety. *Inorganic chemistry* focuses on a small number of atoms (usually one to three) that are held together by ionic bonds (or covalent bonds in the case of water). *Organic chemistry*, which focuses on carbon molecules, is central to all living organisms and is covered in Chapter 4.

Why Matter Matters

To understand biology, you need to know a few core chemistry terms and ideas. You must understand what an atom is and how atoms can be joined together to form compounds or molecules. You must have a general understanding of how ions combine and break apart, forming molecules or providing energy to reactions. You must grasp how reactions can form acids or bases, as well as what effect that has on pH. Lastly, you need to understand how carbon is central to organic chemistry and that organic chemistry is the type of chemistry that goes on in living organisms.

Having a general feel for chemistry will make it easier for you to figure out how electrolytes play an important role in the processes that occur in metabolism. Parts II and III focus on how living things create energy from matter and metabolize substances to continue living. Chemical reactions occur in all of these processes. It's only natural that you understand some chemistry to learn biology. Why do you have to? Because Mother Nature says so. And you know what happens when you don't listen to Mother Nature!

Science is a body of knowledge that is created by studying nature. *Biology* is the study of living things. *Chemistry* is the study of matter, what matter is made of, and how matter can change or be combined. So, because living things are composed of matter, chemistry matters to biology.

What Matter Is

Matter is a material (substance) that takes up space and has mass. It can be a solid, liquid, or gas, and it can be a single substance or a mixture of substances.

Matter has mass

Mass is the term for describing the amount of matter that a material (or substance) has; when the amount of gravity is figured into that value, you get the weight of the material. Therefore, *weight* is determined by the amount of gravitational force acting on a substance that has mass and how much resistance the substance gives to a movement.

Matter takes up space

Space is measured in volume. Because the mass of a substance does not change (although its weight might), the mass of differing materials can be compared by determining how much space (for example, *volume*) each material occupies. This is done using a *unit of mass (m),* which in chemistry is a *gram (g). Volume (v)* usually is measured in *milliliters (ml).* The term used to relate a material's mass to its volume is called *density (D).* The equation for this is $D = m/v$, meaning density equals the amount of mass per unit of volume.

Matter comes in three forms

Matter can be a solid, liquid, or gas. *Solids* have a definite shape and size, such as a person or a brick. *Liquids* have a definite volume (for example, they fill a container), but they take the shape of the container that they fill. *Gases* do not have a definite shape or a definite volume.

The state of matter is changed by adding energy in the form of heat. For example, if you add heat to a block of ice (a solid), it eventually becomes water (a liquid). If you continue to add heat, the water begins to evaporate in the form of steam (a gas).

Matter has two categories

Matter can be a *substance* (akin to an "ingredient") or a *mixture,* which is a combination of different substances. Substances generally are held together by chemical means (such as bonding). Mixtures generally are held together by physical means, and the substances that make up the mixture can be separated by physical means (such as filtering, distilling, sorting, or extracting).

Examples of substances include

- Oxygen gas
- Salt
- Sugar

Examples of mixtures include

- Rocks
- Plants
- Animals
- Concrete

What's the Difference? Elements, Atoms, and Isotopes

All matter is composed of *elements.* If you break down matter into its smallest components, you are left with individual elements. If you are breaking down a *molecule* into its smallest individual pieces, you get elements of the same type. If you are breaking down a *compound* into its smallest individual pieces, you are left with elements of different types. (Molecules and compounds are explained in the next major section.) But, even elements are made up of something: atoms.

One *atom* of an element is the smallest "piece" of matter that can be measured. Of course, atoms are made up of something, called *subatomic particles*, such as quarks, mesons, leptons, and neutrinos. But those subatomic particles cannot be taken away from the atom without destroying the atom. Therefore, the atom is the smallest whole, stable piece of an element that still has all the properties of that element — that is, at least for now, anyway. Who knows, scientists may delve further into atoms and find an even smaller measurable piece of matter. What matters now, though, is that you grasp the concepts of how these pieces relate.

If you are interested in subatomic particles, check out the last link on the timeline at this Web site:

```
www.watertown.k12.wi.us/hs/teachers/buescher/atomtime.asp
```

Elements of elements

When talking about chemistry, the term elements does not refer to water, air, fire, or earth. Instead, *elements* are the "ingredients" that make up the water, air, fire, or earth. To illustrate an element for you, I give you two analogies based on two of my passions: music and cooking.

Think of a basic C chord in music. The note middle C is at the bottom, the note E is in the middle, and the note G is on top. Okay. Think of C as an element, E as a different element, and G as yet another element. If you span an octave by playing middle C and the next higher C note (two of the same notes), you have created a molecule — call it a C molecule. If you play an octave of two E notes, you've created an E molecule. And, if you play an octave of two G notes, you've created a G molecule. Now, if you play middle C, E, and G together, you've created not only a beautiful-sounding chord, but also a compound of elements.

If the music analogy didn't help you, try this one. Think of a recipe for chocolate chip cookies. First, you need to mix wet ingredients: the butter, sugar, eggs, and vanilla. Consider each of those ingredients a separate element. You need two sticks of the element butter. When you combine butter plus butter, you get a molecule of butter. Before you add the element of eggs, you need to beat them. So, when you add egg plus egg in a little dish, you get a molecule of eggs. When all of the wet ingredients are mixed together, you get a compound called "wet."

Next, you need to mix together the dry ingredients: flour, salt, and baking soda. Think of each of those ingredients as a separate element. When all the dry ingredients are mixed together, you get a compound called "dry." Only when the wet compound is mixed with the dry compound is the reaction sufficient to add the most important element: the chocolate chips. Hopefully, this example helps you understand the differences between an element, molecule, and compound. If nothing else, I'm sure it made you want a snack.

"Bohr"ing you with atoms

You can thank the Danish scientist Neils Bohr for coming up with the model of an atom, shown in Figure 3-1. But, actually, the term *atom* was used by Greek philosophers as far back as 450 B.C. Those philosophers knew that matter was made up of tiny building blocks. But it took until fewer than 100 years ago to come up with a model to explain how.

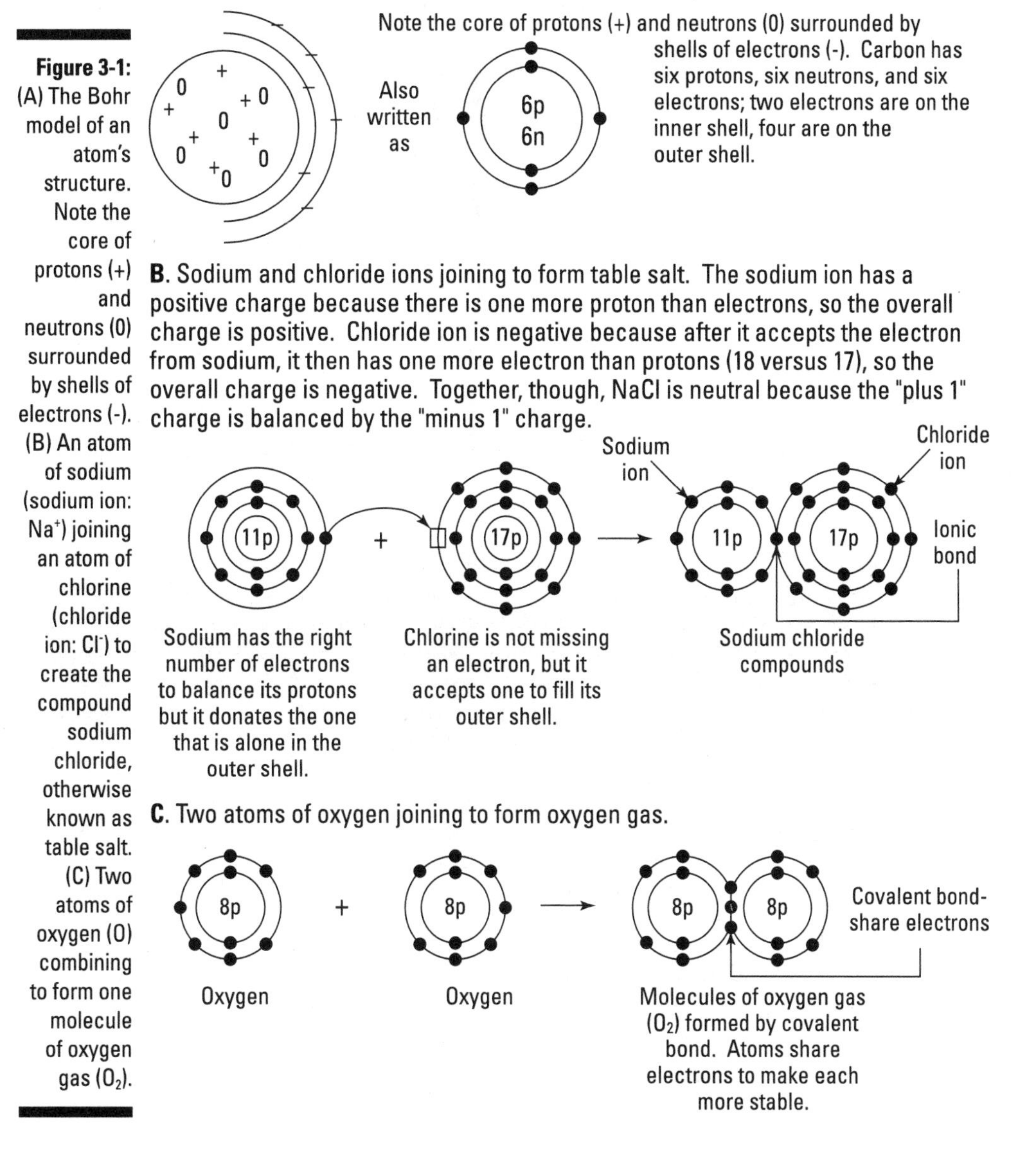

Figure 3-1: (A) The Bohr model of an atom's structure. Note the core of protons (+) and neutrons (0) surrounded by shells of electrons (-). (B) An atom of sodium (sodium ion: Na^+) joining an atom of chlorine (chloride ion: Cl^-) to create the compound sodium chloride, otherwise known as table salt. (C) Two atoms of oxygen (O) combining to form one molecule of oxygen gas (O_2).

Inside the atom: Protons and neutrons

Inside the *nucleus* (core) of the atom are two kinds of *particles* (pieces of matter). They are called *subatomic particles* because they are smaller than the atom (*sub* = less, lower; as in subscript). These subatomic particles (pieces of an atom) include *protons*, which are positively charged, and *neutrons*, which are neutral. Because the protons are positive and neutrons have no charge, the nucleus at the center of an atom is positive, overall.

Although the nucleus of an atom is illustrated as a circle, it does not have a definite shape, like a ball. It is not like a cell in the body. The nucleus of an atom is simply (or not so simply) a conglomeration of positive and neutral particles.

Outside the atom: Electron shells

Atoms are surrounded by electrons (see Figure 3-1), which are negatively charged. Just as batteries need to have the positive and negative poles together to function, an atom is held together (so it can function) by the pull between its positive core (nucleus) and its negative electrons.

Atoms can have several electron shells surrounding its nucleus. The closer a shell is to the nucleus, the less energy the electron needs to be pulled toward the nucleus. However, if electrons want to move from inner to outer shells, energy is required.

The Periodic Table of Elements

All of the known elements are listed in the Periodic Table. The table gets its name from its design. Each row of the table is called a period. Moving across the table horizontally, you go from metals to nonmetals, with heavy metals in the middle. The energy level of the electrons in the outer shell of atoms increases as you read the table from right to left. With more energy in the outer electron shell, there is more of a pull by the atom's nucleus, making the overall size of the atom smaller. Also going from left to right, the ability of the elements to form bases decreases and the ability to form acids increases. The size of the atom increases from top to bottom within each column (called a *family* or *group*).

The periodic law states that the properties of elements are a periodic function of their atomic numbers. Each element gets its *atomic number* from the number of protons in the nucleus of an atom of the element. For example, carbon has six protons in the nucleus of one atom, so, its atomic number is 6. The number of electrons in one atom of an element also is equal to the atomic number (because, overall, atoms are neutral: the positively charged particles are offset by the negatively charged particles one for one).

In some cases, one of the atoms has a more positive charge and the other has a more negative charge. In water, the oxygen is a bit more negative, and the hydrogen atoms are a bit more positive. The opposite charges pull the hydrogen atom of one water molecule toward the oxygen atom of a different water molecule, creating a network of bonds to hold the mixture together. This network of bonds makes it harder to boil or freeze water, which is important because there is so much water in a living organism.

I so dig isotopes

If the number of neutrons changes — that is, if the atom's nucleus gains or loses neutrons — then the nucleus of the atom can decay, but the overall atom still has its chemical properties. But then it is called an *isotope* of the element. Isotopes of an element have the same number of protons in the nucleus of the atom, but they have differing numbers of neutrons.

For example, if you look at the Periodic Table (see Figure 3-2), you'll see the number 6 down and to the left of the symbol C. (For a Web version, check out the site `www.schoolscience.co.uk/periodictable.html`.) The number signifies that the atomic number, and thus the number of protons in the nucleus, of carbon is 6. Then, you will see the number 12.0111 up and to the left of C, which tells you that the *atomic mass* (or atomic weight) of carbon is 12. The atomic mass is determined by adding the number of protons with the number of neutrons that an atom normally contains — that is, when it is most *stable*. Therefore, the most stable isotope of carbon is carbon-12 (^{12}C).

1 H Hydrogen 1.01																	2 He Helium 4.00
3 Li Lithium 6.94	4 Be Beryllium 9.01											5 B Boron 10.81	6 C Carbon 12.01	7 N Nitrogen 14.01	8 O Oxygen 16.00	9 F Fluorine 19.00	10 Ne Neon 20.18
11 Na Sodium 22.99	12 Mg Magnesium 24.31											13 Al Aluminum 26.98	14 Si Silicon 28.09	15 P Phosphorus 30.97	16 S Sulfur 32.06	17 Cl Chlorine 35.45	18 Ar Argon 39.95
19 K Potassium 39.10	20 Ca Calcium 40.08	21 Sc Scandium 44.96	22 Ti Titanium 47.90	23 V Vanadium 50.94	24 Cr Chromium 52.00	25 Mn Manganese 54.94	26 Fe Iron 55.85	27 Co Cobalt 58.93	28 Ni Nickel 58.71	29 Cu Copper 63.55	30 Zn Zinc 65.38	31 Ga Gallium 69.72	32 Ge Germanium 72.59	33 As Arsenic 74.92	34 Se Selenium 78.96	35 Br Bromine 79.90	36 Kr Krypton 83.80
37 Rb Rubidium 85.47	38 Sr Strontium 87.62	39 Y Yttrium 88.91	40 Zr Zirconium 91.22	41 Nb Niobium 92.91	42 Mo Molybdenum 95.94	43 Tc Technetium (99)	44 Ru Ruthenium 101.07	45 Rh Rhodium 102.91	46 Pd Palladium 106.42	47 Ag Silver 107.87	48 Cd Cadmium 112.41	49 In Indium 114.82	50 Sn Tin 118.69	51 Sb Antimony 121.75	52 Te Tellurium 127.60	53 I Iodine 126.90	54 Xe Xenon 130.30
55 Cs Cesium 132.91	56 Ba Barium 137.34	57 La Lanthanum 138.91	72 Hf Hafnium 178.49	73 Ta Tantalum 180.95	74 W Tungsten 183.85	75 Re Rhenium 186.21	76 Os Osmium 190.2	77 Ir Iridium 192.22	78 Pt Platinum 195.09	79 Au Gold 196.97	80 Hg Mercury 200.59	81 Tl Thallium 204.37	82 Pb Lead 207.19	83 Bi Bismuth 208.98	84 Po Polonium (210)	85 At Astatine (210)	86 Rn Radon (222)
87 Fr Francium (223)	88 Ra Radium (226)	89 Ac Actinium (227)	104 Rf Rutherfordium (257)	105 Db Dubnium (260)	106 Sg Seaborgium (263)	107 Bh Bohrium (262)	108 Hs Hassium (265)	109 Mt Meitnerium (266)									

Lanthanides	58 Ce Cerium 140.12	59 Pr Praseodymium 140.91	60 Nd Neodymium 144.24	61 Pm Promethium (147)	62 Sm Samarium 150.35	63 Eu Europium 151.96	64 Gd Gadolinium 157.25	65 Tb Terbium 158.93	66 Dy Dysprosium 162.50	67 Ho Holmium 164.93	68 Er Erbium 167.26	69 Tm Thulium 168.93	70 Yb Ytterbium 173.04	71 Lu Lutetium 174.97
Actinides	90 Th Thorium (232)	91 Pa Protactinium (231)	92 U Uranium (238)	93 Np Neptunium (237)	94 Pu Plutonium (242)	95 Am Americium (243)	96 Cm Curium (247)	97 Bk Berkelium (247)	98 Cf Californium (251)	99 Es Einsteinium (254)	100 Fm Fermium (257)	101 Md Mendelevium (258)	102 No Nobelium (259)	103 Lr Lawrencium (260)

Figure 3-2: The Periodic Table of Elements.

Dating with carbon-14

The isotope carbon-14 (which has 6 protons and 8 neutrons) decays radioactively at a very, very slow rate but steadily over time. Because the rate of decay is steady, if a scientist measures how far the nucleus of the atom is decayed, he or she can date things like bones and fossils. *Dating* means that scientists can determine how long ago an organism was living. Carbon-14 dating is used in fields like *archaeology* (study of past human life), *paleontology* (study of geologic remains such as fossils), and *forensic science* (scientific techniques applied to legal problems).

An isotope of carbon that contains more neutrons than protons is carbon-14 (^{14}C). Like I said in the preceding paragraph, if the number of neutrons is less than or greater than the number of protons in the nucleus of the atom, the nucleus may decay. This decay is *radioactive,* meaning it gives off measurable energy.

Molecules, compounds, and bonds (oh, my!)

Of all the elements in the Periodic Table, living things use only a handful. The four most common elements found in living things are hydrogen, carbon, nitrogen, and oxygen, all of which are found in air, plants, and water. Then, several other elements exist in smaller amounts in living organisms, including sodium, magnesium, phosphorus, sulfur, chlorine, potassium, and calcium. You can see in later chapters how these elements are used in reactions inside the body.

Electrolytes

Most often, the elements such as sodium, magnesium, chlorine, potassium, and calcium circulate in the body as *electrolytes.* Electrolytes are substances that release ions when they break apart in water. When in the "water" of the body, substances such as sodium chloride (NaCl) break apart into the ions Na^+ and Cl^-, which then are used in organs such as the heart or in cellular processes.

Eyes on ions

What's an ion, you ask? *Ions* are charged particles — that is, atoms with a positive or negative charge. Remember that inside atoms there are protons (which are positive) and neutrons (which are neutral), as well as electrons (which are negative) outside it. Ions are positive (+) when they have more protons than electrons; they are negative (-) when they have more electrons than protons.

Just as in any relationship that is breaking up, someone gets a new partner, and the other is left alone. When a molecule of water splits up (that is, ionizes), one hydrogen atom is left alone, and the other hydrogen atom gets the oxygen. However, this situation really is the most equitable situation. The compound H_2O doesn't split into H + H + O. It splits into H^+ (hydrogen ion) and OH^- (hydroxide ion): one positive ion and one negative ion. The hydrogen ion helps to create acids; the hydroxide ion helps to create bases.

Molecules and compounds

When atoms of the same element combine, they form *molecules*. For example, when two atoms of oxygen join together, the result is one molecule of oxygen gas (O_2).

Compounds are formed when molecules of two or more atoms are joined together. For example, because water is a combination of two different elements (hydrogen and oxygen), it is considered to be a compound. Another example of a compound is glucose ($C_6H_{12}O_6$), which combines several atoms of carbon, hydrogen, and oxygen.

Why water is so important

Covalent bonds hold the hydrogen and oxygen atoms together in a water molecule. In covalent reactions, the atoms have an electrical charge (for example, positive or negative), but they do not give or take electrons. Instead, the atoms share electrons, which essentially causes one of the atoms to have a more positive charge and the other to have a more negative charge. In water, the oxygen is a bit more negative, and the hydrogen atoms are a bit more positive. The opposite charges cause the hydrogen atoms of one water molecule to be pulled toward the oxygen atom of other nearby water molecules, creating a strong non-covalent bond to hold the compound together. This strong bond, which is called a hydrogen bond, makes it difficult to boil or freeze water, which is important because a living organism contains so much water. The covalent bond strength means that water is a stable molecule; it is very hard to pull oxygen and hydrogen of a water molecule apart into the atoms. Although water contains covalent bonds, the characteristics of boiling and freezing water are due to its network of hydrogen bonds, not its covalent bonds.

Both molecules and compounds are held together by *bonds*, and bonds can be either *ionic* or *covalent*.

- **Ionic bonds:** *Ionic bonds* hold atoms joined together in an ionic reaction. Ionic reactions occur when ions combine and the atoms involved lose or gain electrons. A simple example of an ionic reaction is the one between sodium (Na^+) and chlorine (Cl^-) to form table salt.

  The sodium ion is written as Na^+, meaning that *Na* is the symbol for the element sodium, and an ion of sodium has one more proton than electrons. Therefore, it is a positive ion with a "plus 1" charge. Remember that opposites attract; so, an ion with a positive charge is naturally attracted to an ion with a negative charge. A chloride ion (an ion of the element chlorine) is written as Cl^-, meaning that *Cl* is the symbol for the element chlorine, and the chloride ion has one less proton than electrons. Therefore, it is a negative ion with a "minus 1" charge. (Note that the number one isn't written out; ions with plus 2 or higher or minus 2 or higher charges have the numbers included.) When the sodium ion and the chloride ion come together, an ionic bond is formed, and the positive and negative charges on both ions are balanced when sodium "gives" an electron to chlorine. That is why the compound table salt is written as NaCl, with no plus or minus signs.

- **Covalent bonds:** *Covalent bonds* are formed when atoms share electrons in a covalent reaction. The term *covalence* refers to the number of electron pairs that an atom shares with another atom. The more electron pairs that atoms share, the more stable they are. And, having stable atoms is a good thing. The bonds that form between the atoms that share electrons are called covalent bonds.

  The word covalence indicates the sharing of *valence*, which comes from the Latin word for strong (as in valor). The atoms share each other's strength, which in essence makes their bonds even stronger. Previously, the *valence number* indicated how many bonds each atom of an element makes with other atoms. But now, the term valence number is used to denote not only how many bonds an atom can make, but also the charge on a molecule made of one or more atoms.

Acids and bases (no, this is not a heavy metal band)

When water splits apart (which doesn't happen to every water molecule all at the same time), hydrogen and hydroxide ions are created. The hydrogen ion (H^+) can combine with negatively charged elements to form acids. *Acids* are molecules that can split apart in water and release hydrogen ions. A common example is hydrochloric acid (HCl). When HCl is added to water, it splits apart into H^+ and Cl^-, increasing the number of hydrogen ions in the water–HCl solution.

Bases are molecules that can split apart in water and release hydroxide ion. The most common example is sodium hydroxide (NaOH). When NaOH is added to water, it splits apart into Na^+ and OH^-. The hydroxide ions can combine with hydrogen ions (therefore, decreasing the number of hydrogen ions in the solution) to form more water. The principles of how acids and bases react in water form the basis of the pH scale.

"Ph"iguring out the pH scale

The term *pH* symbolizes the hydrogen ion concentration in a solution (for example, what proportion of a solution contains hydrogen ions). Describing it in detail involves explaining logarithms, so I'll leave it at this: The pH scale goes from 1–14. A pH of 7 is neutral, meaning that the amount of hydrogen ions and hydroxide ions in a solution are equal. For example, water has a pH of 7 because when water breaks up, the split is equitable into one hydrogen ion for every hydroxide ion.

If a solution contains more hydrogen ions than hydroxide ions, it is said to be acidic, and the pH of the solution is less than 7. The reason it is acidic is because of the properties of acids I explain in the previous section. Remember that if a molecule releases hydrogen ions in water, it is an acid. The more hydrogen ions it releases, the stronger the acid, and the lower the pH value. Table 3-1 shows you the pH of some common substances and may visually help you to figure out the pH scale.

The situation is reversed for bases. If a solution contains more hydroxide ion than hydrogen ion, it is said to be basic, and its pH is higher than 7. Remember that bases dissociate (break apart) into hydroxide ions (OH^-) and a positive ion. The hydroxide ions can combine with H+ to create water. Because the hydrogen ions are used, the number of hydrogen ions in the solution decreases, making the solution less acidic and therefore more basic. So, the more hydroxide ions a molecule releases (or the more hydrogen ions it takes in), the more basic it is.

Table 3-1 The pH of Some Common Substances

Increasing pH (Decreasing Acidity)		*Substances*
0	(most acidic)	Hydrochloric acid (HCl)
1		Stomach acid

(continued)

Table 3-1 *(continued)*

Increasing pH (Decreasing Acidity)		*Substances*
2		Lemon juice
3		Cola, beer, vinegar
4		Tomatoes
4.5		Fish die if water is this acidic
5		Coffee
5.5		Normal rainwater
6		Urine
6.5		Saliva
7	(neutral)	Water, tears
7.5		Human blood
8		Seawater
9		Baking soda, antacids
10		Great Salt Lake
11		Ammonia
12		Bicarbonate of soda
13		Oven cleaner
14	(most basic)	Sodium hydroxide (NaOH)

Buffing up on buffers

In living organisms, blood or cytoplasm are the "solutions" in which the required ions (for example, electrolytes) are floating. That is why most substances in the body hover around the neutral pH of 7. However, the body has a backup system in case things go awry. A *buffer system* exists to help neutralize the blood if excess hydrogen or hydroxide ions are produced. The buffers help to keep the pH in the normal range by "taking up" (combining with) the excess hydrogen or hydroxide ions. If something is wrong with the buffer system, an organism, such as you, can develop *acidosis* if the pH drops too low (blood becomes too acidic) or *alkalosis* if the pH gets too high (blood becomes too basic).

RELATED CHEMISTRY

An important homeostasis reaction

The reaction between bicarbonate ion and hydrogen ion yields (gives, produces) carbonic acid. The reaction that yields carbonic acid also can go the other way, which is why the arrow faces both right and left. Carbonic acid can break apart to supply hydrogen ions when the pH of the blood is too high (meaning it is less acidic, too basic).

$$CO_2 + H_2O \leftrightarrow H_2CO_3$$

$$H_2CO_3 + HCO_{3-} \leftrightarrow H^+$$

The most common buffers in the body are *bicarbonate ion* (HCO_{3-}) and *carbonic acid* (H_2CO_3). Bicarbonate ion is prevalent in the bloodstream. It carries carbon dioxide through the bloodstream to the lungs so that it can be exhaled (see Chapter 5). It also combines with excess hydrogen ion to keep the pH of the blood in the normal range. When bicarbonate ion takes up the extra hydrogen ions, it forms carbonic acid, which keeps the pH of the blood from going too low. However, if the pH of the blood gets too high, carbonic acid breaks apart to release some hydrogen ions, which brings the pH back into balance. The pH of the body is fine-tuned by actions in the kidneys. (See Chapter 9 for an explanation of homeostasis and kidney structure and function.)

Chapter 4

Macromolecules (The Big Ones)

In This Chapter

- Figuring out what organic chemistry involves
- Understanding the structure and function of major macromolecules
- Finding out how sugars, proteins, and nucleic acids are made
- Believing that all fats aren't bad

When you start exploring biology, you'll find that many, many processes are constantly occurring in living organisms. The ability to convert ingested fuel to usable energy is what differentiates a living organism from a dead one. The ingested fuel contains a variety of large molecules (macromolecules) that get broken down as far as possible. When the macromolecules have been broken down into their smallest parts, they can enter the cells, which contain more macromolecules, which are involved in more processes. (Are you picking up on the fact that life is a continuation of cycles?)

This chapter introduces you to some of the major macromolecules found in living things. You encounter these big important molecules wherever the major cell cycles and processes are discussed. So, give them a firm handshake, sit down, and get to know these big, lovable groups.

Organizing Organic Chemistry Basics

In organic chemistry, the focus is on the element carbon. Carbon is central to all living organisms; however, thousands of nonliving things (such as drugs, plastics, and dyes) are made from carbon compounds. Diamonds are carbon atoms in a crystal structure. Diamonds are so hard because the atoms of carbon are so closely bonded together in the crystal form. That same ability to pack closely together makes carbon an excellent structural element in its other forms as well.

One atom of carbon can combine with up to four other atoms. Therefore, organic compounds usually are large and can have several atoms and molecules bonded together. (Chapter 3 covers atoms, molecules, and bonds.) Organic molecules can be large, and they comprise the structural components of living organisms: carbohydrates, proteins, nucleic acids, and lipids.

In this chapter, I touch on some basic concepts of organic chemistry, and I give you information about the structure and function of the structural molecules: carbohydrates, proteins, nucleic acids, and lipids.

Carbon Is Key

In their outer shells, carbon atoms have four electrons that can bond with other atoms. When carbon is bonded to hydrogen (which is common in organic molecules), the carbon atom shares an electron with hydrogen, and hydrogen likewise shares an electron with carbon. Carbon-hydrogen molecules are referred to as *hydrocarbons.* Nitrogen, sulfur, and oxygen also are often joined to carbon in living organisms.

Long carbon chains = low reactivity

Large molecules form when carbon atoms are joined together in a straight line or in rings. The longer the carbon chain, the less chemically reactive the compound is. However, in biology, other measures of reactivity are used. One example is enzymatic activity, which refers to how much more quickly a certain molecule can allow a reaction to occur.

One key to knowing that a compound is less reactive is that its melting and boiling points are high. Generally, the lower a compound's melting and boiling points, the more reactive it is. For example, the hydrocarbon methane, which is the primary component of natural gas, has just one carbon and four hydrogen atoms (CH_4). Because it is the shortest carbon compound, it has the lowest boiling point (-162°C) and is a gas at room temperature. It is highly reactive.

On the other hand, a compound made of an extremely long carbon chain has a boiling point of 174°C (compared to water, which has a boiling point of 100°C). Because it takes so much more for it to boil, it is much less reactive and is not gaseous at room temperature.

Lucky for you, long, generally unreactive carbon chains make up your body. So, even if you feel like you are getting to your "boiling point," it's just your emotions. You really won't boil. And you won't actually melt out in the sun on a hot summer day.

Forming functional groups based on properties

In organic chemistry, molecules that have similar properties (whether they are chemical or physical properties) are grouped together. The reason they have similar properties is because they have similar groups of atoms; these groups of atoms are called *functional groups.*

Chemical properties involve one substance changing into another substance by reacting. An example of a chemical property is the ability of chlorine gas to react explosively when mixed with sodium. The chemical reaction creates a new substance, sodium chloride (table salt). *Physical properties* refer to different forms of a substance, but the substance remains the same; no chemical reaction or change to a new substance occurs. An example of physical properties is ice being an alternate physical form of H_2O. H_2O can take the form of a solid (ice), liquid (water), or gas (steam), but it is H_2O nonetheless.

Some of the properties that the functional groups provide include polarity and acidity. For example, the functional group called carboxyl (-COOH) is a weak acid. *Polarity* refers to one end of a molecule having a charge (polar), and the other end having no charge (nonpolar). For example, the plasma membrane has hydrophilic heads on the outside that are polar, and the hydrophobic tails (which are nonpolar) form the inside of the plasma membrane. The functional group hydroxyl (-OH) is polar and hydrophilic; methyl groups ($-CH_3$) are hydrophobic.

Loading You Up on Carbohydrates

Carbohydrates, as the name implies, consist of carbon, hydrogen, and oxygen (hydrate = water, hydrogen and oxygen). The basic formula for carbohydrates is CH_2O, meaning that there is one carbon atom, two hydrogen atoms, and one oxygen atom as the ratio in the core structure of a carbohydrate. The formula can be multiplied; for example, glucose has the formula $C_6H_{12}O_6$, which is six times the ratio, but still the same basic formula.

But what *is* a carbohydrate? Carbohydrates are energy-packed compounds. They are broken down by organisms quickly, which gives organisms energy quickly. However, the energy supplied by carbohydrates does not last long. Therefore, stores (reserves) of carbohydrate in the body must be replenished frequently, which is why people get hungry every four hours or so. Although carbohydrates are quickly broken down by organisms, they also serve as structural elements (such as cell walls and cell membranes).

Carbohydrates can be monosaccharides, disaccharides, or polysaccharides (see Figure 4-1). Which type a compound is depends on how many carbon atoms it has. For example, *monosaccharides* are simple sugars consisting of three to seven carbon atoms. *Disaccharides* are two monosaccharide molecules joined together; therefore, they have six to 14 carbon atoms. *Oligosaccharides* have more than two but just a few monosaccharides joined together (*oligo-* means "few"). *Polysaccharides* describe carbohydrates formed by a large number of monosaccharides; polysaccharides are very long chains of smaller carbohydrate molecules linked together. Some of those babies are huge.

Note that most of the names of carbohydrates end in *-ose*. Glucose, fructose, ribose, sucrose, maltose — these are all sugars. A *sugar* is a carbohydrate that dissolves in water (water soluble is the technical term), tastes sweet, and can form crystals. Just like, well, the sugar in your sugar bowl.

Making sense of monosaccharides

The most common monosaccharide is glucose, so that is what I am using here to represent all monosaccharides. (Other monosaccharides include fructose and galactose.) Glucose, with the formula $C_6H_{12}O_6$, is called a *hexose* (a six-carbon sugar) because it has six carbon atoms (*hex* = six). There are also *trioses* (three carbons), *tetroses* (four carbons), and *pentoses*, which have five carbon atoms (*pent-* = five).

Although glucose is a hexose having six carbons, there are other hexoses. Other compounds that have the same number of carbons but different structures are called *isomers*. Glucose has three isomers — that is, three different formations of the same chemical formula. These different isomers connect to form disaccharides. How this happens is explained in the section on disaccharides.

Another name for glucose is dextrose because in chemistry, *dextro-* means right, and glucose is a "right-facing" molecule. (I won't get into why in this book.) Fructose (the sugar in fruit) is also a six-carbon sugar molecule, and it also goes by the name levulose (*levo-* means left) because it is a "left-facing" molecule. The difference between these two isomers is the placement of two bonds. Yet glucose does not need to be digested, and fructose must be converted to glucose before it can be used in the body. Also, compared to regular table sugar, glucose is not as sweet, and fructose is sweeter. Both of those property differences hinge on the structural difference between the two molecules because the chemical formula of both glucose and fructose is exactly the same.

Glucose can be a straight-chain compound, or it can exist as two different ring structures that differ only by the placement of the hydroxyl group on the first carbon (see Figure 4-1). These compounds have the same chemical formula, but may not have the same properties because of the way the atoms are oriented in space.

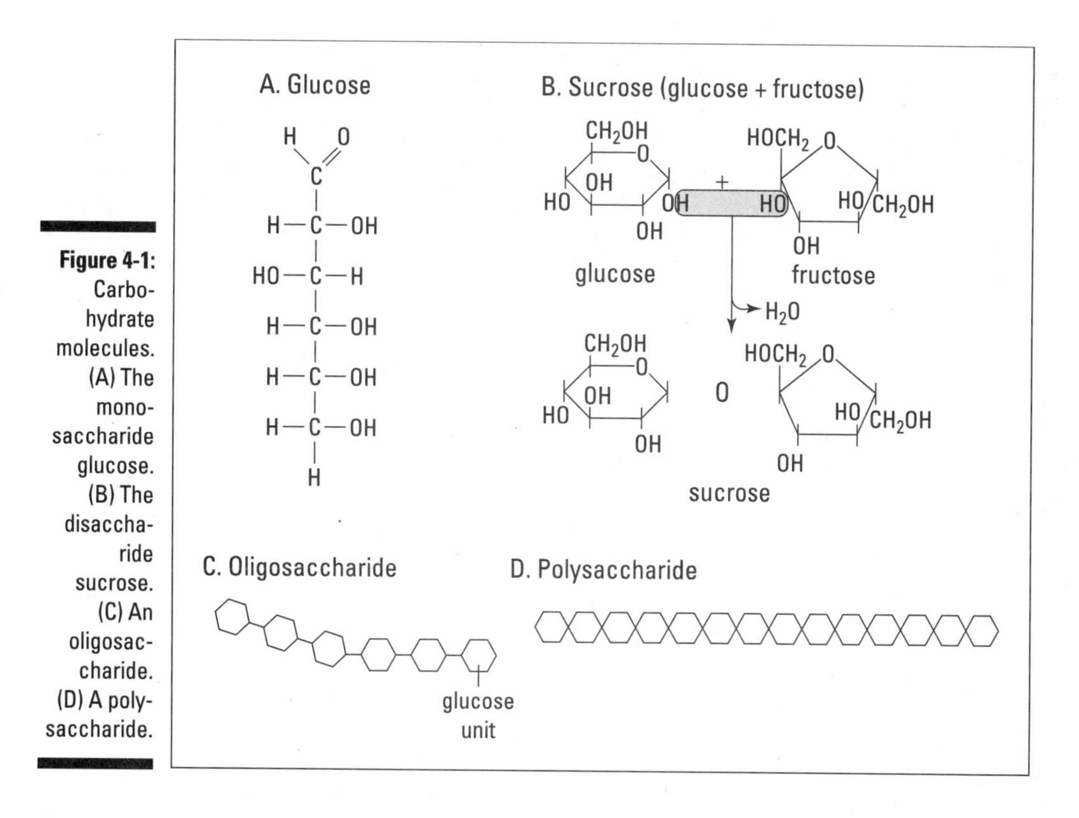

Figure 4-1: Carbohydrate molecules. (A) The monosaccharide glucose. (B) The disaccharide sucrose. (C) An oligosaccharide. (D) A polysaccharide.

Dissecting disaccharides

Disaccharides are carbohydrate molecules formed when two monosaccharides join together. Common disaccharides include sucrose (table sugar), lactose (in milk), and maltose. Sucrose is formed when glucose and fructose join together in a reaction known as a *dehydration synthesis* (also called a *condensation reaction*). Lactose is formed when glucose and galactose join in the same type of reaction, and maltose is two units of glucose joined together.

The term dehydration synthesis may sound technical, but really think about what the words dehydration and synthesis mean. *Dehydration,* as I'm sure you know, is what happens when you don't drink enough water. You dry out because water is removed (but not completely) from some cells, such as your tongue, to make sure that more important cells, such as your blood cells, heart, or brain, continue to function. And, *synthesis* means "making something." If you really think about it, in a dehydration synthesis, something must be made when water is removed. That is exactly what happens. When glucose and fructose get together, a water molecule is given off (removed from the monosaccharides) in the reaction as a *byproduct.*

How atoms are oriented in space

What does "oriented in space" mean? Well, humor me now, and do this: Put your fingers together to make a "tent" with your hands. Then, interlock your fingers so that they form Xs at the second joints in your fingers. Now, slide the fingers on your left hand down to the knuckle joints on your right hand (or vice versa). All ten of your fingers are there in this "finger compound." The "formula" would be two thumbs + two index fingers + two middle fingers + two ring fingers + two pinkies. But, when you slide your fingers, you are changing the positioning of them. You are changing the place or space that they are occupying. So, although there are ten fingers altogether and the "formula" of which fingers (think of your fingers as atoms for now) is unchanged, the way that the fingers exist (are oriented) in space varies. That type of simple difference can make chemicals react differently.

A *hydrolysis* reaction breaks down a disaccharide molecule into its original monosaccharides. When something undergoes hydrolysis, it means that a water molecule splits a compound (*hydro-* = water; *lysis-* = break apart). When sucrose is added to water, it splits apart into glucose and fructose.

Pulling apart polysaccharides

The word *polysaccharides* literally means "many sugars." Polysaccharides are loooooong chains of monosaccharides joined together through the process of dehydration synthesis described in the previous section. When I say these babies are huge, I mean some of them can have thousands of monosaccharide molecules joined together. Starch and glycogen, which serve as means of storing carbohydrates in plants and animals, respectively, are examples of polysaccharides.

Storage Forms of Glucose

Carbohydrates are in nearly every food, not just bread and pasta, which are known for "carbo loading." Fruits, vegetables, and meats also contain carbohydrates. Any food that contains sugar has carbohydrates. And, most foods are converted to sugars when they are digested. So, when carbohydrates from the foods you consume are digested, glucose is the smallest molecule into which a carbohydrate is broken down. Glucose molecules are absorbed from intestinal cells into the bloodstream. The bloodstream then carries the glucose molecules throughout the entire body. Glucose enters each cell of the body and is used by the cell's mitochondrion (organelles of animal cells are discussed in Chapter 2) as fuel. The processes that convert glucose into energy are glycolysis (see Chapter 6), the *Krebs cycle* (see Chapter 8), and oxidative phosphorylation (see Chapter 6).

Once an organism has taken in food, the food is digested, and needed nutrients are sent through the bloodstream. When the organism has used all the nutrients it needs to maintain proper functioning, the remaining nutrients are excreted or stored.

You store it: Glycogen

Animals (including humans) store some glucose in the cells so that it is available for quick shots of energy. Excess glucose is stored in the liver as the large compound called *glycogen*. Glycogen is a polysaccharide of glucose, but its structure allows it to pack compactly, so more of it can be stored in cells for later use. But, you know, if you consume so many extra carbohydrates that your body stores more and more glucose, all your glycogen may be compactly structured, but you no longer will be.

Starch it, please: Storing glucose in plants

The storage form of glucose in plants is *starch*. Starch is a polysaccharide. The leaves of a plant make sugar during the process of *photosynthesis*. Photosynthesis occurs in light (*photo-* = light), such as when the sun is shining. The energy from the sunlight is used to make energy for the plant. So, when plants are making sugar (for fuel, energy) on a sunny day, they store some of it as starch. When the simple sugars need to be retrieved for use, the starch is broken down into its smaller components. They literally save some energy for a rainy day!

A hydrolysis reaction splits sucrose

Disaccharide molecules can be split apart into their monosaccharide molecules by this simple experiment: Mix 1 cup of table sugar into 1 gallon of water until the sugar dissolves. Boil the sugar solution.

By doing so, the sucrose compound has water molecules added to it, and the result is that sucrose splits into glucose and fructose. The chemical equation follows. (The reverse of this equation illustrates the dehydration synthesis that occurs when sucrose is formed.)

$$C_{12}H_{22}O_{11} + H_2O \rightarrow C_6H_{12}O_6 + C_6H_{12}O_6$$

And, did you know that when the starch in a potato is broken down into sugars, that the "sugar" from a potato will actually raise your blood sugar higher and faster than the sugar in ice cream? Now, I'm not telling you to eat ice cream instead of potatoes. Potatoes are a good source of vitamins and minerals, and the skins provide a good dose of fiber. But, hey, ice cream has all that calcium right? Hmmm.

Selling You on Cellulose

Cellulose is a polysaccharide that has a structural role rather than a storage role. In plants, cellulose is the compound that gives rigidity to the cells. The bonds between each cellulose molecule are very strong, which makes cellulose very hard to break down.

Most animals can't digest cellulose because it is so hard to break down. Animals that eat only plants (herbivores) have special sacs in their digestive system to help break down cellulose. (I tell you all about it in Chapter 5.)

Humans can't digest cellulose either. (The proof is in the toilet the day after you eat corn, for example.) Because cellulose passes through your digestive tract virtually untouched, it helps maintain the health of your intestines. One way cellulose helps the intestines is that it clears materials from the intestinal walls, keeping them clear, which may help to prevent colon cancer. Cellulose is the *fiber* (or *roughage*) of which your cereal box says you need more.

Animals have only membranes surrounding their cells. Plants have walls surrounding theirs. Cell walls contain cellulose, and cellulose with its rigid structure gives "crunch" to vegetables when you cut or bite into them. Think celery.

There are some other structural forms of carbohydrates, but usually they are combined with proteins.

Loads of cellulose

Because there are so many plants in the world (think of all the flowers, trees, weeds, grasses, vines, and bushes), cellulose, which is found in every cell of every plant, is the most abundant organic compound on earth.

Building Organisms: Proteins

Without proteins, living things would not exist. Proteins are involved in every aspect of every living thing. Many provide structure to cells; others bind to and carry important molecules throughout the body. Some proteins are involved in reactions in the body when they serve as enzymes. Still others are involved in muscle contraction or immune responses. Proteins are so diverse that I cannot possibly tell you about all of them. This section gives you the basics on their structure and most important functions.

Amino acid chains link information to build proteins

All proteins are made up of *amino acids* (Figure 4-2). Think of amino acids as train cars that make up an entire train called a protein. Proteins are formed by amino acids, which are produced based on the genetic information in a cell. Then, the amino acids that are created in the cell are linked together in a certain order. Each protein is made up of a unique number and order of amino acids. The protein that is created has a specific job to do or a specific tissue (such as muscle tissue) to create.

The structure of amino acids is fairly simple. Each amino acid has an amino group at its core with a carboxyl group and a side chain attached. The side chain (a chemical compound) that is attached determines which amino acid it is (see Figure 4-2).

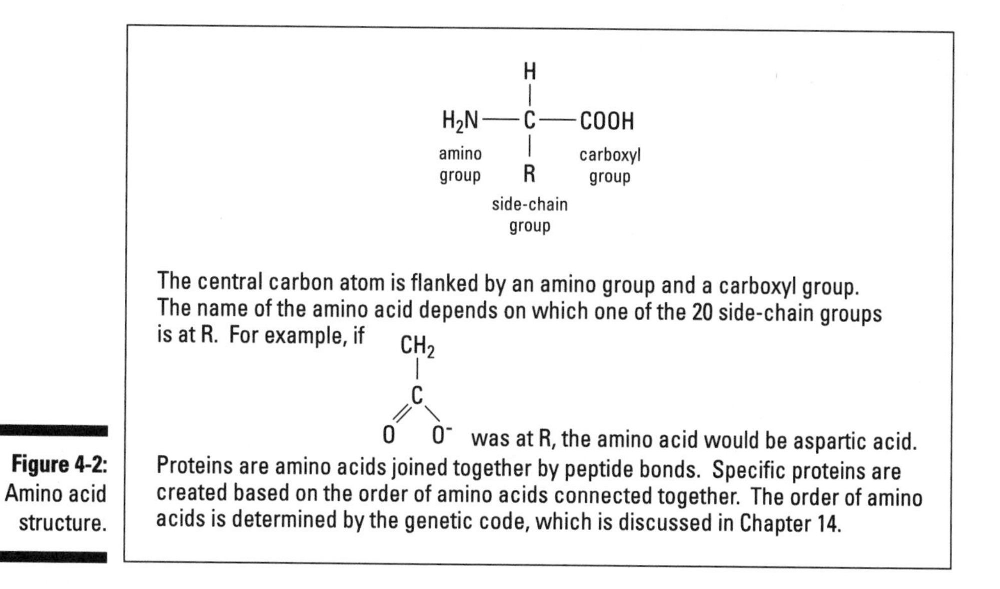

The central carbon atom is flanked by an amino group and a carboxyl group. The name of the amino acid depends on which one of the 20 side-chain groups is at R. For example, if CH_2–C(O)(O^-) was at R, the amino acid would be aspartic acid. Proteins are amino acids joined together by peptide bonds. Specific proteins are created based on the order of amino acids connected together. The order of amino acids is determined by the genetic code, which is discussed in Chapter 14.

Figure 4-2: Amino acid structure.

Speeding up reactions: It's enzymatic, not automatic

An *enzyme* is a protein used to speed up the rate of a chemical reaction. Because they regulate the rate of chemical reactions, they also are called *catalysts*. There are many, many, many different types of enzymes, because for each chemical reaction that occurs, an enzyme specific to that reaction must be made. Metabolic processes don't just automatically happen; they need enzymes. And, a reason that you must consume protein is so that you can make more enzymes so that your processes will occur. Although you won't see every reaction that occurs in the body in this book, you should get a feel for the fact that chemical reactions control the metabolism and life of living things (which explains why a chapter on chemistry appears in a biology book!).

Proteins are long chains of polypeptides, and thus, so are enzymes. However, some enzymes contain parts that are not made up of proteins but assist the enzyme in its function. These are called *coenzymes*. Vitamins often act as coenzymes. The name of an enzyme usually reflects the name of the chemical on which the enzyme acts (that is, the chemical substrate). For example, an enzyme that acts on a fat (fat being the substrate) is called a lipase (remember, *lip-* = fat).

To act on a substrate, an enzyme must contain an *active site*. The active site is the area on the enzyme that allows the substrate and enzyme to fit together (like puzzle pieces). The way that enzymes and substrates fit together is often compared to the way a key fits a lock; the way enzymes kick-start reactions often is referred to as the *lock-and-key model*. Once the substrate and enzyme are connected (the whole thing is now called an *enzyme-substrate complex*), the enzyme can get to work.

So, what is it that enzymes do? Well, when your car is just sitting in your garage, it contains all the necessary things to make it move — motor, gasoline, oil, tires, and so on — yet it still just sits there until a specific action occurs. When you turn the ignition (with the properly fitting key), a spark from a spark plug ignites the fuel, allowing the car to move. An enzyme is akin to the key. Once the "key-ignition complex" is formed, the reaction (sparking the fuel) is able to occur so that the desired end result is achieved (fuel flowing through the engine). In the end, the spark plug loses a spark, and some fuel is used up, but the key remains the same. This is how enzymes work.

During an enzymatic reaction, the substrate is changed during the reaction, and new products are formed during the reaction, but the enzyme comes out of the whole thing unchanged. Then, the enzyme leaves the reaction to form a complex with a different substrate and catalyze another reaction. The products of the reaction continue on in their pathway. Enzymes are able to catalyze reaction after reaction millions of times before they start to wear out. Then, the body creates more enzymes by synthesizing the proper protein chains from the correct amino acids. And, the cycles continue. That's life.

But enzymes can be affected by certain conditions. Well, the enzyme itself isn't as affected as much as the rate at which the enzyme works. If changes occur in the temperature of the cell, concentrations of enzyme or substrate in the cell, or pH of the cell, then the reaction may slow down or speed up. If an organism is too cold, enzymes barely work at all. If the organism becomes way too hot, enzymes can be *denatured*, which means they are permanently changed and are unable to catalyze any more reactions.

In humans, enzymes work best at — surprise! — normal body temperature — 98.6°F. If a person becomes hypothermic (like if he or she falls into a frozen pond and is not rescued for a while), then the metabolic processes start to shut down. Why? Because the enzymes say, "Sorry, we can't work under these conditions!" The enzymes kind of go on strike. The same thing happens in people with hyperthermia (body temperature is way too high), except then the enzymes may quit completely (and are not even tempted by the promise of better pay and benefits).

I include more information about enzymes in Chapter 10.

What I learned about collagen in college

Collagen is the most abundant protein in animals with a backbone (that is, vertebrate animals). It is a fibrous (structural) protein that is found in *connective tissue*, which is all the tissue that joins muscles to bones to allow movement and forms skin that protects the muscle tissue. Connective tissue includes ligaments, tendons, cartilage, bone tissue, and even the cornea of the eye. It provides support in the body, and it has a great capability to be flexible and resistant to stretching (to a point, as I'm sure those of you who have torn cartilage or popped a ligament are well aware).

The lower layer of the skin (called the *dermis*) largely is made up of collagen. When skin is removed from an animal (think about removing the skin from a chicken breast), the collagen allows the skin to be pulled away without tearing the muscle tissue underneath. Collagen (and other fibrous proteins) is arranged in long polypeptide chains that form sheets. If you look at collagen fibers under an electron microscope, it almost looks like those metal bars with a spiral design used in construction to hold concrete walls together.

Just as those metal bars provide support and strength to buildings, collagen does the same between cells and in tissues in your body. And you are loaded with collagen. Between 25 percent and 33 percent of your body weight is comprised of collagen.

Around the hemoglobin (in less than 80 days)

Hemoglobin is an example of the other major type of proteins: globular proteins. *Globular proteins* serve a larger variety of functions than the fibrous proteins. For example, the globular proteins include such useful proteins as enzymes, antibodies, and transport proteins.

These proteins, as their name implies, are globular. And, globs would seem to be more flexible (squishier) than fibers, don't you think? Well, you're right. Many globular proteins can change their shape to fit into very small areas (like an antibody would have to do to go after a virus), cross cell membranes (as a transport protein would have to do), and be involved at the cellular level in chemical reactions (as an enzyme would be).

Hemoglobin is a transport protein found in red blood cells: It carries oxygen around the body. Here's a little information about its structure so you can understand how it performs its function. A hemoglobin molecule is shaped kind of like a 3-D four-leaf clover without a stem. Each leaf of the clover represents a certain chain of protein. In the center of the clover, but touching each protein chain, lies a heme group. At the very center of a heme group is an atom of iron.

When gas exchange occurs between the lungs and a blood cell (you can find details about this in Chapters 5 and 7), it is the iron that binds (attaches to) the oxygen. Then, the iron-oxygen complex releases from the hemoglobin molecule in the red blood cell so that the oxygen can cross cell membranes and get inside any cell of the body. However, the atom of iron and the hemoglobin are not used just once. The iron and hemoglobin usually carry carbon dioxide back to the lungs and deposit the CO_2 there so it can be exhaled (see Chapter 7). The hemoglobin remains in the same red blood cell for its entire "life." When the red blood cell it calls home is ready to die, the iron either is recycled and gets picked up by another red blood cell to be incorporated into another hemoglobin molecule, or it is excreted as cellular waste and gives color to your feces. That's another story you have to go to Chapter 9 to read.

And, being globular, hemoglobin molecules do indeed change shape. In people with *sickle cell anemia*, which is an inherited disorder, the red blood cells become shaped like a sickle (that is, thin and curved) instead of round like normal red blood cells. The change in shape occurs after defective hemoglobin molecules lose their oxygen (but not their iron) and polymerize — that is, join together with other molecules to make a larger molecule. Because people with this disorder have red blood cells that are shaped differently, their blood cells cannot pass through their blood vessels very easily. Extreme pain occurs when blood and oxygen cannot get to certain areas of the body, and also as a result, tissues are damaged or die.

Sickle cell anemia is caused by a single mutated amino acid in the protein chains that make up hemoglobin. One amino acid! And, you know what causes that amino acid change? A gene. Man, the power is in the genes! But genes are made up of nucleotides in the nucleic acids, so the power really is at an even more minute level. And those nucleotides affect everything in life (see Chapters 11 and 14).

Links of Life: Nucleic Acids

Until as recently as the 1940s, scientists thought that genetic information was carried in the proteins of the body. They thought nucleic acids, which then were a new discovery, were too small to be significant. Then, James Watson and Francis Crick figured out the structure of a nucleic acid, and scientists eventually found things were the other way around: Nucleic acids created the proteins!

Nucleic acids are large molecules that carry tons of small details: all the genetic information. Nucleic acids are found in every living thing — plants, animals, bacteria, viruses, fungi — that uses and converts energy. Just think about that fact for a moment. It really is awesome when you realize that every single living thing has something in common. People, animals, plants, and more all are connected by genetic material. Every living thing may look different and act different, but deep down — and I mean way deep down in the nucleus of cells — living things contain the same chemical "ingredients" making up very similar genetic material.

There are two types of nucleic acids: *DNA* (which stands for deoxyribonucleic acid) and *RNA* (which stands for ribonucleic acid). Nucleic acids are made up of strands of *nucleotides*, which are made up of a base containing nitrogen (called a *nitrogenous base*), a sugar that contains five-carbon molecules, and a phosphoric acid. That's it. Your entire genetic composition, personality, maybe even intelligence hinges on molecules containing a nitrogen compound, some sugar, and an acid.

The *nitrogenous bases* are molecules either called *purines* or *pyrimidines.*

Purines include

- Adenine
- Guanine

Pyrimidines include

- Cytosine
- Thymine (in DNA)
- Uracil (in RNA)

Deoxyribonucleic acid (DNA)

DNA contains two strands of nucleotides arranged in a way that makes it look like a twisted ladder (called a *double helix*). Figure 4-3 shows you what a DNA molecule looks like.

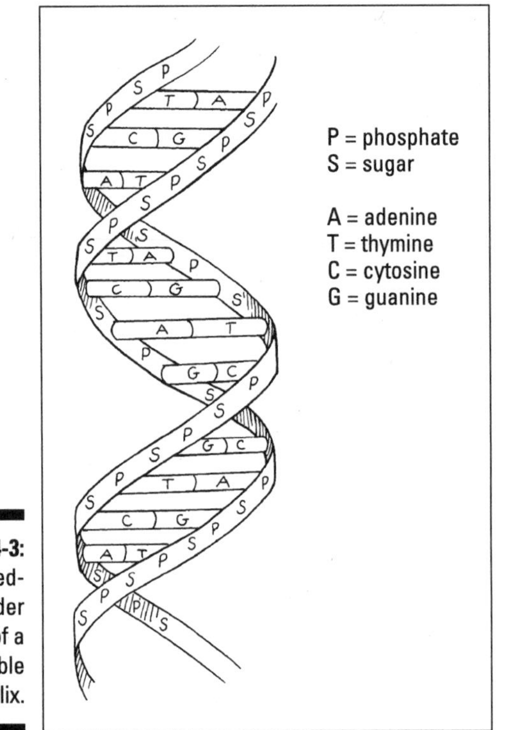

Figure 4-3: The twisted-ladder model of a DNA double helix.

The nitrogenous bases that DNA builds its double-helix upon are adenine (A), guanine (G), cytosine (C), and thymine (T). The sugar that is in the composition of DNA is 2-deoxyribose.

If you refer to Figure 4-3, you can see that adenine is always paired with thymine (A-T), and guanine is always paired with cytosine (G-C). These bases are held together by hydrogen bonds, which form the "rungs" of the "twisted ladder." The sides of the ladder are made up of the sugar and phosphate molecules.

Certain sections of nitrogenous bases along the strand of DNA form a *gene*. A gene is a unit that contains the genetic information or codes for a particular product and transmits hereditary information to the next generation. But genes are not found only in reproductive cells. Every cell in an organism contains DNA (and therefore genes) because DNA also codes for the proteins that the organism produces. And proteins control cell function and provide structure. So, the basis of life happens in each and every cell. Whenever a new cell is made in an organism, the genetic material is reproduced and put into the new cell. (You can find details about this in Chapter 11.) The new cell can then create proteins within itself and also pass on the genetic information to the next new cell.

The order of the nitrogenous bases on a strand of DNA (or in a section of the DNA that comprises a gene) determines which amino acid is produced. And the order that amino acids are strung together determines which protein is produced. Which protein is produced determines what structural element is produced within your body (such as muscle tissue, skin, or hair) or what function can be performed (such as if hemoglobin is being produced to transport oxygen to all the cells).

Every cellular process and every aspect of metabolism is based on genetic information and the production of the proper proteins. If the wrong protein is produced (such as in sickle cell disease), then disease occurs. (I give you examples in Chapter 14.)

Ribonucleic acid (RNA)

The nitrogenous bases that RNA uses are adenine, guanine, cytosine, and uracil (instead of thymine). And, the sugar in RNA is ribose (instead of 2-deoxyribose). Those are the major differences between DNA and RNA.

In most animals, RNA is not the major genetic material. Many viruses — such as the human immunodeficiency virus (HIV) that causes AIDS — contain RNA as their genetic material. However, in animals, RNA works along with DNA to produce the proteins needed throughout the body.

For example, RNA has three major subtypes: messenger RNA (mRNA), transfer RNA (tRNA), and ribosomal RNA (rRNA). All three of those subtypes are involved in protein synthesis. (Get the details in Chapter 14.)

Fats Aren't Forbidden: The Truth About Lipids

In addition to carbohydrates, proteins, and nucleic acids, you need one more type of large molecule to survive. Yet, if you are like most people, you try to avoid it in your diet. Fats. Or, if you call them by their technical name — lipids.

Lipids are nonpolar molecules, which means their ends are not charged (like a water molecule is — you know how water can conduct electricity?). Because they are nonpolar and water is polar, lipids are not soluble in water. That means the lipid molecules and water molecules do not bond or share electrons in any way. The lipids just float in the water without blending into it. I'm sure you've heard the old adage, "oil and water don't mix." Well, oil is a liquid lipid. Butter and lard are examples of solid lipids.

Three major types of lipid molecules exist.

- **Phospholipids:** These lipids are made up of two fatty acids and a phosphate group. These are the type of lipids used in the cell membranes of every cell in every animal. These lipids have structural functions. They aren't the type that are floating around bloodstreams and clogging arteries.
- **Steroids:** These lipids have four connecting carbon rings and a functional group that determines which steroid molecule it is. These lipid compounds generally create hormones. Cholesterol is a steroid molecule that is used to create hormones, such as testosterone and estrogen. (See Chapter 10 for more details.) So, yes, for a healthy sex life and other important functions, you need cholesterol. But, cholesterol is transported around the body by other lipids. If you have too much cholesterol floating in your bloodstream, that means there is an excess of fats carrying it through your bloodstream, and that can cause trouble. The fats and cholesterol molecules can get stuck in your blood vessels, leading to blockages that cause heart attacks or strokes. I go into this in more detail in Chapter 5.
- **Triglycerides:** Triglycerides are made up of three (tri-) fatty acid molecules and a glycerol molecule. These are the typical fat molecules. They are formed from an excess of glucose; after the liver stores all the glucose it can as glycogen, whatever is remaining is turned into triglycerides. The triglycerides float through your bloodstream on their way to be deposited into adipose tissue. Adipose tissue is the soft, squishy, visible fat you can see on your body. Adipose tissue is made up of many, many molecules of fat. And, the more fat molecules that are added to the adipose tissue, the bigger the adipose tissue (and the place on your body that contains it) gets.

When you use up all your stored glucose (that doesn't take long, sugars "burn" quickly in aerobic conditions), your body starts breaking down glycogen, which primarily is stored in liver and muscle. Liver glycogen stores can typically last 12 or more hours. After that, your body starts breaking down adipose tissue to retrieve some stored energy. That is why aerobic exercise (and enough exercise to use up more calories than you took in that day) is the best way to lose fat. Notice I didn't say "pounds" here. Pounds measure everything in your body's composition: fat tissue, muscle tissue, and bone, along with the water, your organs, skin, and some incidental stuff.

Can you turn fat into muscle? No!

People do not "turn fat into muscle." That saying is such a myth, or more likely a misunderstanding on the part of the people who say it. Fat and muscle, as you now know, are two different types of tissues. Fatty (adipose) tissue is made from lipid molecules; muscle tissue is made from proteins. When you "lose weight," you do not know for sure whether you are losing proteins from muscle tissue or bones, or if you are losing fatty tissue, or, for short-term weight loss, if you are just losing water. You may have an idea based on the way your clothes fit or the way you feel. But muscle tissue is much more dense than adipose tissue; it weighs more than fat. So, when you begin an exercise program, you could theoretically be losing some fat, but building muscle tissue (which incidentally burns fat, too) and not seeing a budge on the scale (or be frustrated by an increase!). Overall, the change that is occurring is a healthy thing, but it takes time for your body to adjust. After a while, you notice your number of pounds going down, once the new muscle tissue starts burning fat better, the addition of new muscle tissue slows down, and the loss of adipose tissue continues.

Lifting weights (or doing any anaerobic exercise) to "lose weight" may help build some muscle tissue, which will help to burn some fats, but without aerobic exercise, the process of using up stored body fat is a long one. If you decrease the amount of calories you take in, and you perform aerobic exercise to burn off excess glucose (so that it is not stored as fat) plus use up fat that is already stored, and you include some weight training to build muscle tissue to help use fats even more efficiently, you will see results. Be patient.

But, more importantly, make sure that your goal is realistic and healthy. For example, an insurance company weight chart says that I should weigh about 135 pounds (I am 5'7"). However, I recently went to a nutritionist who determined my body mass through *electrical impedance*. During this very quick procedure (less than 5 minutes), you lie down on a table, have a few electrodes hooked at various spots on your body (the foot is one place I remember), and then a very small electrical current is sent throughout your body (so small that you don't feel a thing). The current "bumps" into your body parts and tissues (actually your body parts *impede* the flow of the electrical current, which gives the name of the procedure), and information is retrieved as to how many pounds of bone and muscle you contain. In my body, there are 133 pounds of bone and muscle (I am dense, in the physical sense, anyway), so a weight of 135 is pretty unrealistic for me. What a relief to know that I'll never be able to weigh what I did in high school or college; I had been shooting for an unattainable goal! Now (after having three babies and actually growing a few inches during adulthood), my body is different, and I can aim for a more realistic weight that is healthy for me at *this* point in my life — 165 pounds. Now *that's* a load off!

Don't forget. The ultimate advice if you are putting together all this information about nutrition and what does what in your body, and planning how you can improve your health is to check with your doctor first before you start any program.

Combinations of Three Macromolecules: Blood Group Antigens

Blood group antigens are carbohydrates that are attached to proteins or lipids. An *antigen* is a substance foreign to the body that causes an immune response. An *immune response* occurs when *antibodies*, which are proteins in your immune system, are summoned to attack an antigen.

When you say you are *blood type A*, what you are telling people is that the cells in your body make antibodies only to type B antigens. The A-type surface antigens on the cells are not recognized. These surface antigens can be attached to the surface of your blood cells (more specifically to the plasma membrane surrounding the cells) or to proteins or lipids anywhere in your body. That means that your body makes antibodies against type B antigens. (If your blood type is positive or negative, that refers to the Rh factor.) So, in essence, your body kills off the cells containing type B antigens, allowing type A to be dominant. You can receive type A blood or type O blood and can donate blood to those with type A or type AB.

If you are *blood type B*, the situation is reversed. Your cells have type B antigens attached, so your body makes antibodies against only type A. Once the type A antigens are kept at bay, your blood cells "show" type B as the dominant type. You can receive type B or type O blood, and you can donate to those with type B or type AB blood.

If you are *blood type AB*, your cells do not make antibodies against type A or type B surface antigens. Therefore, you can receive blood from a donor with any blood type (*universal recipient*), but you can donate blood only to other people with type AB blood.

If you are *blood type O*, your cells make antibodies against both type A and type B antigens. This means that if you need blood, you can only receive more type O blood. But, you can donate your blood to anybody; thus, you are a *universal donor.*

Please donate blood

Type O blood is the most common. In the United States, 49 percent of African-Americans have type O blood, 45 percent of Caucasians, 40 percent of Asians, and 79 percent of Native Americans. So, most of you out there have type O blood; please share it with the rest of us. Blood supplies are low. You are the universal donors — donate you should!!

Part II
Living Things Need Energy

In this part . . .

Living things are like little engines. Engines need fuel, burn fuel, and produce wastes. So do living things. It's one of the important factors that determine whether an organism is alive or not. This part of the book shows you how plants and animals get the fuel they need to stay alive. Then, you take a good look at what they do with it.

In these chapters, you find out all about respiration and how it differs from actual breathing; why your body needs carbohydrates, proteins, and fats; and how plants create their own fuel from minerals, water, and sunlight. You look at the structure of plants and see how they move water up their stems to get their nutrients to their cells. And you come to understand what is actually going on in your gut to get out what little nutrients exist in that snack you're munching on.

Chapter 5

Acquiring Energy to Fill the Tank

In This Chapter

- Finding out how animals get their food
- Looking at food chains
- Taking a look at how animals mechanically break apart food
- Getting proper nutrition
- Deciphering chemical digestion
- Figuring out how breathing works

Just as you need to put gas in your car's engine and air in your tires so that your car can move, you need to put food and oxygen into your body so that it runs. And you are not alone. Every human, as well as every other living organism, needs to "fill their tanks." Within the organism's body, food is broken down into nutrients. Nutrients are converted to energy, which sustains life. Animals ingest their food and breathe in their oxygen; plants absorb their food and have a special way of "breathing." This chapter explains how.

How Animals Get Nutrients and Oxygen

Humans may think that they only have to drive to the supermarket, pull up to a drive-in window, or stand at the front door and wait for a delivery person to "get food." That may be true in the literal sense, but in biology terms, acquiring nutrients is a biochemical process, as is breathing.

The hunt for food

The biochemical process all starts with a signal from your empty stomach. That signal to your brain starts you on the search for food. Millions of years ago, early humans would go hunt animals for meat or gather nuts, fruits, and berries. They would walk every day on a search for sustenance, much like herds of animals do. They would graze most days and feast when they had killed an animal for meat. Once nomadic tribes began settling in one place, hunting continued, and farming was born. People started growing their own food, which required tilling, hoeing, planting, digging — in other words, work. People put energy into acquiring food for energy.

Now, conveniences have greatly minimized the energy people put into getting their food. However, many humans are taking in more food than they need, which has some devastating results. People are no longer nomadic. They do not wander in search of food. They do not stalk and attack animals on a regular basis (some perhaps just once a year or so, and not without a license). Other animals (with the exception of dogs, cats, and other pets who have also come to expect convenience) still work at getting their food.

Heterotrophs on the hunt

Heterotrophs are animals that essentially feed on other living organisms. Heterotrophic organisms are the opposite of *autotrophic organisms*, which can use simple inorganic substances and sunlight to make the organic compounds that they need to survive. Plants are an example of autotrophs.

Heterotrophs cannot make their own organic compounds. They must obtain organic compounds from other living things that contain organic compounds. There are three classes of organisms that do this:

- *Herbivores* consume only plants and get their organic compounds from the plants. Examples of these animals include deer, cows, and other grazing herd animals.
- *Carnivores* eat only other animals. The animals that they eat have already eaten plants, so the carnivores get their organic compounds from the animal tissue and the digested plant material inside those unfortunate animals. Examples include lions and tigers.
- *Omnivores* eat anything. These animals (including you, me, and all other humans) consume plants and other animals. Vegetarians who consume only plant-based foods also need proteins found only in animal tissues. Other omnivores include bears, which eat plant-based materials, as well as fish or smaller animals.

You are part of the food chain (luckily, you're near the top)

Food chains provide a visual example of how energy is transferred throughout the universe. I say "universe" here rather than world, because the sun is involved. The sun is the starting point of energy in food chains because the sun provides energy that is used by plants when they make food for themselves (remember, they are autotrophic). However, plants not only provide energy for themselves, but also for some heterotrophs. Thus, a food chain begins.

In a simple food chain, a *producer* makes the "food" that provides the energy, and a *consumer* uses it. For example, when herbivores and omnivores consume plants, they acquire the food that was produced by the plants, and that the plants had acquired energy from the sun. Sounds kind of like "The House that Jack Built," eh?

In a more complex food chain, several producers of energy may be in the chain, as well as several levels of consumers. When an omnivore or carnivore consumes another omnivore or carnivore, the energy inside of the prey (the animal that was eaten) is obtained from smaller omnivores, carnivores, or herbivores and passed to the predator (the animal that did the eating).

However, the consumer does not acquire all of the energy in the food that the producer made. When food is digested by a producer (say, you), some of the material you consumed is converted to energy used within your body. The excess is excreted as feces (go ahead, say "eeww"). Excreted waste is not lost energy; the energy is just in another form that is useable by different organisms (like bacteria, earthworms, dung beetles, and so on.). But, it is not useable by the next higher level in the food chain. In fact, the longer a food chain is, the less energy that the higher consumers actually acquire, which is why food chains aren't all that long to begin with. You see, all reactions are not 100 percent efficient, and much energy is lost as heat, too.

Suppose that you are on a safari in Africa. You and your fellow travelers roast an antelope for dinner over a campfire. You go to sleep with a belly full of antelope meat, but awake in the middle of the night to the sounds of a lion consuming one of your fellow campers. (Go ahead, say "eeww" again.) Prior to going to sleep, the unfortunate camper used the tent's toilet facilities to excrete feces (and maybe do a crossword puzzle). Some of the energy that he consumed, as well as some of the energy he produced, was lost from his body as feces; some also was lost through heat and sweat (probably when he saw the lion coming). So, when the lion consumed him, the lion did not get the full amount of energy that was in his system before his untimely demise. However, the energy in the feces, as well as the energy he lost through heat and perspiration, is not lost to the universe. The energy remains in the universe; it just isn't harnessed by the lion. Instead, the feces (and the dead camper's remains) provide energy for the microscopic organisms that decompose it, and the rotted organic material gets into the earth, which then gives energy back to plants. And, the cycle continues; the circle of life goes on.

Food chains are discussed again in the part of the book about ecology and ecosystems (Chapter 17).

Sample food chains

Food chains can be simple, with just one producer and consumer, or they can be complex. However, food chains usually do not contain more than four or five "links" in the chain. If they did, energy sources would become too depleted to sustain the higher organisms. Sunlight is the primary provider of energy. Plants are primary converters of this energy to organic food, and herbivores usually are the primary consumers. The more complex food chains have two or more consumers.

Simple Food Chain:

Sunlight ---> Grasses (producers) ----> Cows (consumer)

Complex Food Chain:

Sunlight ---> Tree (producer) ----> Giraffe (consumer 1 eats the leaves from the tree) ----> Lion (consumer 2 eats the giraffe) ---> Hyena (consumer 3 eats the lion when it is old and dying)

Breaking Down the Breaking Down of Food

Different groups of animals eat different foods, but what do animals physically do with the food they take in? The different methods all make up *mechanical digestion*, which occurs from the time the animal consumes the food until it enters the stomach. This is opposed to *chemical digestion*, which is what happens once chewed or ground-up food is flooded with enzymes and acids to break it down further.

Ruminating over ruminants

Ruminants are mammals that can break down cellulose. Humans have one stomach that fills with hydrochloric acid and enzymes to help break down food. Ruminants, such as cows, have a stomach with several compartments. Cows are used as the model here to help you understand how ruminants break down food.

Cows, like all herbivores, eat grasses and other plant material. Plants contain cellulose, which is very hard to digest (even for an herbivore). So, when a cow swallows some grass, the chewed grass first enters the compartment of the stomach called the *rumen*. The rumen contains a salty solution and bacteria that helps to break down the cellulose. Cows then regurgitate (spit up) the material from the rumen, called *cud*, back into their mouths. They "chew their cud" to help break down the cellulose even further. The cud is swallowed again, and it re-enters the rumen. This cycle repeats as necessary until the material is broken down far enough to be churned up and passed into the true stomach. From there, digestion continues through the intestines and excretory system of the cow. And, I'm sure you've experienced the end result of the entire process — methane gas and cow manure!

The daily grind of a gizzard

Gizzards are a sac-like structure in animals that help to break down food. Many animals have teeth to help tear apart or chew food. But some animals don't chew. Do chickens have lips? No. Do ducks have teeth? No. Birds generally swallow things whole, including stones. Yes, stones. The stones that are swallowed end up in the gizzard to help grind the other food that the bird eats. The walls of the gizzard are extremely muscular, and when the walls of the gizzard rub back and forth, the contents of the gizzard get ground up. Earthworms also have gizzards.

The truth about teeth

Not all teeth are for chewing, but all teeth are for maceration. *Maceration* is the action of physically breaking down food into pieces. Chewing is a grinding action that only herbivores and omnivores do.

A lion's huge dagger-like teeth are intended for killing an animal and tearing its flesh. The lion then swallows the chunks of meat whole. No chewing involved. Because lions often don't chew, they do not have many grinding teeth (like molars). Herbivores, on the other hand, have many grinding teeth, which are flat, as well as incisors (think scissors), which can clip grasses and plants. Humans and other omnivores have a combination of these kinds of teeth: canine teeth for tearing food, incisors for biting off pieces, and molars for grinding up food.

What Is There to Eat? Nutrition and Wise Choices

It's a shame that humans have such sensitive taste buds. If humans didn't have such an evolved sense of taste, maybe they would be like other animals and just eat what is part of their natural diet and eat it only when they are truly hungry. But, alas, food tastes good, and humans often are tempted to put really cheap fuel in their systems. Would you do that to your car on a regular basis? Or, do you use the premium stuff to make sure your car's engine doesn't knock and ping?

If you want to keep your bodily systems from knocking and pinging, follow the recommendations made by the United States Department of Agriculture (USDA) in the Food Pyramid in Figure 5-1. The Food Pyramid serves as a way of visualizing the proportion of items from different food groups that should make up your diet. Table 5-1 gives you an idea of what exactly one serving of some different foods is considered to be.

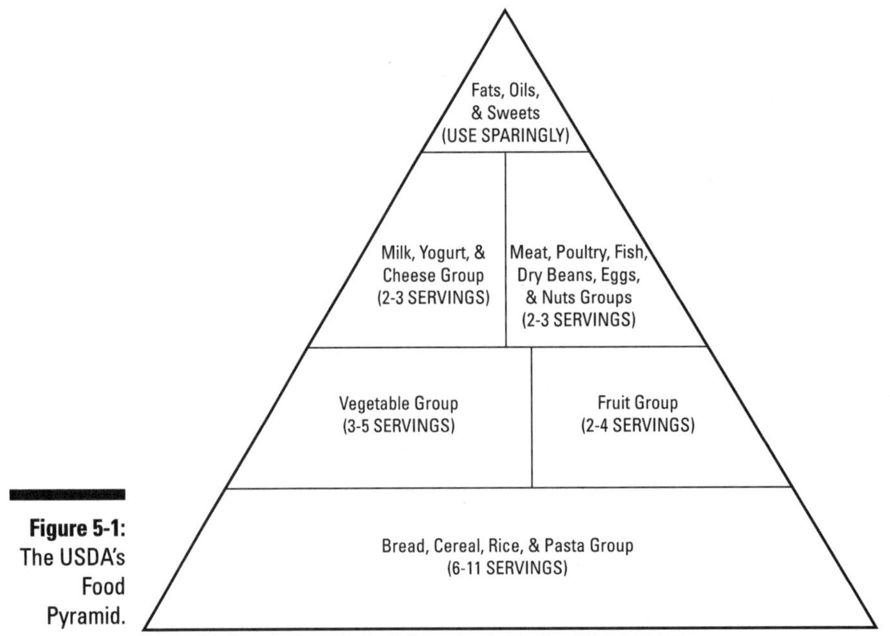

Figure 5-1: The USDA's Food Pyramid.

Source: U.S. Department of Agriculture/U.S. Department of Health and Human Services

Table 5-1 What's a Serving, Anyway?

Food Group (from Food Pyramid)	*Recommended Serving Size*
Bread	1 slice 5 to 6 small crackers
Cereal	1 ounce ready-to-eat cereal ½ cup cooked cereal
Rice, pasta	½ cup cooked rice or pasta
Vegetables	1 cup raw leafy vegetables ½ cup chopped raw vegetables ½ cup cooked, chopped vegetables ¼ cup vegetable juice

Food Group (from Food Pyramid)	Recommended Serving Size
Fruits	1 medium piece of fresh fruit (apple, banana, orange, peach) ½ cup cooked or canned, chopped fruit ½ cup fruit juice
Milk products	1 cup milk 1 cup yogurt ½ ounce natural cheese 2 ounces processed cheese
Meat	2 to 3 ounces cooked lean meat
Fish	2 to 3 ounces cooked fish
Poultry	2 to 3 ounces cooked lean poultry
Dry beans	½ cup cooked dry beans
Eggs	1 egg
Nuts, seeds	2 tablespoons peanut butter
Fats, oils, sweets	No specific amount; use very little

Source: *The Food Guide Pyramid (Washington, D.C.: International Food Information Council Foundation, U.S. Department of Agriculture, Food Marketing Institute, 1995).*

Counting and cutting calories

There may be four food groups, but all foods consist of three main nutrient groups: carbohydrates, proteins, and fats. Each of these is converted to usable energy, which is measured in *calories*. Calories basically are a unit of measuring heat energy. It takes 1 calorie to raise the temperature of 1 gram of water 1 degree Celsius (opposed to Fahrenheit). The calories that you count, however, are really called *kilocalories*. *Kilo-* means 1,000, so kilocalories are larger units than calories. It takes 1 kilocalorie to raise the temperature of 1 kilogram (1,000 g) of water 1 degree Celsius. The energy of a kilocalorie is used to measure the amount of energy that can be provided by foods. From here on out, though, I use the term calorie instead of kilocalorie in keeping with what you are used to reading on food labels.

Each gram of carbohydrate contributes 4 calories, as does each gram of protein. But each gram of fat is more than twice as calorie dense: 1 gram of fat contributes 9 calories. It seems pretty obvious that if you eat fewer grams of fat, you take in fewer calories. But, for overall health, you need carbohydrates, proteins, and fats in the right proportions. The guideline is 60 percent of your calories from carbohydrates, 30 percent of your calories from fats, and 10 percent of your calories from proteins.

Finding calorie and nutrient info

Plenty of charts and books are available that list the calorie and nutrient content of specific foods. Check out *Nutrition For Dummies*, 2nd Edition, by Carol Ann Rinzler (Hungry Minds, Inc.), which explains further why you need the major nutrients and then gives you several pages of nutrient data for common foods at the end of the book. The American Diabetes Association makes a chart readily available. Even fast-food restaurants usually have a poster of nutrition information hanging up by their counters these days.

If you want to start making sure that the fuel you put into your body is premium, ask your doctor, nutritionist, or even pharmacist if he or she has a listing of calorie and nutrient content. Or, if you have easier access to a computer than to a health-care professional, you can search through the USDA's Nutrient Database at `www.nal.usda.gov/fnic/cgi-bin/nut_search.pl`. Got that?

The optimal number of calories you need to maintain a healthy weight is based on how much energy you need when you are resting and how much energy you use when you are working. This amount varies for people of different sexes, ages, activity levels, and weights. You can determine the amount of calories you need just to cover the basics of what your body does all day long (breathing, heart beating, digesting, cellular processes, etc.). This amount is called the *basal metabolic rate* (BMR). To determine your BMR, divide your weight (in pounds) by 2.2 to convert it to kilograms. Then, divide your weight (in kg) by the values in Table 5-2 to determine your BMR. Remember, this number does not account for your activity level (such as if you workout two days per week or six days per week, take leisurely strolls around your neighborhood once in a while, run marathons, garden, play tennis, golf regularly, or just lay in a hammock reading biology books).

Table 5-2 A Few Equations to Determine Basal Metabolic Rate

Males	*Equation*
10–18 years old	(17.5 x weight in kg) + 651
18–30 years old	(15.3 x weight in kg) + 679
30–60 years old	(11.6 x weight in kg) + 879
>60 years old	(13.5 x weight in kg) + 487

Females	*Equation*
10–18 years old	(12.2 x weight in kg) + 746
18–30 years old	(14.7 x weight in kg) + 496
30–60 years old	(8.7 x weight in kg) + 829
>60 years old	(10.5 x weight in kg) + 596

Source: *The National Research Council, Recommended Dietary Allowances (Washington, D.C.: National Academy Press, 1989).*

So, if your body requires 2,000 calories per day, it means that those 2,000 calories provide enough energy for all the metabolic functions that occur to keep you alive and functioning well. If, however, you take in 3,000 calories, the extra 1,000 calories are stored as fat in your adipose tissue. But, if you take in a few hundred calories less per day, you will lose 1 pound for every 3,500 calories you do not consume.

Whether you need to lose weight is based more on your body mass index (BMI). To determine your BMI, first you divide your weight (in pounds) by your height (in inches) squared. Then, multiply the result of that math work by 705. The value you arrive at is your BMI. A value of approximately 21 is a great goal. If your value is higher than 28, you have a definite increased risk of diseases such as diabetes and heart disease and should consider lowering your weight (since you can't increase your inches to lower the BMI).

After millions of years of not having food readily available 24 hours per day, humans have developed a mechanism for storing energy that can be used during times of low food intake. Therefore, your body does not give up extra potential energy easily. It packs energy-rich fat on to your hips, thighs, abdomen, and buttocks. If you continue to take in more calories than you use, you will gain weight. It is much easier for your body to create fat than it is to use it.

Every 3,500 extra calories equals 1 pound of fat. Therefore, if you need to lose 20 pounds to be at your healthy weight, that means you need to burn (use up) 70,000 calories worth of energy. Since most exercise uses 350 to 500 calories per hour, that's about 8 to 10 hours of exercise per pound. And, while you are exercising to use up those extra calories, you cannot be taking in more than you need, or your weight will stay the same. If you prefer to do less exercise, you can take in less food (fewer calories) than you need, so that your body uses up some of its stored energy (thereby nipping away at the 70,000 extra calories you are sitting on).

Living organisms follow the rules of physics and chemistry, and the human body is no exception. Everything in the universe (including you and your calories) is governed by two laws of thermodynamics. *Thermodynamics* is a specialized field of study that focuses on the changes between heat and energy (*thermo-* = heat; *dynamics* = change).

- **The first law of thermodynamics: Energy cannot be created or destroyed.** This law applies to chemical processes, including the ones controlling your metabolism. It means that the total energy of a system (like your body) and its surroundings stays the same, although the form of the energy may change, or the system may gain or lose energy to the surroundings. So, if you have excess calories to burn, the energy from those calories may produce heat energy inside you (making you sweat as you exercise them away), which you release to your surroundings. But, overall, the amount of heat between you and your surroundings stays the same. When you are trying to lose weight, you are just trying to move the energy that is stored within you to your surroundings. When you gain weight, energy from your surroundings (like the ice cream in your freezer) moves from the surroundings to your system.
- **The second law of thermodynamics: *Entropy* is a state of matter that is a measure of randomness, and randomness in the universe increases when possible.** This law means that molecules will keep moving randomly, until they reach a stable condition. There is a classic experiment involving gas in two cylinders. The gas molecules keep moving from one cylinder to the next, until the amount of gas between the two cylinders is equal. This simply indicates that organisms, cells, molecules, atoms — everything of this universe — wants to be balanced and does not stop trying to achieve balance, even if that means that things are chaotic (random) for a while.

All of this math and physics should make one thing clear: The key to weight loss simply is using more energy than you take in. A fad diet is not necessary. You can eat what you want, as long as the numbers work out right at the end of the day. However, for optimum health, you should not get your 2,000 needed calories from ten doughnuts and then call it a day. Your body needs certain nutrients to function properly. Carbohydrates, proteins, and fats supply organic molecules needed to convert or create energy. Table 5-3 lists the required vitamins and minerals that you need.

Table 5-3 Vitamins and Minerals Needed by the Body

Vitamins	*Minerals*	*Trace Elements*
Water-soluble	Major minerals	Zinc
Vitamin B	Potassium	Chromium
Vitamin C	Calcium	Selenium

Vitamins	Minerals	Trace Elements
Fat-soluble	Sodium	Iron
Vitamin A	Chloride	Copper
Vitamin D	Sulfur	Iodine
Vitamin E	Magnesium	Manganese
Vitamin K	Phosphorus	Fluoride
		Molybdenum

Mining for information on minerals and vitamins

Minerals are inorganic molecules that are part of the earth. Because plants grow in the earth, some minerals are in plants, too. And, because animals eat plants, animals have minerals in their systems, too, but very small amounts.

Vitamins are organic molecules that occur in all living things naturally. (Really, vitamins do not come only in pill form.)

Almighty minerals

The body doesn't need a whole lot of minerals, but some are essential for proper functioning of your body. These essential minerals are called *major minerals*; minerals that are needed only in very small amounts are called *trace elements*.

If you look at the Periodic Table of the Elements in Chapter 3, you find the major minerals and trace elements on that chart. The elements are nonliving substances found in the earth or the earth's atmosphere, and therefore in living things. Elements such as carbon, oxygen, nitrogen, and hydrogen make up most of the organic compounds such as carbohydrates, proteins, and fats. But some of the other "inorganic" elements play important roles in the body, too.

For example, phosphorus can be found in the bonds holding strands of DNA together in all living things. Potassium, sodium, magnesium, and calcium help to regulate pretty important functions such as heart beats, fluid balance, and muscle contraction.

Vital vitamins

Vitamins are the little alarm clocks of the body. They regulate the building of tissues and cells, they help in metabolism, and they promote healing and prevent diseases. They allow the details of the body to get done. Vitamins are made up of the same chemicals as carbohydrates, proteins, and fats, but they do not need to be broken down to be used. They do not provide energy. Vitamins are extremely small, so some protein molecules (which are large) usually carry them around the body to make sure that their important jobs don't go unperformed.

Fat-soluble vitamins (vitamins A, D, E, and K) need to be "dissolved" in fat molecules (phospholipids to be exact) for the cells to be able to use them. The phospholipids carry the vitamins through the bloodstream and into cells. *Water-soluble vitamins* (the B vitamins and vitamin C) often act as or with enzymes, which speed up reactions. Vitamins A, C, and E are also considered to be antioxidants (see Chapter 15).

Craving (knowledge about) carbohydrates

Carbohydrates are compounds of carbon, hydrogen, and oxygen that quickly supply the body with energy. Carbohydrate molecules, such as sugars, "burn" fast. Did you ever hold a marshmallow over a campfire too long and see how fast it is consumed by flames? That is because glucose is quickly converted to energy, and fire is a very visible form of energy (it provides heat and light energy).

Glucose and other carbohydrates supply quick, short-term energy. Fat that is created to store extra glucose molecules supplies long-term energy, which is why you don't start to "burn fat" until 20 minutes or more into an aerobic workout. First, your body quickly burns off the glucose that is readily available in your cells. Then, it starts breaking down fat molecules and converting them into glucose for fuel.

Glucose is the most important carbohydrate molecule. You can acquire glucose directly from foods containing carbohydrates (such as breads, pasta, sweets, and fruits). However, your body also creates glucose when it breaks down proteins and fats. (Don't worry; I explain how in Chapter 6.)

Glucose is converted to usable energy by the metabolic process called glycolysis, the end product of which is pyruvate. Pyruvate then enters the Krebs cycle, which produces the energy compound adenosine triphosphate (ATP). This process takes place in the mitochondria of each cell in your body. (Mitochondria are described in Chapter 2; the details of the Krebs cycle are available in Chapter 8.) Therefore, the energy that glucose creates is measured in molecules of ATP, which is the "currency" of living organisms. ATP is "spent" when you are using energy, and created from the fuel you put into your body.

When there is more fuel in your body than is needed to meet current energy requirements, your body stores some fuel from carbohydrate as glycogen. Plants store their extra fuel as starch, or they create more structural molecules of cellulose. Glycogen, starch, and cellulose all are long chains of glucose (and other sugar) molecules called polysaccharides.

Proteins: You break down their chains; they build yours

Living organisms are made of proteins, so they must acquire protein to replenish their supply. Every muscle, cell membrane, and enzyme is made from proteins. So, to create more muscle fibers, new cells, and other elements that help your body run, you need to take in protein. In essence, you need protein to make protein.

Proteins are made from amino acids. Nine amino acids must be obtained from the foods you eat, so they are called essential amino acids. (It is essential that you consume foods containing those nine amino acids.) Humans can synthesize 11 amino acids from a variety of starting compounds that are not necessarily derived from amino acids themselves. (Because these are made in the body, they are called nonsessential amino acids because it is not essential that you consume foods containing them.)

When you think of protein sources, you probably think of meat. There's a good reason for that. Your muscle fibers are made of protein, right? Well, so are the muscles of all animals. And, when you eat meat, you are consuming the muscle tissue of another animal. Whether it is beef, chicken, turkey, pork, or fish, the muscle tissue contains protein. Beans, nuts, and soy all are good sources of protein, but the protein is plant protein, not animal protein.

Protein structure: A mean old acid has a wave of excitement

What is the difference between plant and animal protein? The answer lies in the structure of protein.

A *protein* is a long chain of polypeptides, which is made when amino acids are joined together by a peptide bond. Picture an unhooked pearl necklace lying in a straight line. Each pearl represents an amino acid, and the thread between each pearl represents the peptide bond. The entire necklace represents the entire protein. When proteins are broken down by digestion to just the amino acids, there is no significant difference between plant and animal protein.

Polypeptides

Polypeptides means "many peptides" (*poly-* = many). But, what's a peptide? A wave of excitement? No. A peptide forms when two or more amino acids join together. A dipeptide forms when two amino acid molecules join together. A tripeptide, as you might expect, is a chain of three amino acids. Peptide bonds link amino acids together when a condensation reaction (a reaction in which water is produced) occurs between the amino group (NH_2) and a carboxyl group (COOH).

Amino acids (sounds like "a mean old acid") are organic compounds that consist of an amino group (NH_2), a carboxyl group (COOH), and a specific side chain (R). The *side chains* are certain combinations of atoms that have different chemical properties. Altogether, there are 20 different side chains, which create 20 different amino acids (see Chapter 4). The amino acid that is produced depends on which side the chain joins the amino and carboxyl groups.

The 20 different amino acids can be combined in a variety of ways to produce different proteins. Do you know what designates which amino acid should be produced and in what combination the amino acids should be joined? Your genes. The information contained in your genetic material provides the "recipe" for the proteins that your cells are to make at certain places in your body at certain times. I explain how in Chapter 14. For now, just understand that the order of amino acids in the polypeptide determine what purpose the protein serves.

And know that plants produce all the amino acids they need, whereas humans produce only 11 of the 20 amino acids. The reason for this is that plants are rooted — they don't walk around. They don't hunt. They can't drive to the grocery store. They need to be able to make (synthesize) all of the amino acids required for their protein synthesis because they have no outside sources of protein (and thus amino acids). Humans (and other animals), however, can fetch other sources of protein; therefore, you and I acquire amino acids by breaking down dietary sources of protein and also by making some.

The amino acids that the human body can make itself are called *nonessential amino acids*. The ones that humans need to acquire from food are called *essential amino acids*, because it is essential that humans get these through diet so that all required proteins can be synthesized (see Table 5-4). So, if you are ever in a nutritional foods store, don't go buying any of the amino acids in the right-hand column; your body already makes those. And, if you eat a balanced omnivorous diet (with definite emphasis on the word "balanced"), you should be getting enough of the amino acids in the left-hand column.

Table 5-4 Essential and Nonessential Amino Acids for Humans

Essential Amino Acids	*Nonessential Amino Acids*
Arginine*	Alanine
Histidine	Asparagine
Isoleucine	Aspartate
Leucine	Cysteine
Lysine	Glutamate
Methionine	Glutamine
Phenylalanine	Glycine
Threonine	Proline
Tryptophan	Serine
Valine	Tyrosine

**Arginine is a nonessential amino acid synthesized from glutamate. Humans don't make enough arginine for growth, however, so it is sometimes considered an essential amino acid. This fact about arginine explains why sometimes you may see ten nonessential amino acids listed, and other times you may see 11 listed.*

So, the answer to "What is the difference between plant and animal protein?" is this: Because animals acquire essential amino acids plus make their own nonessential amino acids, animal protein is *complete*. Plant proteins, however, are called *incomplete* because they do not contain enough of some of the amino acids that humans need. This is the reason that vegetarians combine certain foods. For example, beans contain low amounts of the amino acid methionine. Rice contains low amounts of the amino acid lysine. But, if you eat rice and beans together, you enhance the amino acid content and make the protein more complete.

Protein functions that make you function

Proteins run nearly every metabolic process in your body, and they are part of the structure of every cell in your body. Here are a few examples.

Outwardly, the protein keratin makes up the outer layers of your skin (your epidermis), your nails, and your hair. One reason that you need to take in protein on a daily basis is because these external structures never stop growing. Whereas humans store fat and glucose in their bodies, the body really doesn't have excess protein lying around. So, when there is a protein deficit due to the fact that protein is needed constantly, protein is removed from places in the body where it is being used. For example, people with anorexia, who do not consume enough food, eventually start breaking down the muscle fibers, such as from their heart, when protein is needed.

Inside, muscle tissue is loaded with protein, and bones contain protein, too. Red blood cells contain *hemoglobin*, which is a compound made of *heme* (contains iron and carries oxygen) and *globin* (a protein). *Immunoglobulins* are protein structures created by your immune system that serve as antibodies to fight bacterial and viral invasions within your body.

Proteins also combine with other substances in the body to perform specific functions:

- *Lipoproteins* are a combination of lipids (fats) and proteins that carry cholesterol throughout the body.
- *Glycoproteins* are a combination of carbohydrates (sugars) and proteins that are found in cell membranes and mucous of the digestive tract, as well as in the extracellular matrix (see Chapter 2). They also play roles in the determination of blood type and cell recognition, which is important in the development of an embryo.
- *Phosphoproteins* are a combination of phosphoric acid and proteins that create the main protein in milk: casein. The phosphorylation of proteins, especially enzymes, is a major way of regulating their activity.

One of the most important functions of proteins, however, is when they act as enzymes.

Enzymes are proteins that serve in chemical processes, such as those that occur during digestion. Enzymes serve as *catalysts* — that is, they help to speed up a reaction, but are not used up or changed during the reaction. There are six major types of enzymes:

- *Ligases*, which join two molecules together
- *Lyases*, which split two molecules apart
- *Hydrolases*, which split two molecules apart when water is added
- *Isomerases*, which create isomers (different chemical structures that have the same chemical formula)
- *Oxidoreductases,* which catalyze oxidation (electron is donated) reactions and reduction (electron is accepted) reactions
- *Transferases*, which transfer chemical groups from one compound to another

You can find more information about the structure and function of enzymes in Chapter 4 and in Chapter 10.

Fats: Yes, you need some (but don't overdo it)

This section on wise nutritional choices wouldn't be complete without including information about fats. Although fat seems to be taboo these days, it's not fat that makes people fat. Consuming more fuel than is burned leads to the production and deposit of fatty tissue, whether the fuel comes from fats, proteins, or carbohydrates. A person can "get fat" eating only "fat-free" foods, which can be high in calories due to the replacement of fat with sugar. If you consume more calories than you use in a day, and you do this day after day after day, you will add fat to your body no matter what you eat.

Your body needs fats, though, believe it or not. Fats are used to make tissues and hormones, and fats insulate your nerves (just as wires are insulated — did you ever hear the expression that somebody's nerves were frayed?).

Fat is a source of stored energy, it gives your body a shape, it reduces heat loss by insulating your organs and muscles, and it cushions your body and organs (much like shock absorbers). It also helps to keep your tissues moist, your joints anointed, and your body from swelling up when it rains. You know how oil and water don't mix? Well, oil is a liquid fat, and the oils that your skin produces ooze out of your pores to coat your skin. The oils coating your skin serve as a moisture barrier: It keeps you from dehydrating on a dry, windy day (when water would be drawn from your skin and into the atmosphere) and keeps excess moisture from entering your cells. Imagine if every time you took a shower, the water you washed with actually entered your skin and got into your cells! Fats help to keep you from becoming a water balloon.

Every cell membrane contains fats (lipids), which help to separate what is going on metabolically within the cell from the fluid outside the cell. And sometimes what is going on inside the cell involves fats, too. *Hormones* are products made during a metabolic process that occurs inside cells. For example, the hormone insulin, which controls the level of glucose in your blood, is produced by cells in the pancreas. And, lipid molecules are part of hormones.

Fats are combinations of *fatty acids* and *glycerol* molecules. Fatty acids are carbon-hydrogen chains with an acid on the end; glycerol is a carbon-hydrogen compound containing hydroxyl groups (-OH).

Yes, fats are important nutrients in your body, but there are good fats and bad fats. Foods have three kinds of fats: triglycerides, phospholipids, and sterols. The body also makes a compound from protein and fats, called lipoprotein, which is involved in the transport of cholesterol through the body.

Triglycerides

As its name implies, a *triglyceride* is a combination of three fatty acid molecules and a glycerol molecule. This is the most common type of fat; it is the form that travels through your bloodstream. Triglycerides are what you burn up for fuel after carbohydrate stores are used up, what you use to store energy as adipose tissue, and what your doctor can measure during a blood test. Basically, an elevated triglyceride level indicates that you are consuming excess fuel, namely carbohydrates (sugars), and that your body has created more triglycerides, which are on their way (via your bloodstream) to being stored as fat.

When a triglyceride molecule is being created, three water molecules are given off: one each for the fatty acid that combines with the hydroxyl groups of glycerol. This type of reaction is called a dehydration synthesis because water is removed from the "ingredients" of the reaction (dehydration) when another compound is made (synthesis). When triglycerides are broken apart to produce energy, water is taken into the reaction to split the triglycerides into the three fatty acid molecules and glycerol. This type of reaction is called a *hydrolysis* reaction because water splits apart molecules (*hydro-* = water; *-lysis* = split).

Phospholipids

These molecules are made up of lipids and phosphate (ions of phosphorus), and they carry hormones and vitamins through your bloodstream, as well as in and out of cells. The vitamins that they carry (vitamins A, D, E, and K) are called *fat-soluble vitamins* because they blend into and are carried by this type of fat. I explain how hormones are created and what they do in Chapter 10.

Sterols

I'm pretty sure that you've heard of cholesterol. Well, cholesterol is a sterol molecule — surprised? These types of fat are made of lipids and alcohol. They contain no calories because they do not produce energy. Instead, these molecules are used to build other molecules, such as hormones and vitamins. Cholesterol is a starting compound from which steroid hormones — that is, the sex hormones estrogen and testosterone, are synthesized. Cholesterol also helps your nerves to send impulses to your brain and receive messages from your brain. And, cholesterol is part of the two-layered lipid membrane that surrounds each and every cell.

So, you *need* cholesterol; it has several necessary functions in the body. However, you, being the animal that you are, create most of the cholesterol that your body needs. Cholesterol is produced in the livers of animals (including yours). That is why only animal products contain cholesterol and not plant-based products. I always find it funny when I see a bag of potato chips boasting "No Cholesterol!" or "Cholesterol Free!" Potatoes do not have any cholesterol to begin with. But, if the oil they are fried in is animal based (such as lard), they do soak up loads of cholesterol. If the potato chip manufacturer used vegetable oils (such as peanut oil, corn oil, and so on) to fry the spuds, then the potatoes remain cholesterol free (but far from fat free). But, it's not like the potato chip maker went to great pains to remove cholesterol for you.

If you have blood drawn to have your cholesterol measured, and your result is 215, that means that there are 215 milligrams (mg) of cholesterol in every tenth of a liter of blood (a deciliter; dl). You know what a liter bottle looks like, so if you divide that amount by 10, that is the amount of blood I'm talking about. Then, 1,000 mg equals 1 gram, so 215 mg is approximately one-quarter of a gram. (If you need help with metric to standard conversions, see Appendix B.) If your cholesterol level is higher than 250 mg/dl, you probably need to take steps to lower it. However, the levels of the types of fat that carry cholesterol around through your bloodstream are more telling of potential heart disease than is the actual cholesterol level.

Lipoproteins and your risk for heart disease

Lipoproteins are compounds made from a fat (*lipo-* = fat, as in lipid) and protein. Their job is to carry cholesterol around your body through the bloodstream. Your body can produce four types of lipoproteins:

- High-density lipoproteins (HDLs)
- Low-density lipoproteins (LDLs)
- Very low-density lipoproteins (VLDLs)
- Chylomicrons

Sometimes in the news you read or hear about HDL being "good" cholesterol and LDL being "bad" cholesterol. However, HDL and LDL, as you now know, are lipoproteins, not cholesterol molecules. They just attach to and transport cholesterol. Here is what is "good" and "bad" about the lipoproteins.

Chylomicrons are very small, newly created lipoproteins that fall into the VLDL category. VLDLs have very little protein and a lot of fat. (Fat is less dense than protein, like fat "weighs" less than muscle.) As VLDLs travel through your bloodstream, they lose some lipids, pick up cholesterol, and become LDLs. The LDLs deliver the cholesterol to cells in your body that need it, but along the way, VLDLs and LDLs can squeeze through blood vessel walls. While doing that, the cholesterol can get stuck to the wall of the blood vessel, causing deposits (plaque) to form. If enough cholesterol gets stuck, an artery may get clogged, which means blood cannot flow through. If that happens, a heart attack or stroke may occur. So, although LDLs help the body by transporting cholesterol, if you have too many of them, the cholesterol may start to block blood vessels, which increases your risk of heart disease, heart attack, and stroke.

Differentiating heart attack and stroke

A *heart attack* (technical name is *myocardial infarction*; *myo-* = muscle, *cardi-* = heart) occurs when blood flow through a blood vessel in the heart is blocked. This causes a small area of heart tissue to die (an *infarct*); a massive heart attack causes a larger area of heart tissue to die, which may be fatal.

A *stroke* occurs when blood flow through blood vessels in the brain is blocked. This causes a lack of oxygen to a certain area in the brain, which damages the brain tissue. The symptoms that occur during a stroke depend on what area of the brain was damaged. Some people lose the ability to speak fluently, others have paralysis on one side of the body, and still others are left with very little function.

HDLs, on the other hand, are the lipoproteins that contain more protein than lipid, which makes them more dense and gives them their name. Because they are more dense, they cannot squeeze through the blood vessel walls, so they shuttle cholesterol right out of the body. They are not able to deposit cholesterol in blood vessels, because they cannot get into them, so they do not increase the risk of heart disease, heart attack, or stroke. Ideally, you want to have more of these dense little guys floating in your blood than you want the LDLs or VLDLs.

Aspiring to Be Respiring: How Animals Breathe

Okay. Stop reading for a minute and just focus on your breathing. You breathe in, and your shoulders lift up and your chest expands. You breathe out, and your shoulders and chest move downward and inward. Okay. Start reading again. But realize that you don't have to think about moving your shoulders and chest to get air in and out of your body. It's still happening, isn't it? In this section, I explain why animals have to breathe and how they go about doing it. Take a deep breath, now.

All animals must exchange gases between themselves and their environment on a continual basis throughout every moment of their lives. In simple animals, the process of gas exchange is simple; it may occur between the surface of the animal and the environment. But in more complex animals, more complex systems of gas exchange have evolved; air from the environment must be processed in the respiratory system.

Respiration is the entire process of taking air in to the system, exchanging needed gases for unnecessary gases, using the needed gases and releasing the waste form of gases. In more complex animals, breathing covers the first step of respiration. *Breathing* is the physical action of taking air in to the system and releasing gaseous waste.

There are four types of gas exchange systems:

- *Integumentary exchange*, which occurs through the skin
- *Gills*, which exchange gases in water environments
- *Tracheal systems*, which are used by insects
- *Lungs*, which are found in land animals

Integumentary exchange

The *integument* is the skin or surface of an animal. This may not be skin like on you and me, but it is the outer membrane of the animal. Very small animals and a very few larger animals that live in moist environments use this type of gas exchange. Worms are an example.

Earthworms have capillaries right under their "skin." As the worms move through the soil, they loosen the soil, which creates air pockets. The worms take in oxygen from the air pockets and release carbon dioxide right through their outer surface. However, to be able to exchange gases directly with their environment, earthworms must stay moist.

You know how worms get flooded out of the ground when it rains and end up all over your driveway and sidewalk? Well, they head right back into the soil as soon as they can (and not just because they are potential bird food). Otherwise, their outer surface dries out, and they can no longer take in oxygen and get rid of carbon dioxide. If that happens, they die. (This is also the premise behind why pouring salt on a slug stops it in its slimy tracks. The salt dehydrates its outer surface, which prevents it from exchanging gases.)

Going over gills

Animals that live in water have *gills*, which are extensions of their outer membranes (see Figure 5-2). The membranes in gills are very thin (usually just one cell thick), which allows gas exchange between the water that flows over them. Capillaries connect to the cells in the gills so that gases can be taken in from the water and passed into the bloodstream of the aquatic animal. Also, gaseous waste can diffuse from the capillary into the cells of the gill and pass out into the watery environment.

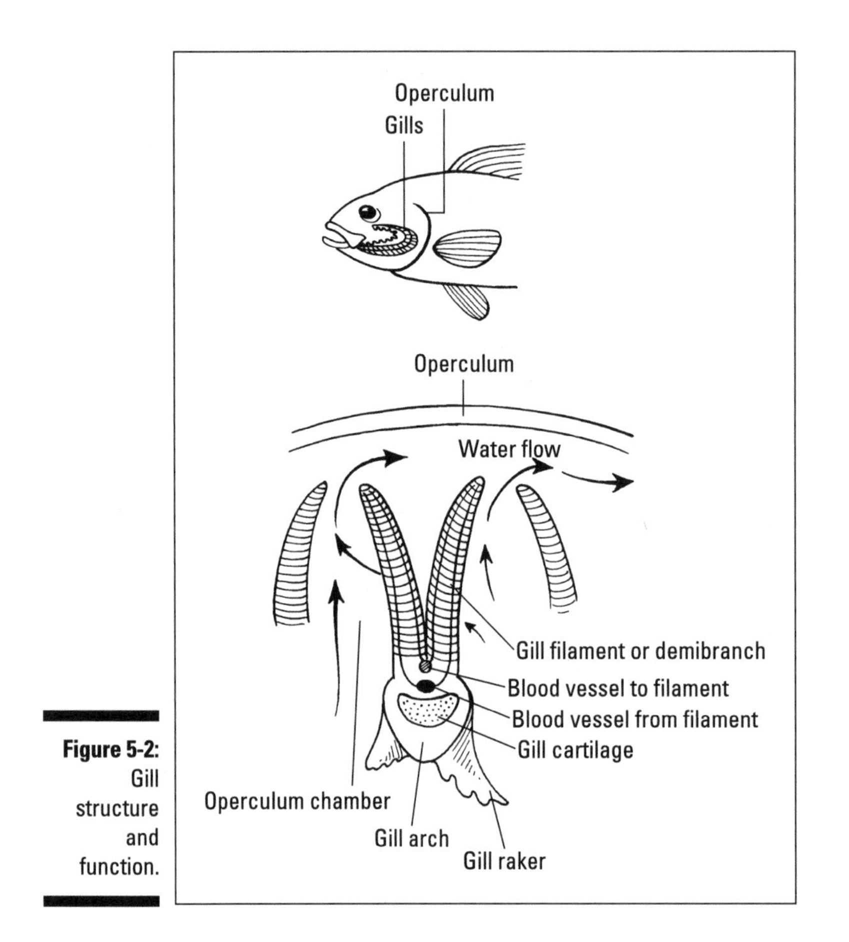

Figure 5-2: Gill structure and function.

Gills may look different on the wide variety of marine life, but they function essentially the same. Fish gills may be the most familiar, but creatures such as lobsters and starfish also have gills. The gills of a lobster are feathery; those in a starfish surround the bumpy projections on the bottom of its surface. The gills are very delicate and must be protected. The protective gill coverings on aquatic organisms are as diversified as the animals.

In fish, the gills are membranous filaments covered by a flap called an *operculum*. The fish opens and closes the flaps protecting the gills by opening and closing its mouth. After water enters the mouth, the water is forced through the gills and then out the back of the operculum. As the water passes over the gills in one direction, the blood inside the capillaries inside the gills is moving in the other direction. Oxygen from the water diffuses into the capillaries in the gills; carbon dioxide diffuses out of the capillaries in the gills. Once the oxygen is in the capillaries, it can be transported around the body of the fish so that all cells in its body acquire the needed gas.

Tracheal exchange systems

Some insects have air tubes that open to the outside of their body. This network of tubes is called a *trachea*; the holes that open to the outside surface are called *spiracles*. (In humans, the trachea is a tube that carries air down into the lungs.)

In a tracheal exchange system, oxygen diffuses directly into the trachea, and carbon dioxide exits out through the spiracles. The oxygen and carbon dioxide do not need to be carried through a circulatory system, because the tracheal system pervades all parts of the insect's body. The cells of the body exchange air directly with the tracheal system. It would be as if you had noses all over your body (akin to spiracles), and your trachea ran the length of your body with branches reaching every area within you. Wouldn't leave much room for organs and muscles to move you around, would it? I guess if humans could fly, it would be a different story.

Earthworms, fish, and some insects get their oxygen by diffusion rather than a breathing process. However, some insects (such as bees and grasshoppers) combine a breathing process with a tracheal system. Bees and grasshoppers contract muscles to pump air in and out of their tracheal systems. A grasshopper contains air sacs on some of the air tubes in their tracheal system. The bags "pump" (like fireplace bellows) after pressure from muscles is applied. Sounds a little like what happens in larger land animals, doesn't it? This could be considered an evolutionary link. (I explain evolutionary links in Chapter 16.)

Longing for lung info?

Well, here it is. Do you know what lungs are? They're the opposite of gills. Gills extend out off an organism (called an *outgrowth*), and lungs are internal ingrowths of the surface of the body (also called *invaginations*). For animals that live in water, gills suffice because the environment is moist. However, for animals that live on land (and breathe in air), the lungs are inside the body so that they can stay moist. That is also why the only outside openings of your respiratory system are your nose and your mouth — both of which are small compared to the rest of your body. With the openings to the respiratory system being small, it minimizes the chance of water evaporating from the system. (You know how your nose dries out in winter when the air is very dry? Imagine if your lungs dried out; you would not be able to exchange gases.)

Lungs may be different shapes and sizes in various land animals, but they function essentially the same as they do in humans. I am going to use humans as the model for the mechanics of the lungs so that you might gain a better understanding of yourself.

Humans have a pair of lungs that lie in the chest cavity (Figure 5-3); one lung is on the left side of the trachea, and the other is on the right. The trachea is the tube that connects the mouth and nose to the lungs. Inside the lungs, the trachea branches off into *bronchi*, which branches off into *bronchioles*. (If you turn the respiratory system upside down, it looks like a tree.) There is also a muscle lying underneath the lungs called the *diaphragm*. The *ribs* surround the chest cavity to protect the lungs (and heart) and to assist in the motions of breathing.

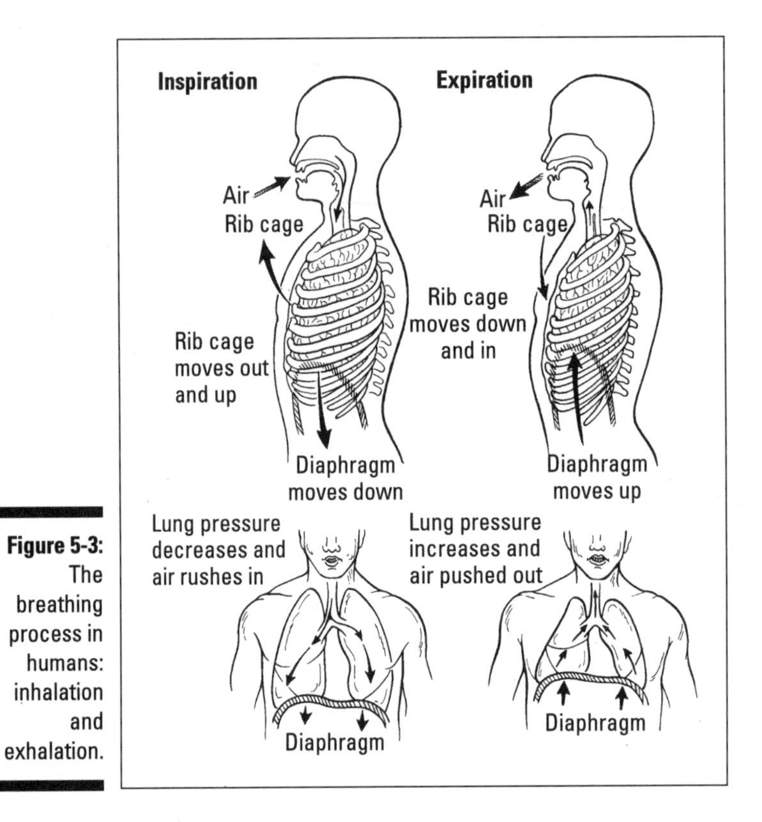

Figure 5-3: The breathing process in humans: inhalation and exhalation.

When you inhale (breathe in), air enters through the nostrils (the openings to the nose), and flows through the nasal cavity. Inside the nasal cavity, hair, cilia, and mucus trap dust and dirt particles, purifying the air that enters the lungs. Occasionally, you must cough and either spit or swallow to move the trapped particles out of your throat. If you swallow dirt and it enters your stomach, it is simply digested and excreted.

The other motion that happens when you breathe in is that the diaphragm muscle contracts (becomes smaller and moves down), which allows your ribcage to move upward and outward. Because the lungs have more room when your chest is expanded, they open up. (Think of a balloon filling up.) The opening up of the lungs means there is more room in the lungs, so air rushes in (or is kind of sucked in) to fill the space. When your lungs fill, the air passes through all the branches of the bronchi into the tiniest air sacs, which are called *alveoli*. The alveoli are the structures where oxygen and carbon dioxide exchange.

Alveoli are one-cell thick. Capillaries are one-cell thick. Capillaries surround the alveoli, so diffusion is a fairly simple process and a very short trip for the gas molecules across the two cell membranes. When blood travels around the body, the oxygenated blood is in the arteries and arterioles, and the deoxygenated blood is carried back to the lungs (to pick up more oxygen and release carbon dioxide) in veins and venules. Capillaries are the blood vessels that bridge the gap between arterioles and venules.

The blood vessels in order of decreasing oxygen content are arteries, arterioles, capillaries, venules, and veins.

At each of the 300 million alveoli in the lungs, oxygen is at a higher concentration than it is in the capillaries surrounding the alveoli. In the capillaries, there is more carbon dioxide than oxygen. So, remember that diffusion occurs from an area of higher concentration to lower concentration (kind of an attempt to even things out)? What do you think happens? The oxygen in the alveoli diffuses across the alveoli's membrane and the capillary's membrane and into the capillary. The carbon dioxide diffuses across the capillary's membrane and the alveoli's membrane into the lung.

Once oxygen is in the capillary, it is snatched up by hemoglobin, and the red blood cells transport oxygen throughout the body. (The blood in your arteries is bright red because of the combination of oxygen and hemoglobin.)

Every cell in your body needs oxygen for its metabolic processes, so oxygen is vital to proper functioning. Without it, you die. So, breathe deeply and make sure that those blood cells carry it to every nook and cranny of your body.

When the red blood cells release oxygen around the body, they simultaneously pick up carbon dioxide being excreted by the cells in tissues around the body. Hemoglobin also carries carbon dioxide, but the combination between the two creates a dark purple color, which you can see in the veins near your wrists.

In the alveoli, the hemoglobin compounds in the red blood cells release the carbon dioxide. This makes hemoglobin ready to pick up oxygen, and allows carbon dioxide to be excreted during *exhalation* (breathing out).

When you exhale, the diaphragm muscle relaxes and moves back up. This action causes the ribcage to move downward and inward, minimizing the size of the lungs. This bellows-type movement increases the pressure inside the now-smaller lungs, which forces (or pulls) air up out of the lungs. In the exhaled air is carbon dioxide that was deposited by the red blood cells.

How Plants Acquire Their Energy

Just as animals ingest food and drink liquids, plants must get food into their systems in order to continue living. Plants create energy for animals to use, so they must replenish their nutrients. And plants breathe, in a way. They don't have spiracles, gills, or lungs, but they do exchange gases. Instead of plants taking in oxygen and giving off carbon dioxide, they do the reverse. They take in the carbon dioxide that all the animals give off, and they give off oxygen for all the animals to use. Pretty cool design, isn't it?

Presenting basic plant structure

Before I get into how plants get their nutrients, I better give you some information about the structure of plants. Otherwise, you might not know what I'm talking about. And I'd hate that.

Basically, plants have a root system, a stem or trunk, branches, leaves, and reproductive structures (sometimes flowers, sometimes cones or spores, and so on). Figure 5-4 shows you the basic structures. Most plants are *vascular*, which means they have a system of tubules inside them that carry nutrients around the plant, just as your vascular system (or *cardio*vascular system if you include the heart) has arteries, veins, and capillaries. Vascular plants are differentiated from plants such as algae, which do not have a vascular system. Most vascular plants are *seed plants*, and seed plants are most often the model used in biology and botany textbooks and the type of plant with which you should be most familiar.

There are two main types of seed plants: *gymnosperms* (conifers, which produce pinecones) and *angiosperms* (flowering plants). Of the 500,000 different species of plants, more than 300,000 are flowering plants. Flowering plants are divided by how many cotyledons they have. *Cotyledons* are the tissues that provide nourishment to a developing seedling. Flowering plants can be *monocotyledons*, which means they have one cotyledon, or *dicotyledons*, which means they have two cotyledons.

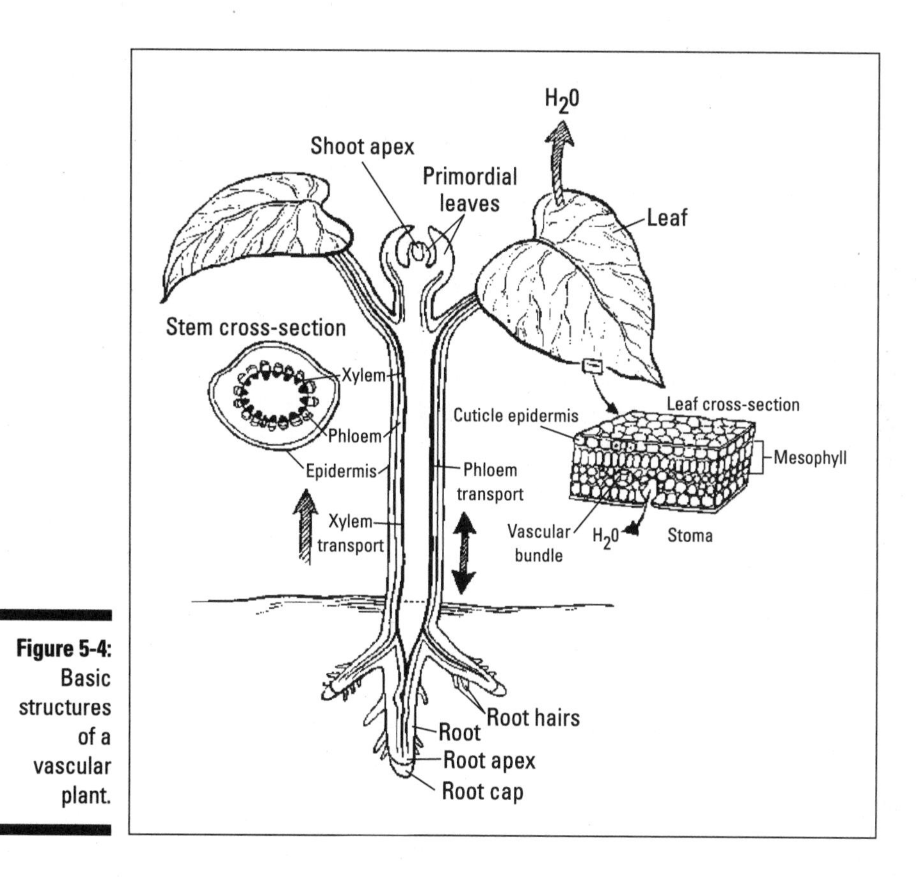

Figure 5-4: Basic structures of a vascular plant.

Three types of *plant tissues* exist:

- *Vascular tissue,* which consists of *xylem* (pronounced "zi-lem") and *phloem* (pronounced "flow-em") — the main tubes through which nutrients are transported
- *Dermal tissue,* which includes the outer cells (epidermis), guards cells surrounding a stoma, and special cells found on the outer surface of plants, such as hair cells (yes, some plants have hair) or cells that cause a stinging sensation (ever bump into nettle?)
- *Ground tissue,* which consists of three types of cells:
  - *Parenchyma cells* are the most common ground cells. They are involved in many basic cell functions including storage, photosynthesis, and secretion.
  - *Collenchyma cells* have thick cell walls and are involved in mechanical support
  - *Sclerenchyma cells* are similar to collenchyma cells, but their walls are even thicker.

Plant cells differ from animal cells in that they have a cell wall. Animal cells have a cell membrane as the outer membrane of a cell; plant cells also have a cell membrane, but they have the additional, more rigid cell wall surrounding it. I give you details about the organelles in animal cells in Chapter 2, but here is a list of what you'd find in a typical animal cell. Then, compare it to what you'd find in a typical plant cell.

Animal cells have these organelles and suborganellar structures:

- Centrioles
- Cell membrane
- Nucleus and nucleolus
- Mitochondria
- Golgi apparatus
- Small vacuoles
- Lysosomes
- Endoplasmic reticulum
- Ribosomes

Plant cells have all the organelles an animal cell has, plus:

- Cell wall (cellulose in cell wall provides structure and rigidity)
- Large vacuole (for storage of large molecules of starch)
- Chloroplasts, which contain chlorophyll (the green pigment in plants)

The reason that plant cells have the extra structures is that plants harness and store their energy in a different way than animals do. First of all, remember that plants get their energy from the sun, whereas the start of the animal food chain is a plant? Animals cannot harness energy from the sun directly. The plants harness energy from the sun using chloroplasts, and they store the energy they convert into carbohydrate molecules in the cell wall and vacuoles.

Making energy from the ultimate energy source

Photosynthesis is the process by which plants convert energy from the sun. It is the process that allows plants to create organic molecules that they use as fuel (remember that plants are autotrophic, meaning "self feeders"?). Here is how it works.

The molecules of chlorophyll contained in the chloroplasts (which are scattered throughout each cell) absorb energy in the form of light from the sun. Some plants need more sunlight than others, but all need at least a little.

Instead of taking in oxygen and breathing out carbon dioxide like animals do, plants take in carbon dioxide from the atmosphere. Plants absorb water from the ground up through their roots — I'm sure you've seen the effects of too little water and sunshine on a houseplant before. During photosynthesis, the energy from the sun splits the water molecules into hydrogen and oxygen. The oxygen molecules are given off by the plant and emitted into the atmosphere, so that you and I can breathe and create energy. Molecules of ATP are created within the plant cell. These reactions are called *photochemical* or *light reactions* because they require light to occur.

Enzymes within the plant then catalyze the combination of hydrogen and carbon dioxide to create a carbon compound that is called an *intermediate*. An intermediate is a compound used to continue a process to create a different compound. In plants, the intermediate is called *phosphoglyceraldehyde* (it might be easier to remember PGAL). PGAL goes on in the process to produce glucose, which the plant uses as fuel to survive. These reactions are called *carbon-fixation reactions* (or *dark reactions* to differentiate them from the light reactions above) because atoms of carbon are "fixed"; that is, they are put into stable compounds that can be used purposefully instead of just floating around the cell aimlessly.

When the plant has created more glucose than it needs to sustain life, it combines glucose molecules into larger carbohydrate molecules called starch. The starch molecules are stored within the large vacuoles in the plant cells. When necessary, the plant can break the starch molecules down to retrieve glucose for energy or to create other compounds, such as proteins (plants use proteins to carry electrons used in photosynthesis), nucleic acids (to create DNA), or fats.

Flowin' through the xylem and phloem

Plants undergo photosynthesis to produce energy for themselves (and ultimately humans). Light and water are needed to perform this process. But, how do the plants get the water and light into their cells? I'll tell you.

Tissues called the xylem and the phloem usually are found together in what are called *vascular bundles*. Both types of tissue conduct substances up through the root and stem of a plant. The xylem conducts water and minerals from the soil; the phloem "flows" sugar molecules.

Photosynthesis

The equation used to explain photosynthesis essentially is the reverse of the reactions that occur in animals during glycolysis. Glycolysis is the splitting apart of glucose by water to produce carbon dioxide and energy; photosynthesis is the splitting apart of water by energy to produce glucose from carbon dioxide with oxygen given off.

$$6\ H_2O + 6\ CO_2 + \text{light} \rightarrow C_6H_{12}O_6 + 6\ O_2$$

water + carbon dioxide + light → glucose + oxygen

All plant cells have a cell wall, but cells in the xylem have an additional cell wall to give them extra strength (helps to avoid a blowout of water through the stem). *Vessel elements* are specialized cells in the xylem that form columns called *vessels*. Water passes through holes at the ends of each vessel element, and continues up through the entire vessel column.

Phloem tissue contains cells called *sieve-tube elements*, which connect in columns called *sieve tubes*. Each sieve-tube element has a pore on the end of it, through which the cytoplasm from one sieve-tube element can "touch" the cytoplasm of the next sieve-tube element. This structure allows the fuel that the plant makes in the leaves to pass through and nourish the rest of the plant. This process is called *translocation* — look for details about it in Chapter 6.

Transporting water from cell to cell

Plants have two ways of moving water from outside the root toward the inside of the root to the xylem and phloem tissue. Water can flow between the cell walls of adjacent cells. Think of this area as a hallway. Or water can flow between cells through tubes connecting the cytoplasm of each cell, much like people can walk through doors of adjoining rooms.

Water moves through these areas by one of three mechanisms:

- *Capillary action* causes liquids to rise up through the tubes in the xylem of plants. The action results from *adhesion*, which is caused by the attraction between water molecules and the walls of the narrow tube. You know how there is always a few drops of liquid along the walls of a straw? It's caused by adhesion between the two *different substances* of water and plastic. The adhesion forces cause water to be pulled up the column of vessel elements in the xylem and in the tubules in the cell wall.

- *Osmosis* is responsible for drawing water from the soil into the xylem cells in the root. The action occurs from two forces. First, with the xylem continuously moving water up out of the root, the pressure in the root is high, especially in small, low-to-the-ground plants (like grass). This force is evident in grass early in the morning when you can see tiny drops of water at the tip of the blades. (This isn't dew; dew is caused when moisture out of warm, humid air condenses during the cooler night temperatures and settles all over the grass.) The second force involved in this process is the high concentration of minerals inside the root as opposed to the ground. In an attempt to equalize the concentration, the plant moves minerals up through the xylem, which forces water up as well.
- *The cohesion-tension theory* describes how the majority of water moves through plants. *Cohesion* is similar to adhesion except it involves *like* substances instead of different substances. Cohesion causes the water molecules to be attracted to each other in a column (kind of like a singles bar for "unattached" H_2Os). Once the water molecules become attached, they fill the column in the xylem and act as a huge single molecule of water. As water evaporates from the plant (a process called *transpiration*), some water is "pulled" out of the plant. Well, since the water is one big column, the rest of the water gets pulled up, too. And, that necessitates more water being pulled into the plant. This mechanism is called "bulk flow of water."

The inspiration for transpiration

Transpiration is the technical term for the evaporation of water from plants. As water evaporates from the leaves (or any part of the plant exposed to air), it creates a tension in the leaves and tissues of the xylem. Because plants lose water through openings in the leaves called *stomata* (singular = stoma), they must regain water. Therefore, the inspiration for transpiration is the loss of water. The loss of the medium that carries necessary minerals inspires the plant to pull more water in from the ground.

Mighty important minerals (to a plant, anyway)

You may have read that minerals move through the water being pulled up through a plant. Have you wondered which minerals they are? Well, to satisfy your curiosity, I've devised a handy table of which minerals plants need and what happens if plants don't get them (Table 5-5). You should clip this table out, stick it in your wallet, and keep it with you at all times (definitely kidding). Then, find out what plants do with their minerals and how they digest their nutrients in Chapter 6.

Table 5-5 Important Minerals for Plants

Mineral	*Effect of Deficiency*
Boron	Leaves on ends die and fall off early; plant growth is stunted; flowers and seeds usually aren't produced
Calcium	Leaves roll and curl, roots are poorly developed and may look gelatinous
Chlorine (a gas)	Tips of leaves wilt and blacken
Copper	Terminal shoots wilt and die; color of leaves is faded
Iron	White marks in veins; bleaching of leaves, tips of leaves look scorched
Magnesium	Veins look green, but leaf tissue looks white or yellow, brittle; leaves may wilt, fall off, or die
Manganese	Same as magnesium but stems yellowish green and often hard
Molybdenum	Leaves are light yellow and may not grow
Nitrogen (gas)	Stunted growth; leaves light green, then yellow, then dry out and fall off
Phosphorus	Stunted growth; leaves sometimes look purplish; stems thin
Potassium	Leaves pale green or streaked yellow, and wrinkled between veins
Sulfur	Leaves light green to yellow; stems thin
Zinc	Leaves die; white streaks between veins in older leaves

Chapter 6

Using Energy to Keep the Motor Running

In This Chapter

- Looking at digestive systems in different animals
- Running through the human digestive system
- Seeing how plants break down glucose and use and create energy
- Following reactions of photosynthesis, photophosphorylation, and photolysis
- Going through the glycolytic pathway

Once an organism ingests (or absorbs) food, the organism's digestive system immediately starts breaking down the food to release the nutrients. In this chapter, you can find information on the digestive system of humans and other animals, as well as how plants form and break down glucose starting with energy from sunlight.

Down the Hatch: Types of Digestion

Here's something for you to digest: different types of digestive systems. I start with the most simple design and work my way up to the complex human system. Ready?

The smallest digestive system can be found in organisms made up of just several cells. However, the system, in which specialized cells surround a digestive cavity (gut), gets the job done. The cells acquire nutrients that the other cells in the organism can use for processes that keep the organism alive.

As organisms increase in complexity, they expend more energy and thus require more nutrients. In keeping with their more complex metabolism, their digestive systems are more complex. However, the process of getting nutrients into the entire organism happens at the cellular level; that is, the cells lining the digestive tract are responsible for taking in the nutrients and passing them on to the rest of the body, regardless of the size of the body. Mice digest, and so do elephants. The process in both is fairly similar.

Intracellular digestion

Materials can enter cells in four different ways. These are methods of *intracellular digestion* (meaning that digestion occurs inside the cells/organisms).

- *Active transport:* This method requires that energy (in the form of adenosine triphosphate, ATP) be used to move nutrients across the plasma membranes separating the cells of the digestive system and into the other cells of the organism
- *Diffusion:* This method relies on simple movement of molecules from where the concentration of nutrients is high (such as in the environment of bacteria in a compost heap) to an area of lower concentration of nutrients (such as into the bacteria).
- *Phagocytosis:* This method involves an organism (or a cell) engulfing solid nutrients. The cell surrounds the material that it is going to "eat," pulling the nutrients inside it and forming a *food vesicle*. The food vesicle connects with a specialized cellular organelle called a *lysosome*. The lysosome contains enzymes that can digest the solid material in the food vesicle. The nutrients are released from the solid material and then absorbed through the membrane of the food vesicle and into the rest of the cell.
- *Pinocytosis:* This method is just like phagocytosis, except that instead of solid material being engulfed, liquid droplets are taken inside the cell, forming a pinocytotic vesicle (instead of a food vesicle).

The word *phagocytosis* means "cells eating" (*phago-* = eating; *cyto-* = cells). Do you have a mental picture of Pac Man? *Pinocytosis* means "cells drinking" (*pino-* = drink; *cyto-* = cells).

Extracellular digestion

In some organisms, digestion occurs outside the cells of the digestive system; when it does, it is referred to as *extracellular digestion*. Some organisms that "eat" their food this way include fungi and parasites. Some of these organisms digest organisms that already are dead and decaying; some feed off of living organisms.

Did you ever take a walk through the woods and notice a dead tree with a fungus growing on it? The fungus is "eating" the tree, much like bacteria help to decompose dead animals. (I'm sure you've experienced the stench of road kill on a hot summer day when you have your windows rolled down.) Plants (specifically fungi) that survive by eating decomposing organic matter are called *saprophytes* (*sapro-* = rotten; *phytes* = plant).

And just how does a fungus "eat" a tree? It secretes enzymes that break down the cellulose in the tree bark. This form of digestion occurs outside of the fungus, though, so it is extracellular. The nutrients released when the bark is decomposed are absorbed and used by the fungus.

Although a fungus is classified as a plant instead of an animal, it does not have the capability that plants have of producing its own organic material for fuel. Fungi have to get their nutrients from outside sources. Sometimes the fungus lives on the outside of the source, and sometimes it grows inside an organism.

The Ins and Outs of Digestive Systems

Medium-sized and larger animals (in comparison to those multicellular organisms) have more appropriately sized digestive systems for their bodies. This section takes a look at some different types of digestive systems and how they function in breaking down food into nutrients that are needed by the animal.

Incomplete versus complete digestive tracts

Of the animals that you can see without a microscope, the ones with the most primitive digestive system are animals that have a gut with just one opening that serves as both mouth and anus (yuk!). An example of an animal with this type of *incomplete* digestive tract is the jellyfish.

Increasing in complexity are animals that have gut tubes, where foods are digested and nutrients are absorbed, with a mouth at one end and an anus at the other. Although simple, this type of system is *complete*. An example of an animal with this type of system is an earthworm.

So, what's the benefit of a complete digestive system? Well, for one thing, the mouth doesn't serve as the anus! Seriously, though, a "flow-through" system allows food that was just *ingested* (taken in) to be *digested* (broken down) before it gets *egested* (excreted out) along with materials taken in earlier. An organism with a complete system does not have to take in food constantly to replace food that is egested before nutrients could be acquired.

Continuous versus discontinuous feeders

Animals that must "eat" constantly because food is taken in and then pushed out soon afterward are called *continuous feeders*. Most of these animals either are permanently attached to something (such as clams or mussels) or are *very* slow moving.

Clams also represent a group of animals called *filter feeders*. These animals siphon water and filter out food particles using their two valves. Because of that feature, they also are called *bivalves; bi-* meaning "two" and, well, valves meaning valves. (Okay. So a clam is a continuous, bivalved filter feeder. Are you picking up on the fact that scientists just *love* to categorize things? If this type of organization turns you on, check out Appendix A!) One valve opens to siphon the food from the water, and then the other valve opens to release the filtered water. This happens continually throughout the organism's life — kind of like breathing, for you and me.

Animals that are *discontinuous feeders* consume larger meals and store the ingested food for later digestion. These animals generally are more active and somewhat nomadic. The ability to "eat and run" serves a preying carnivore well. Otherwise, an animal such as a lion would need to be hunting and eating constantly, which would exhaust the lion and also cause the lion to be out in the open savannah much longer, taking the chance that he could be prey for another predator. (Yes, lions have predators, too. Although they are near the top of the food chain, there really is no one animal that claims a top spot. Just as you and I are near the top, the possibility exists that during a beautiful hike through the Rockies, a hungry grizzly bear with cubs in tow might take the top spot and knock us down a link on the food chain!)

Although you might find yourself snacking and grazing constantly throughout a day, you really are a discontinuous feeder (or should be). This means that you can consume food rapidly, but digest it gradually and then (in theory) not have to eat again for several hours.

However, you and all the other animals that are discontinuous feeders must have a place in the body to store the food. In humans, this organ is the *stomach*. In birds, insects, and worms, this organ is called the *crop*. In rodents, food is stored in pouches in their cute, little cheeks.

An Inside Look at the Human Digestive System

You know that the mouth is where you put your food, but did you realize it was part of your digestive system? Well, it is. The act of chewing (the technical term for it is *mastication*) is the first step in digesting. Your body must break down food into smaller and smaller pieces so that the nutrients contained in the food can be released from the food and used by your body. After all, the true purpose of eating and digestion is to gain nutrients to keep your body functioning.

Mmmm. . . . Your mouth is a busy place

During *mechanical digestion*, the teeth start breaking down the food into smaller bits, but the rest of your mouth gets involved, too. Your taste buds detect the chemicals that make up the food you are eating — such as carbohydrate, protein, fat — so that the appropriate enzymes are produced and secreted throughout your digestive system.

Your saliva contains an enzyme that starts breaking down long carbohydrate molecules as soon as you put food in your mouth. *Salivary amylase* is the enzyme that begins splitting apart the bonds between glucose molecules in a long chain of starch. You know how you salivate just before you are about to eat something? That's the effect of your eyes or nose sensing something delicious, sending the message to the brain that you are about to open your mouth and take a bite, and your mouth's way of getting ready by producing saliva containing salivary amylase. Try to be aware of this happening the next time you eat. Or, you may discover that it is happening right now, just because you are *reading* about salivating! (Hey, your brain is still getting the message "salivating!")

The mouth also secretes in the saliva a potent lipase called *salivary lipase*. It is secreted continuously and accumulates in the stomach between meals. Between 10 percent and 30 percent of dietary fat is hydrolyzed in the stomach by this enzyme.

Once your teeth have chewed, your taste buds have sent on the information about what you are eating, and the enzyme in your saliva starts breaking apart starches, you are probably ready to swallow. Your tongue pushes the chewed food to the back of your throat, and you swallow the food down your esophagus. Saliva makes this process much easier, too.

Learn the Heimlich maneuver!

If you ever "swallowed something wrong" and began to cough and choke, it was because the swallowed food entered the *trachea* instead of the esophagus. The trachea is also known as the windpipe, and it carries air that you breath in through your nose down into your lungs. Both the trachea and esophagus are tubes in your pharynx (throat). The esophagus connects to your stomach. If food gets into the trachea by mistake (that is, "goes down the wrong pipe"), your body's natural reaction is to cough it up to get it out of there. Food that plugs up the trachea or gets into the lungs (called *aspirated* food) can prevent the lungs from functioning properly and cause a person to suffocate. A lack of oxygen for just a few minutes can lead to brain damage or death. So, chew thoroughly and swallow carefully! And, learn the *Heimlich maneuver* so that you can help someone who might be choking. You never know. Somebody else who is reading this book may do so and help *you* some day.

Plop. Your swallowed food just dropped into your stomach. Now what? Salivary amylase stops breaking apart the starch molecules, and the enzymes in the stomach take over.

Breaking the chain

The enzyme pepsin starts breaking down proteins into smaller chains of amino acids (the building blocks of proteins) when food drops into the stomach. When pepsin starts acting on the proteins in swallowed food, a hydrolysis reaction occurs, which breaks apart the long protein chains (polypeptides) into smaller pieces (dipeptides or peptides).

But here's an interesting question: If pepsin breaks down proteins, why doesn't it destroy the proteins that make up the tissues in your digestive tract?

Well, when pepsin is secreted, it is in its inactive form called *pepsinogen*. Because it is inactive, it does not damage the cells that make it. When it is in the cavity of the stomach, pepsinogen is converted to its active form, pepsin, by losing a few dozen of its amino acids. The lining of the stomach is not affected by pepsin because pepsin acts only on proteins (mostly collagen, found in animal tissues), and the stomach is covered with mucus (a substance made from fats) that protects the protein-containing tissues — that is, unless you have a stomach ulcer. (Check out the sidebar "Stomach ulcers.")

Stomach ulcers

The lining of the stomach can be eroded by digestive enzymes. This condition is called an ulcer. About 20 years ago, the medical community thought that stomach ulcers were largely caused by stress, worry, frustration, and other negative emotions. However, several years ago, a bacteria called *Helicobacter pylori* was found to cause an infection in the stomach lining. Because the mucous lining of the stomach is inflamed by the bacteria, the stomach acids and enzymes secreted during digestion are able to eat away at the proteins in the tissue of the stomach wall. Ulcers can be painful, and if they bleed or cause a hole (perforation) through the wall of the stomach, they can become a medical emergency. Luckily, because most ulcers are actually caused by *H. pylori*, physicians often can treat the condition with an antibiotic instead of surgery, as was common not too long ago.

Once the stomach has continued to churn up the food particles, and the stomach acid has started breaking down the peptide bonds in proteins, the whole goopy substance is squirted into the top of the intestines. The *pyloric valve*, which is the "gate" between the stomach and small intestine, is opened occasionally by the *pyloric sphincter* muscle to allow a little of the stomach contents into the duodenum of the small intestine.

The long and winding road

Don't let the word "small" fool you. The small intestine is much longer (10 feet long) than the large intestine (5 feet long). The term "small intestine" refers to the fact that this part of the intestines is narrow in diameter; the large intestine is wider in diameter, but shorter in length.

When partially digested food enters the small intestine, it mixes with *bile* and *pancreatic juice*. These substances are secreted by the liver and pancreas, respectively, into the small intestine to help digest fats and carbohydrates. Bile salts emulsify fats; that means they help to liquefy the fats. *Lipase*, an enzyme that is part of the pancreatic juice, breaks apart fat molecules into fatty acids and glycerol. *Pancreatic amylase*, also in the pancreatic juice, continues to break down carbohydrates into disaccharides. Enzymes called — get this — *disaccharidases* break apart the disaccharide molecules into monosaccharides that can be absorbed through the cells lining the small intestine.

So, at this point in digestion, the carbohydrates and fats in your food are broken apart to their simplest forms in the small intestine. Only the proteins need to be broken down further.

There are two more enzymes in the pancreatic juice that help digest proteins. Like pepsin, their inactive forms are secreted, and they become activated in the cavity of the small intestine. The inactive forms are *trypsinogen* and *chymotrypsinogen. Trypsin* and *chymotrypsin* are the active forms that do the work of breaking apart peptide fragments. Once the peptides are broken down to small chains, *aminopeptidases* finish them off by breaking apart the peptides into individual, absorbable amino acids.

Finally, after several hours in the digestive system, the carbohydrates, fats, and proteins all are in their smallest components: monosaccharides (such as glucose), fatty acids and glycerol, and amino acids. Now they can be used by the body. But to be used by the entire body, they have to leave the digestive system.

Absorbing the Good Stuff, Passing the Junk

Nutrients that the body can use are absorbed into the cells lining the small intestine. The rest of the material that cannot be further digested or used passes onto the large intestine.

Worthy nutrients remain in your system

Sugars such as glucose that were gleaned from carbohydrates, as well as amino acids that previously made up proteins in your food, pass directly into the cells of the small intestine by active transport. Remember, this means that energy — in the form of adenosine triphosphate (ATP) molecules — is used to move the sugar and amino acids into the intestinal cells. *Capillaries*, which are the smallest type of blood vessels, surround the outside of the small intestine.

By way of *capillary exchange*, the sugars and amino acids get into the bloodstream. Capillary exchange serves as a trading system. The small intestine sends the beneficial nutrients into the capillaries of the circulatory system, and the capillaries dump off the cellular garbage collected from around the entire body (for example, dead blood cells) into the small intestine so that they can continue on through the excretory system.

The sugars and amino acids that are now inside the capillaries get shuttled through the bloodstream to the liver. However, the products of fat digestion get coated with proteins and are then called *chylomicrons* (see Chapter 5).

Instead of being carried through the bloodstream, the chylomicrons get transported through the lymph system, which deposits lymph fluid into veins near the heart.

Callin' on the colon to dump the junk

Once nutrients have passed out of the small intestine, whatever material remains continues on to the large intestine (the *colon*). Here, most of the water contained in the leftover material gets reabsorbed back into the body. An error in this absorption results either in constipation (too much water is absorbed) or diarrhea (not enough water is absorbed). After the water is reabsorbed, the waste materials compact into a solid (feces).

The large intestine absorbs ions (such as sodium) into its cells from the material passing through it. Sodium ions are necessary for many cellular processes, such as the active transport of materials across cell membranes. The large intestine also collects (from the bloodstream) ions to be excreted, helping to regulate the amount of ions in the body. If the amount of ions in your body (also called electrolytes) is not in the normal range, serious effects occur. For example, if your level of sodium and potassium electrolytes is abnormal, the ability for muscles to contract properly or for nerves to send impulses correctly is affected, and that can affect your heartbeat, possibly causing a heart attack.

Why you *must* wash your hands

Although the bacteria that produce vitamin K in your intestines are helping to keep you healthy, they can be extremely harmful if they are anywhere in your body other than your intestines. One of the bacteria that live in your colon is *Escherichia coli* (*E. coli*). There have been numerous news reports during the past few years about deaths and illnesses related to foods contaminated with *E. coli*, but those problems were caused by a strain of *E. coli* (called O157:H7) that does not normally live in the intestines. However, any strain of *E. coli* that is ingested can cause diarrhea and vomiting. Serious *E. coli* contamination also can contribute to sepsis (bacteria in the bloodstream, traveling around your body causing infections elsewhere), which can lead to coma and death. The No. 1 way that *E. coli* gets into food is from dirty hands. When you wipe yourself after defecating, some of the bacteria that were excreted with your feces can easily get onto your hands. If you don't wash your hands to remove any bacteria (*E. coli* is not the only bacteria that live in your intestines!), and then you pick up food with your hands and eat it, you may ingest the bacteria and make yourself sick. And, if you don't wash your hands, but then touch someone else (or someone else's food), you could make them very sick. So, *please* wash your hands after going to the bathroom; if not for your sake, then at least for everyone else out here!!

Several types of bacteria call the large intestine home. During digestion, some of the bacteria produce vitamin K, which humans need but cannot produce. This necessary product from these friendly little beneficial bacteria is absorbed through the lining of the large intestine.

When digestion and absorption are finished, the body has what it needs (or what it has to work with at the time), and the feces pass out of the colon into the rectum. The rectum is like a holding tank. When the rectum is full, you feel the need to *defecate* (meaning remove fecal material). This feat, signaling the end of the digestive process, is performed through the anus.

Back to the liver

Sugars and amino acids get transported to the liver for a very specific reason. The liver is like the quality control department in this factory called your body. Blood flows through the liver, which can detect any abnormalities in blood levels of various substances and start to correct them.

For example, the liver can detect the level of glucose in the blood. If the level is too high (hyperglycemia), the liver removes some of the glucose from the blood and turns it into glycogen to store it. If excess glucose is still in the blood after the liver has made enough glycogen, the liver switches its metabolic process to storing the extra glucose as fat. The fat molecules are then carried by the bloodstream and deposited around the body. (You know where they are!)

If the level of glucose in the blood is too low, the liver breaks down some of its stored glycogen back into glucose and puts the glucose into the blood. If all of the glycogen stores are used up, the liver starts to break down some stored fats to get some glucose. However, if fat stores are used up (such as during starvation), the body starts to break down amino acids to get the carbon-oxygen-hydrogen molecules the body so desperately needs. To get the amino acids, however, proteins (such as muscles) in the body are broken down. Remember that the heart is a muscle, so eventually starvation leads to death.

Although sugar seems to be such an evil (because an excess of it is stored as dreaded fat), just the right amount of glucose needs to be in the bloodstream. Why? Because glucose is the main fuel for your brain, and I show you how important your brain is in Chapter 10.

Plants Digest, Too

If you have an idea of what animals do with the foods they take in, I'll tell you what plants do before it's time to *leave*. I can nip this kind of humor in the *bud*, if you want. But these puns *stem* from mom, and she can go on for *flowers*. . . . Okay. I'll just plant the *seeds* of knowledge in you. (Don't forget to water them and give them plenty of sunlight!)

How Plants Absorb Nutrients and Create Fuel

Plants have roots that stick down into the earth. The roots pull water, which has nutrients dissolved in it, up from the ground. (Chapter 5 provides more information about this if you need it.) A few special forces cause the water to move up the stem of the plant through the specialized tissue called the xylem:

- **Osmosis:** *Osmosis* uses the difference in concentrations of nutrients between the soil and the root to move water (and nutrients) into the plant. More minerals and nutrients are in the center of the root, which is an area called the *stele* or *vascular cylinder* (higher concentration), than are in the outside of the root (lower concentration). One reason is because the water and nutrients keep moving toward the center of the root to the xylem, which is a tube that then sends the water and nutrients up the root and into the stem. During osmosis, water moves from an area of lower concentration to the area of higher concentration.
- **Capillary action (adhesion):** Once the water and nutrients are inside the xylem, adhesion and cohesion continue to move the water up through the plant. Adhesion occurs when the water molecules cling to the xylem tissue (kind of an "opposites attract" thing). Adhesion provides the force to pull water up the sides of the tube in the xylem.
- **Cohesion-tension:** *Cohesion* occurs when water molecules stick to each other. Cohesion causes the water in the tube of the root and stem to become one long column of fluid and nutrients. As water evaporates from the plant into the atmosphere (called *transpiration* in plants but *respiration* in animals), the column of water continues to move up to fill the space left by the water molecules that were "pulled out" of the leaves upon evaporation. This force of water evaporating from the leaves is called *bulk flow,* and it is ultimately caused by the sun's energy. Why? Because the sun heats the water inside the leaves, causing the water to evaporate (much like a pot of water on a stove produces steam as water evaporates — you just don't see steam coming from plants).

Photosynthesis and transpiration

When the sun heats the water in plant leaves, it not only results in transpiration, but it causes photosynthesis to occur as well.

Photosynthesis is the biochemical process of energy from the sun splitting water molecules inside the plant and combining with carbon dioxide molecules so that a hydrolysis reaction occurs, creating molecules of glucose that the plant can consume as fuel and oxygen that animals can use in their bodies.

The energy from the sun is "turned into" energy inside the plant. (Then, the energy in the plant is transferred to the animal that eats it, and so on, and so on. See the info on food chains in Chapter 5.)

The equation for photosynthesis looks like this:

$$\text{Light} + 6\ H_2O + 6\ CO_2 \rightarrow C_6H_{12}O_6 + 6\ O_2$$

I gave you the basics of photosynthesis in Chapter 5, but I'm going to give you a few more details here about how this process happens. The process of photosynthesis is such an important process because it is the basis of all food chains. Without plants undergoing photosynthesis, animals would not be able to harness the energy from the universe or obtain oxygen to breathe.

The devastation of the rain forests is such an issue because without those trees and plants working as a virtual oxygen factory, humans are removing a key ingredient to their survival, as well as to the survival of other animals. The energy and oxygen that plants provide for animals can come from nothing else.

Plants and animals work together: Decaying animals and feces provide the organic matter in the soil that plants need, and plants provide carbohydrates and oxygen that animals need.

"Phyl"ling you in on chlorophyll

To get deeper into photosynthesis, you have to put on your chemistry hat for a while. Ready? Inside the cells in a leaf are molecules of chlorophyll, which is a pigment. A *pigment* provides color because it absorbs light at every wavelength except the certain color you can see. Do you remember the mnemonic device "ROY G. BIV" or "VIBGYOR" from junior high science? Those memory aids give you the colors in a full spectrum of light: red, orange, yellow, green, blue, indigo, violet.

In plants, you see mostly green. The pigment molecule *chlorophyll a* absorbs every wavelength of light except green. The green light rays are reflected off the leaf and picked up by your eye, which allows you to see the green. (Starting to see the light?) But, in autumn, when the *chlorophyll a* molecules are used up, you start to see the leaves in a new light, so to speak. The pigment molecule *chlorophyll b* absorbs light at every wavelength except red, orange, and yellow, so those are the colors you see when chlorophyll b is absorbing the light rays from the sun.

Dancin' down the electron transport chain

When I say a leaf absorbs energy in the form of light from the sun, what I really mean is that the energy is absorbed by the electrons in the atoms of the chlorophyll molecule. These tiny little electrons are so excited to be part of the plant that they just bounce all over the place. (Imagine a toddler who was just informed he's going to a toy store, and he can pick out something.)

While doing the happy dance, the electrons give off energy and then bounce into another pigment molecule, where they are incorporated and thrilled again to the point where they give off more energy. This cycle is called the *electron transport chain,* and it continues until the electrons bump into the "bouncers": chlorophyll a P_{680} and P_{700}. These "bouncers" don't like rowdy little electrons at this dance club, so they restrain them: that is, chlorophyll P_{700} absorbs electrons into *photosystem I*; chlorophyll P_{680} absorbs electrons into *photosystem II.*

Photophosphorylation elation

Photophosphorylation (I know, it's a long one) is the name for the chemical process that occurs when plants make that familiar energy compound ATP from the smaller molecule adenosine diphosphate (ADP) and a molecule of inorganic phosphate (P_i) — the *-phosphorylation* (addition of phosphorus) part — plus energy from light (the *photo-* part). Photophosphorylation describes the reactions of photosynthesis that use energy acquired from the absorption of light (whether the light is from the sun or a fluorescent bulb in your kitchen) and convert it to energy that is later used in the dark reactions (called the Calvin-Benson cycle; see the section called, "Making sugar in the dark").

These reactions don't really occur in a circular-type cycle, but you can start (and end) at photosystem II. Look at Figure 6-1 to follow the reactions.

Oxidation-reduction reactions

Oxidation-reduction (redox) equations describe the transfer of energy between atoms. In most chemical reactions, energy is transferred. Some molecules lose electrons, some gain electrons. In Chapter 2, I explain how electrons try to fill the outer shell of an atom. If the outer shell is not filled, the atom is charged because it is, simplistically, creating energy by looking for an electron to fill its outer shell. The more energy an atom has (the more electrons it is missing), the higher its valence number (that is, +1, +2, etc.).

A substance that gains electrons decreases its valence number, so it is said to be *reduced.* The substance that loses electrons is said to be oxidized, and its valence number increases. In a redox reaction, the substance that gains electrons is the oxidizing agent. The substance that loses the electrons is the reducing agent.

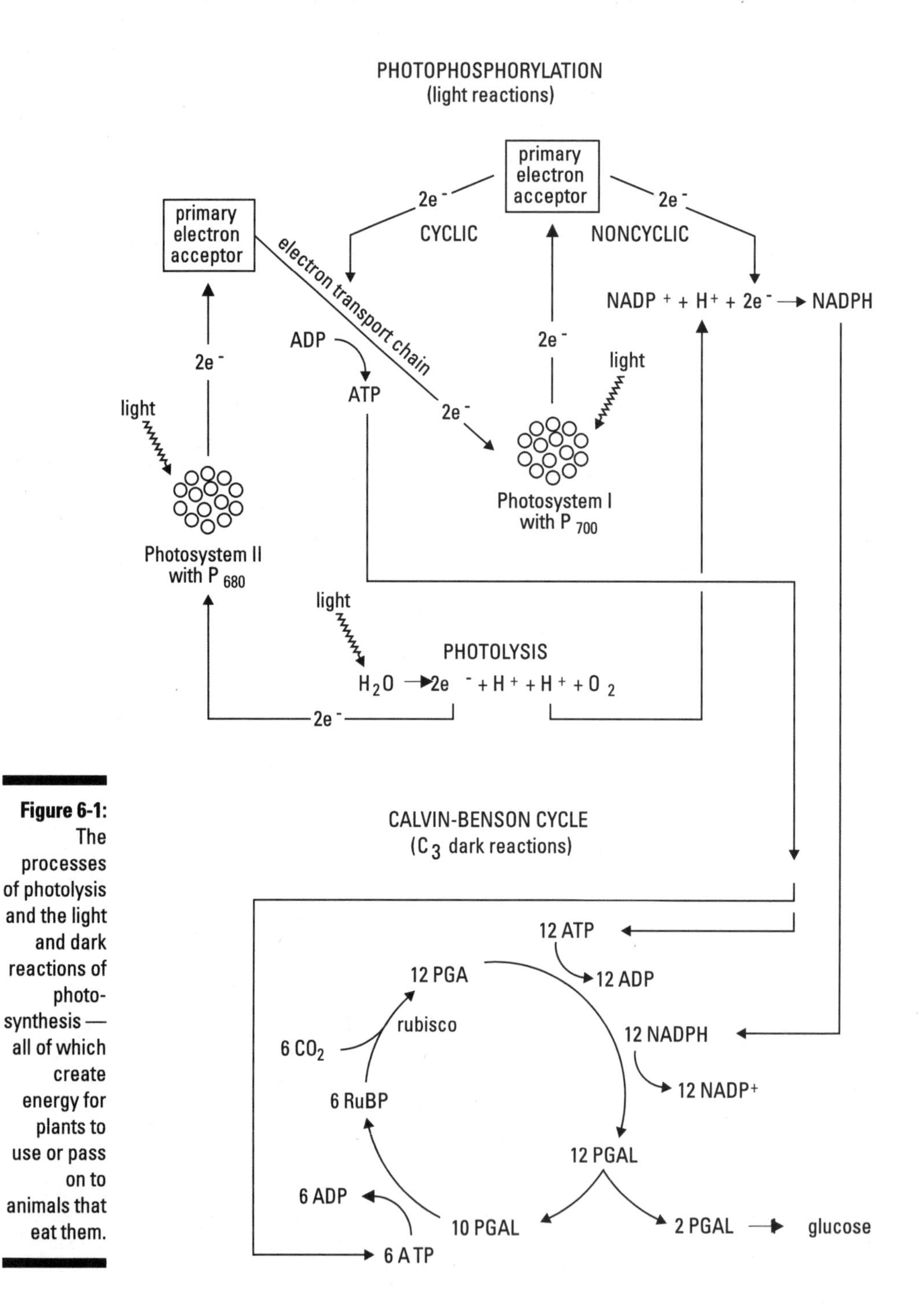

Figure 6-1: The processes of photolysis and the light and dark reactions of photosynthesis — all of which create energy for plants to use or pass on to animals that eat them.

- When light strikes the chlorophyll molecules in photosystem II, two electrons become energized and are sent to the primary electron acceptor.
- The primary electron acceptor welcomes the tiny electrons with open arms — that is, it accepts them — and sends them down the electron transport chain.
- The electrons are passed from one carrier protein to another in the electron transport chain. Some of the carrier proteins contain the metal iron, such as cytochrome and ferredoxin.
- The two original electrons lose some energy as they get passed down the electron transport chain. (Think of how babies wear out when passed from relative to relative.) The energy that they lose is used to *phosphorylate* (that is, add the phosphate molecule to ADP) the reaction that creates ATP. (*Note:* This process is how plants "harness" energy from sunlight. The energy they take in from light is turned into energy [ATP] that they can use to create their fuel [glucose]. The process of photophosphorylation is the first step in creating that fuel.)
- At the end of the electron transport chain lies photosystem I. When light energy strikes the chlorophyll molecules in photosystem I, the electrons are sent to a second primary electron acceptor. (Don't get confused, although I completely understand if you wonder why they didn't call this one the *secondary* electron acceptor! Sorry. I'm just telling you about this pathway; I didn't make this stuff up.)
- The second primary electron acceptor again welcomes the excited electrons with open arms and sends two of them up through another electron transport chain.
- At the end of this short electron transport chain, the two electrons have more energy than they do after going through the first, longer transport chain. So, they combine with a hydrogen ion (H^+) and a molecule of *oxidized* nicotinamide adenine dinucleotide phosphate ($NADP^+$) to form *reduced* nicotinamide adenine dinucleotide phosphate (NADPH).
- NADPH is a coenzyme packed with plenty of energy. So, although NADPH gets its energy by taking the two electrons that started way back at the beginning of this process, the NADPH passes on that energy when it is used to start the reactions of the Calvin-Benson cycle. (See the section "Making sugar in the dark: The Calvin-Benson cycle," later in this chapter.)

Splitting the light: Photolysis

You know that plants need water. The water molecules that a plant takes in are split by light energy to allow the reaction that creates glucose to occur. The reaction is a hydrolysis reaction (splitting water), but photolysis (splitting by light) is part of the equation, too.

Don't drown your plants!

Plants take in carbon dioxide and give off oxygen. But plants do need oxygen. They need it in the soil around their roots. The oxygen keeps the soil aerated, which allows the minerals to be dissolved into the water and absorbed through the root hairs. If the soil is waterlogged — which happens if you give your plant too much water — the water molecules displace the oxygen molecules, and the soil compacts. That means it is too heavy for the nutrients to move through it and get into the roots. This damages or kills the plant. The process is very similar to how people drown. If you were to dive into a lake and hit your head, knocking yourself unconscious so that you couldn't resurface, the water you would be lying in would displace the oxygen from your lungs, preventing you from being able to breathe or get oxygen to your brain and body cells. This damages or kills you in much the same way that overwatered plants drown in their own pots. So, be kind to your plants and don't overwater them. Oh, yeah, and don't dive into unknown waters, okay?

What happens is this: The energy from light (like the sun) splits a water molecule into two hydrogen ions and one-half of an oxygen (O_2) molecule. The creation of these ions releases two electrons. And it is those two electrons that initiate the whole process of phosphorylation. Didn't you wonder where photosystem II got the two electrons from in the first place?

Aren't you also wondering why those two electrons go to photosystem II first instead of photosystem I? Good question. Well, because the chlorophyll in photosystem II absorbs light at 680 nm, it would make sense that the electrons would go to the lower spot on the light spectrum. It may just be that photosystem I at 700 nm was discovered first, even though it is not the first place those electrons go.

So, now, you should be starting to see how water and chlorophyll — both of which are inside leaves — mix with sunlight to create fuel for the plant. But wait, there's more. At this point, the plant has created the energy-rich molecule NADPH, but it has not yet created glucose (the ultimate energy molecule). That happens in the dark reactions. (Sounds spooky, doesn't it?)

Making sugar in the dark: The Calvin-Benson cycle

The Calvin-Benson cycle, (also known as C_3 photosynthesis), is a series of reactions that occur during photosynthesis but are not initiated by light. These reactions use the energy created during photolysis and photophosphorylation.

After this cycle is performed six times, one molecule of glucose is created. Here's how it happens. You can refer to Figure 6-1, too, to follow the cycle, which is broken down into the following types of reactions:

- Carboxylation
- Reduction
- Regeneration
- Carbohydrate synthesis

Overall, the entire Calvin-Benson cycle can be put into an equation:

$$6CO_2 + 18ATP + 12NADPH + H^+ \rightarrow 18\ ADP + 18P_i + 12\ NADP^+ + 2\ glucose$$

But, don't worry. In the following sections, I break down this equation for you into smaller chunks.

Carboxylation: Fixin' the carbon dioxide

A *carboxylation* reaction is one that takes carbon dioxide, which is an unreactive inorganic molecule, and uses it to create an organic molecule that is involved in metabolic reactions.

In plants, carbon dioxide is taken from the air and, through the series of reactions in the Calvin-Benson cycle, used to create glucose. Specifically, one molecule of carbon dioxide combines with one molecule of ribulose bisphosphate (a sugar-phosphate molecule abbreviated RuBP) to create two molecules of *phosphoglycerate* (PGA), which is a molecule containing three carbon atoms. The fact that a three-carbon molecule is created explains why the Calvin-Benson cycle also is called C_3 photosynthesis: C_3 stands for three carbons. Because this entire process needs to occur six times, the overall equation is:

$$6\ CO_2 + 6\ RuBP \rightarrow 12\ PGA$$

Reduction: Packin' in the energy

Remember the ATP and the NADPH molecules that were created during photophosphorylation? Well, this part of the cycle is where they come in during photosynthesis. This step also explains why these are the "dark" reactions. ATP and NADPH bring the energy into these reactions of photosynthesis; energy from light is not used directly here — it was already converted to the ATP and NADPH molecules!

At this point in photosynthesis (after six turns of the Calvin-Benson cycle), the 12 PGA molecules from the carboxylation reaction combine with 12 ATP molecules (from 12 photophosphorylations) and 12 NADPH molecules (from 12 photolysis and photophosphorylation reactions). What all these molecules create is one big energy-dense molecule called *phosphoglyceraldehyde* (PGAL).

The energy to create PGAL is taken from the molecules of ATP and NADPH. Remember that ATP is formed from ADP and P_i; NADPH is formed from $NADP^+$ and H^+. So, when the ATP and NADPH are broken apart to release energy, they also release ADP, P_i, and $NADP^+$. But don't worry. Mother Nature would not let energy go to waste. The ADP, P_i, and $NADP^+$ are recycled back through the steps of photophosphorylation.

Regeneration: Using the second product to create the first

Back in the carboxylation reaction of the Calvin-Benson cycle, RuBP is used to create PGA. Then, in the reduction reaction, PGAL is formed. In this regeneration reaction, the PGAL is converted to RuBP so that the entire process of photosynthesis can keep occurring. I told you Mother Nature would not let energy go to waste. You can't get much more efficient than the organism creating its own food, as well as the products from which it forms fuel! Can you imagine how efficient your car's engine would be if it used not only gas but could also use the products created when gas is broken down to re-form gas? You'd never have to pull up to the pumps (and price increases wouldn't be such an issue)!

For this regeneration reaction to occur, more energy in the form of ATP is used. The overall equation is:

$$10 \text{ PGAL} + 6 \text{ ATP} \rightarrow 6 \text{ RuBP}$$

Right about now, you should be asking yourself what happened to the other two PGAL molecules.

Carbohydrate synthesis: Finally, making the sugar

If you were following along closely and keeping tabs on the numbers, you would see that in the reduction reaction a total of 12 PGAL molecules were created (after six Calvin-Benson cycles, of course). Then, in the regeneration reaction, only ten PGAL molecules were used to create RuBP. What happened to the other two PGAL molecules?

I won't leave you in suspense anymore. Look at Figure 6-1. See the point in the circle where 12 PGAL is labeled? An arrow splits off the circle labeled 2 PGAL, and that has an arrow pointing toward glucose. That means that this is the point in photosynthesis where the glucose is made! Yeah! Finally!

The Calvin-Benson cycle is also called C_3 photosynthesis because a molecule containing three carbon atoms is created? (It's called PGA.) Well, eventually PGAL is created, but that also contains only three carbon atoms. And, if you put two of those three-carbon molecules together, you create a six-carbon molecule, right? Right. And glucose is a six-carbon molecule.

In plants, other six-carbon monosaccharides like fructose also can be formed in this step. Then, the monosaccharides can be combined to form disaccharides such as sucrose or polysaccharides such as starch and cellulose.

Respiring Plants Break Down Glucose

Plants do not have digestive systems like animals do. But they do break down (digest) fuel to create useable energy like animals do. Why, you ask? Well, to grow, plants need to make more cells. Energy is needed to do that. Energy also is needed for the most basic metabolic processes in a plant, such as to continually run the reactions of photosynthesis, which ironically, convert more energy. Fuel is found not only in sugars such as glucose, fructose, sucrose, and starch, which are used throughout the life of the plant, but also in the fats in seeds, which are used to help the plant get started.

You know how animals not only take in food and water but also oxygen to survive? In animals, the process of taking in oxygen and giving off carbon dioxide and water is called *respiration*. Plants undergo respiration, too, except that they take in carbon dioxide and give off oxygen and water.

The entire process of *photosynthesis* (including both light and dark reactions) is summed up like this:

$$\text{Energy (light)} + 6\ H_2O + 6\ CO_2 \rightarrow C_6H_{12}O_6 + 6\ O_2$$

The entire process of *respiration* can be summed up like this:

$$C_6H_{12}O_6 + 6\ O_2 \rightarrow 6\ CO_2 + 6\ H_2O + \text{energy}$$

Notice anything here? Notice how respiration is the reverse of photosynthesis? It is. In plants, the creation of glucose and the breaking down of glucose happen continuously, and both major processes happen in each and every cell. There are no organ systems in plants, so the breaking down of fuel is not separated from where it is used. This design eliminates the need for a circulatory system, per se, as well as the need for a circulatory pump like a heart. And it eliminates the need for an excretory system. Because plants are so efficiently designed, they really do not create much waste. Their "waste" consists of oxygen and water, which they give off as products of this entire process of respiration.

Where it happens

In plants, the steps of respiration are fairly similar to respiration in animals (see Chapter 5). The steps include glycolysis, the Krebs cycle — which is also called the tricarboxylic acid (TCA) cycle or the citric acid cycle) — and the respiratory chain.

Glycolysis is the process of breaking down glucose. This process happens in the cytoplasm of the cell. In plants, glycolysis occurs through the glycolytic pathway (most often) or the oxidative pentose phosphate pathway. The glycolytic pathway is important because along the pathway, the substances that are produced (called *intermediates* because they are between the original substance being degraded — glucose — and the final product, which is pyruvic acid) are then used to form other important structural substances in the plant.

Once pyruvic acid is produced, it crosses into the mitochondria of the plant and starts the Krebs cycle. The rest of the process of respiration occurs in the mitochondria as well. After the Krebs cycle is completed and high-energy molecules are created, the energy is passed through a chain of events called the respiratory chain. At the end of that chain, oxygen and water are released.

How it happens

Take it slowly through these pathways. Feel free to re-read if necessary. No one is right behind you with a case of road rage. And, be sure to look at Figure 6-2 to follow what's going on.

Glycolytic pathway (glycoloysis)

This pathway turns one molecule of glucose into two molecules of pyruvic acid, two molecules of NADH, and two molecules of ATP. (Yes, if you look at Figure 6-2 you see four molecules of ATP. But, if you look really closely, you can see that this part of respiration also uses two molecules of ATP.) A total of ten chemical reactions are in this pathway; I decided to spare you the details here. But it is important to note that some of the reactions in this pathway are *reversible*. That means that a plant can make glucose by running through this same pathway backwards.

Krebs cycle

This cycle is a major biological pathway because it occurs in both plants and animals, which is why you see it mentioned so often in a biology book. If your goal is to learn just one bit of biochemistry, may I suggest this cycle?

The Krebs cycle is part of aerobic respiration. At the beginning of aerobic respiration, the pyruvic acid created from glucose in the glycolytic pathway has a molecule of NAD^+ (an electron carrier) added to it to get things moving. The reaction causes the release of carbon dioxide and the high-energy molecule NADH, and then the product of the reaction is formed — acetyl coenzyme A (acetyl CoA). In this case, acetyl CoA is a carbohydrate molecule that starts the Krebs cycle in motion.

Here's how it works:

With the addition of water and acetyl CoA, oxaloacetic acid is converted to citric acid. With the loss of water, citric acid changes to *cis*-aconitic acid. More water is taken in, and *cis*-aconitic acid becomes *iso*-citric acid.

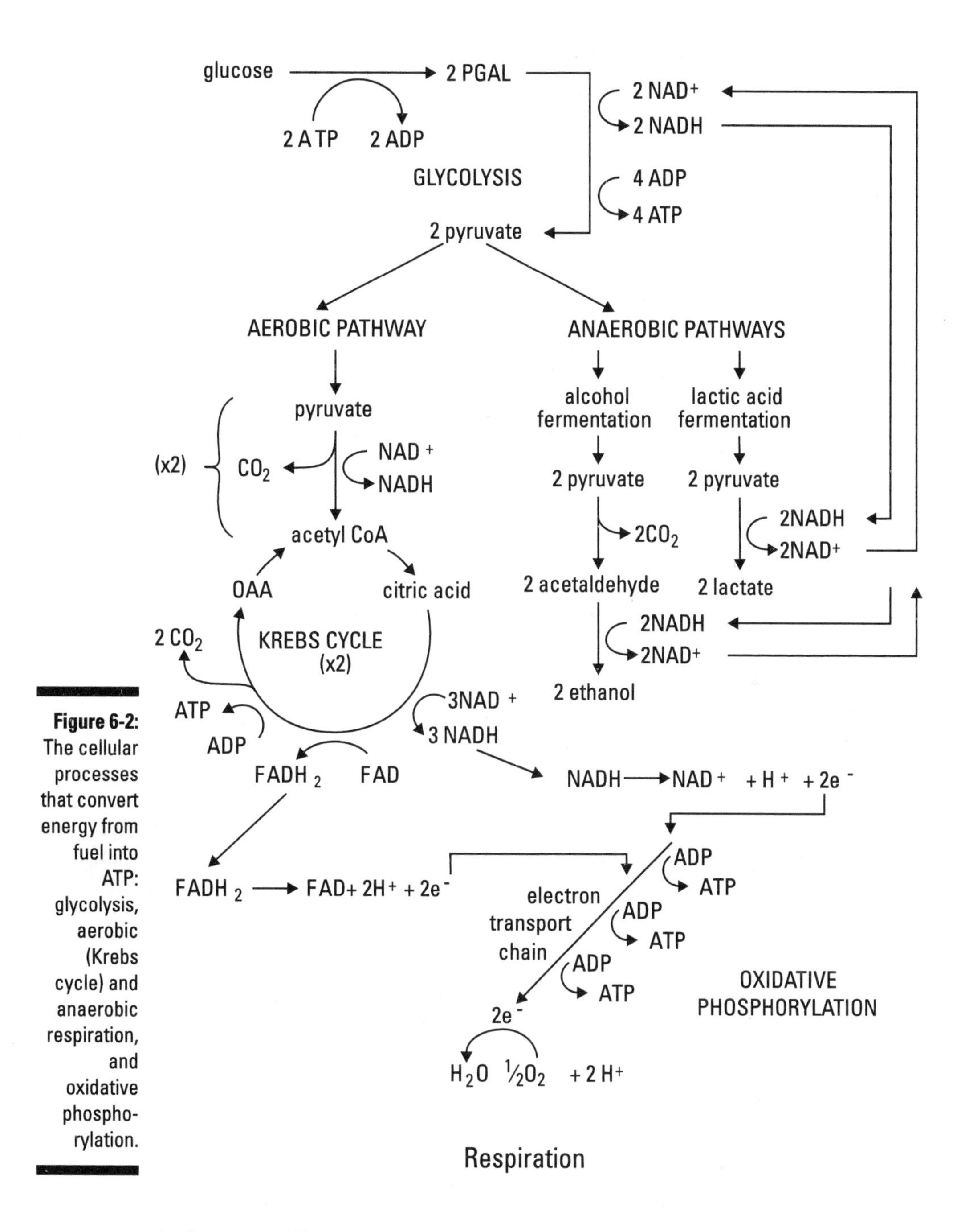

Figure 6-2: The cellular processes that convert energy from fuel into ATP: glycolysis, aerobic (Krebs cycle) and anaerobic respiration, and oxidative phosphorylation.

At this point, NAD^+ joins in, converting *iso*-citric acid to α-ketoglutarate; the reaction gives off carbon dioxide and NADH. The α-ketoglutarate converts to succinyl-coenzyme A when NAD^+ and coenzyme A are added. Carbon dioxide and NADH are given off in this reaction. Succinyl CoA is joined by guanosine diphosphate (GDP) and an inorganic phosphate molecule (P_i) to form succinic acid. Coenzyme A and guanosine triphosphate (GTP) are given off.

Succinic acid (or succinate) is converted to fumaric acid (fumarate) when oxidized flavin-adenine dinucleotide (FAD) is added. FAD is an electron carrier like NAD^+, and it is also considered to be a nonprotein coenzyme. That means it helps to pass on the energy to keep the reactions moving so that the ultimate goal can be reached. FAD is reduced to $FADH_2$ in this reaction.

At this point in the cycle, more water is added to fumarate, which converts the fumarate to malic acid. NAD^+ joins the cycle again and converts malic acid to oxaloacetic acid. NADH is given off.

At the end of one pass through the Krebs cycle, you have the following amounts of energy-rich molecules:

- Three molecules of NADH
- One molecule of $FADH_2$
- One molecule of ATP

Sure, one molecule of ATP equals one molecule of ATP. But how many ATP molecules do three molecules of NADH equal? The answer to this profound question lies in the respiratory chain and oxidative phosphorylation.

Respiratory chain

Each electron carrier produced during the Krebs cycle does its job well. NADH and $FADH_2$ — the electron carriers — are produced when their oxidized partners (NAD^+ and FAD, respectively) become reduced. If you read the sidebar about oxidation-reduction reactions earlier in this chapter, you're ahead of the game right now. If you didn't read it, here it is in a nutshell: When a substance is reduced, it gains electrons; when a substance is oxidized, it loses electrons. (If you want to know why it sounds backwards, head back to the sidebar.) So NADH and $FADH_2$ are compounds that have gained electrons, and therefore, energy.

In the respiratory chain, oxidation and reduction occur over and over. The purpose is to pass electrons. (The respiratory chain can be referred to as an *electron transport chain.*) For example, the reduced electron carrier NADH enters the respiratory chain and adopts an oxidized state. That means that it is available to become oxidized and lose an electron. Of course, that is precisely what happens, and NADH becomes NAD^+. The NAD^+ immediately adopts a reduced state, meaning that it is able to pick up an electron and become reduced. Because it "picks up" an electron, it is called an electron carrier. The picked-up electron is passed to the second electron carrier, and the process continues repeatedly until the final electron carrier is reached. At the end of the respiratory chain, oxygen is the final electron acceptor, and it becomes reduced to water.

Oxidative phosphorylation

As NADH and $FADH_2$ pass through the respiratory chain, transporting electrons, they themselves lose energy. I guess carrying electrons just sucks the energy right out of them! The energy that they lose is used to add phosphorus to ADP to create ATP. And creating ATP is the ultimate goal of breaking down fuel to generate energy.

For each NADH molecule that is produced in the Krebs cycle, three molecules of ATP can be generated. For each molecule of $FADH_2$ that is produced in the Krebs cycle, two molecules of ATP are made.

But remember that two molecules of NADH are produced when pyruvate is converted to acetyl CoA just before the Krebs cycle begins. Those NADH molecules are equivalent to three molecules of ATP.

And two molecules of NADH are produced in the glycolytic pathway. Those NADH molecules are equivalent to two molecules of ATP.

So, overall, one molecule of glucose going through the entire process of aerobic respiration can produce 36 molecules of ATP. But what happens if respiration is anaerobic?

If there's no oxygen: Anaerobic respiration

Sometimes animals are low on oxygen (after really hard workouts, long runs), but metabolism never stops. Cells continually are using and creating energy to keep a body working. During this situation, anaerobic respiration occurs to give the organism enough energy to acquire more oxygen. Anaerobic respiration in animals follows the pathway of *lactic acid fermentation*. In plants, anaerobic respiration follows the pathway of *alcohol fermentation*.

The purpose of both lactic acid fermentation and alcohol fermentation is to generate NAD^+ so that the glycolytic pathway can continue. If the supply of NAD^+ runs out, glycolysis cannot occur, and ATP cannot be generated.

Alcoholic fermentation

Sometimes humans really appreciate when plants can no longer go through glycolysis. Then, instead of breaking down sugars by the normal glycolytic pathway, the plant material ferments. The alcohol that they produce during fermentation is the source of alcohol in beer (fermenting barley) and wine (fermenting grapes).

Here's what happens: The pyruvate molecule that normally is converted to acetyl CoA at the start of the Krebs cycle during aerobic respiration instead produce acetaldehyde and carbon dioxide (bubbles in champagne and beer). The acetaldehyde uses NADH to produce ethanol (the alcohol). NADH is used because this process is happening under anaerobic conditions. Without oxygen, the respiratory chain cannot function, and NAD^+ cannot be produced.

In the process of alcoholic fermentation, NAD^+ is released when the ethanol is produced, and the NAD^+ can be used to allow glycolysis to continue. Unfortunately, if alcoholic fermentation continues long enough (that is, if oxygen does not become available), the alcohol that is produced kills the plant, and glycolysis has no need to continue.

Lactic acid fermentation

In animals, if oxygen is not available, lactic acid fermentation occurs to allow glycolysis to continue. To generate some NAD^+ to put through the glycolytic pathway, pyruvate (from the beginning of the cycle) is converted to lactic acid when NADH is added to the reaction. The lactic acid is stored in muscle tissue until oxygen becomes available. (Lactic acid also causes the pain you feel usually two days after lifting weights, doing dozens of abdominal crunches, or otherwise fatiguing a muscle group.) Once oxygen is available, lactic acid is broken down to release energy, although it provides less energy than aerobic respiration produces.

Because the lactic acid is stored up until oxygen is available, the animal is kind of behind the eight ball, so to speak, when it can get oxygen. Therefore, lactic acid fermentation is said to create *oxygen debt* because once oxygen is available, it must first be used to break down the lactic acid before it can be used in aerobic respiration.

Part III
Living Things Need to Metabolize

In this part . . .

If an organism were able to take in food and oxygen, but then its body couldn't do anything with the fuel, the organism would still die. Organisms need to process the fuel and get it to all of its cells. Nutrients must be extracted from food, and the nutrients and oxygen must be transported to the cells. At the cellular level, processes occur to get the nutrients and oxygen across the cell's borders. All these processes require energy, and as with any type of energy that is used, waste is created. The wastes must also be removed from the organism or its internal environment becomes akin to a toxic waste dump.

You read about how animals and plants get nutrients into each of their cells. Here, you find information about the circulatory system and capillary exchange in animals, as well as translocation and transpiration in plants. Then, you examine how the organisms get rid of the waste. You also take a tour of the digestive system and see how wastes are excreted by humans, fish, worms, insects, and plants. Better get out that plastic suit and rubber gloves!

Chapter 7

Taxi! Transport of Nutrients

In This Chapter

- Piecing together how the bloodstream carries nutrients around the body
- Tracing the path of blood through the heart and the blood vessels
- Finding out how capillaries and cells exchange nutrients and waste products
- Seeing how plants transport water and nutrients

Can you imagine if you continued to consume foods and products but never threw anything away? Your house would soon resemble a garbage dump. The smell would be overwhelming, and the condition would be unsanitary.

Can you imagine if you and your neighbors put garbage into garbage bags, but the garbage man never came and picked them up? Garbage would be fodder for vermin and would seep into the ground. The neighborhood would be unsightly, to say the least, and the stench of rotting organic matter would permeate the entire area. Yuck. Unhealthy, right? Right.

Can you imagine if you and your neighbors were totally disabled so that none of you could go to the store for food, nor could you contact anyone to bring it to you? Once you used up all the food in your house, you would be left to whither and starve. There would be no more food available to you and no prospect of it coming in the future.

Think of an organism as a neighborhood, and each cell in your body as a single dwelling full of neighbors. If "food" (nutrients) cannot be delivered to each home in the neighborhood, the "neighbors" (the cells) would eventually die. Or, if nutrients were available but there was no garbage service, conditions in each "home" (cell) and eventually the entire "neighborhood" (organism) would become unsanitary and unhealthy. This is why an organism needs to make sure that every cell it is made of is able to get nutrients and dispose of waste. Circulatory systems see to it that the basic services of delivery and pick-up are taken care of.

After an organism ingests food, the organism breaks down the food into the smallest possible molecules, which serve as nutrients. However, those nutrients cannot just sit in the digestive system, or the rest of the body would starve. The nutrients culled from the food the organism takes in must be transported around the body to each and every cell. In this chapter, you explore how transport happens in different organisms.

Circulating Circulation Information

The *circulatory system* is the method of transport in plants and animals. Animals must have nutrients and oxygen reach every cell in their body. Plants must have nutrients and carbon dioxide reach every cell in their "body." And, both plants and animals must have waste products removed from their systems. The circulatory system is how these things are carried around the living organism. Animals can have one of two kinds of circulatory systems: open or closed.

Open circulatory systems

This type of system is found in animals such as insects and some mollusks (snails, clams). Inside these animals is an open cavity called a *hemocoel* into which a blood-like fluid called *hemolymph* is pumped (*hemo-* = blood). A heart does the pumping, and it has holes called *ostia* through which the hemolymph is pumped in and out. The hemolymph carries the oxygen and nutrients, and when it fills the hemocoel, the tissues of the organism are flooded with the fluid. No vessels holding in the fluid here. Simple, yet effective.

Closed circulatory systems

This is the type of system you are personally familiar with. Closed systems are said to be closed because they have vessels that contain the fluid — in these animals, blood. In small animals, such as insects, the nutrient- and oxygen-filled fluid does not have far to travel to bathe all the tissues of the organism. But, in larger animals, the organisms are too big for diffusion to work from a fluid-filled cavity to all tissues and cells.

You do not have an open cavity inside your body that gets filled with blood every time your heart pumps. You have a network of "highways" that perform the transportation and keep the blood from seeping out. In animals, each blood vessel in the network is responsible for transporting nutrients and oxygen to cells and for removing wastes and carbon dioxide from cells. The blood vessels are the arteries, arterioles, capillaries, venules, and veins. Besides humans, all other vertebrates have closed circulatory systems, as do birds and some invertebrates, such as earthworms and squids.

Traveling the Highways and Biways: How Blood Circulates

The concept of circulation would be so easy to explain to you if it were possible to make a beating heart be still. But, as you are well aware (albeit subconsciously), the heart beats constantly throughout your life. If it stops, you are said to be in cardiac arrest, or you are pronounced dead. So, although I strongly discourage anyone from stopping their heartbeat, in this chapter I need you to imagine that the heart is not actually pumping continually. Consider each step that I tell you about to be a snapshot of what is going on in the body at that millisecond in time.

Ya gotta have heart

Hearts may come in different sizes and shapes, but their function remains the same in organisms that have them: They pump fluid throughout the circulatory system.

The heart of an insect

You may not be able to hear the heartbeat of a grasshopper or a cricket, but rest assured, it's in there. Insects have open circulatory systems (see the section earlier in this chapter).

A wormy heart

Although you may think earthworms are insects, they are not. Technically, they are called *annelids*. These little guys and gals have closed circulatory systems. As in people, the blood of earthworms is red because it contains *hemoglobin*, which is the iron-containing molecule that carries oxygen. The fact that earthworms have a closed circulatory system and red hemoglobin-containing blood explains why you had to dissect one in junior high school. However, earthworms are a bit more simple in design than humans. (Okay, a lot more simple.)

In humans, hemoglobin is contained in red blood cells; in earthworms, hemoglobin just floats in the plasma. Plasma is the fluid base that human red blood cells float in; it serves as a liquid medium of transport. In humans, blood vessels include arteries, arterioles, capillaries, venules, and veins, all of which traverse the entire body. In earthworms, however, there is one dorsal blood vessel (top side; think dorsal fin on a shark — you know, the one that sticks up out of the water), one ventral (bottom side) blood vessel, and a network of capillaries. The heart of an earthworm is a series of muscular rings near the thicker tip of the worm. The worm's heart pumps blood away from the heart through the ventral blood vessel. From there, it oozes into all the capillaries to reach all the cells of the worm, and then it travels back to the heart through the dorsal blood vessel.

Something's fishy with this heart

Fish have hearts, too, although the system that their hearts keep going is quite simple. The blood of a fish follows a single loop through the body of the fish — kind of like a racetrack compared with a network of highways.

A fish has a large vessel called the *ventral aorta*, through which blood leaves the heart when it pumps. The ventral aorta carries the blood to the *gills*, and the blood then passes through the capillaries along the gills to pick up oxygen. When the blood picks up oxygen, the blood then is said to be *oxygenated.* From the gills, the oxygenated blood immediately flows into the *dorsal aorta*, which carries the oxygenated blood to the rest of the capillaries in the fish. This part of the loop is called the *systemic circulation.* Once the systemic circulation is completed, the blood returns to the heart.

The circulatory system in a fish is simple and effective, but because the blood never returns to the heart after being oxygenated at the gills, the blood pressure of a fish is quite low. One pump of the heart must move the blood through the entire system. The blood needs to become oxygenated, deliver nutrients, and pick up waste all in one pump. This works for fish, but not for higher vertebrate animals, which generally have higher metabolic needs and therefore degrade food into nutrients and produce waste much more quickly.

Lub-dub, lub-dub: The beating of a human heart

Human hearts, as well as the hearts and circulatory systems of some other mammals, are complex. Because these higher animals are larger, they need to have a higher blood pressure to get the blood circulated throughout their entire larger bodies. *Blood pressure* is a force that sends the blood through the circulatory system.

Humans and other mammals have *two-circuit circulatory systems*: one circuit is for *pulmonary circulation* (circulation to the lungs; *pulmo-* = lungs), and the other circuit is for *systemic circulation* (the rest of the body). Pulmonary circulation allows blood to pick up oxygen in the lungs (and dispose of carbon dioxide). But, then the oxygenated blood needs to go back to the heart to be pumped through the rest of the body via systemic circulation. More specifically, pulmonary circulation delivers deoxygenated blood to the lungs so that it can become oxygenated and then delivers oxygenated blood back to the heart. When oxygenated blood returns to the heart, it gets pumped through the systemic circulatory system, which carries blood to all the cells in the body.

The human heart (Figure 7-1) has *four chambers*: *two ventricles*, each of which is a muscular chamber that squeezes blood out of the heart and into the blood vessels, and *two atria*, each of which is a muscular chamber that drains and then squeezes blood into the ventricles. The two atria reside at the top of the heart; the two ventricles are at the bottom. And, the heart is divided into left and right halves, so there is a left atrium and left ventricle, as well as a right atrium and right ventricle.

The reason that the heart is divided into halves is because of the two-circuit system. The right side of the heart can pump blood to the lungs, while the left side of the heart pumps blood to the rest of the body. Blood goes in both directions on each and every pump.

The cardiac cycle: Path of blood through the heart

Every minute of your life, your heart pumps about 70 times. Every minute of your life, your heart pumps the entire amount of blood that is in the body — 5 liters, which is equivalent to 2½ big bottles of soda. And, the heart is only as big as a clenched adult fist! It's small but strong. And, it never stops working from the time that it starts to beat when humans are nothing but wee little embryos in their mother's wombs until the moment they die. The heart doesn't even get an entire second to rest. It beats continually every 0.8 seconds of your life and "rests" for only 0.4 seconds between every pump.

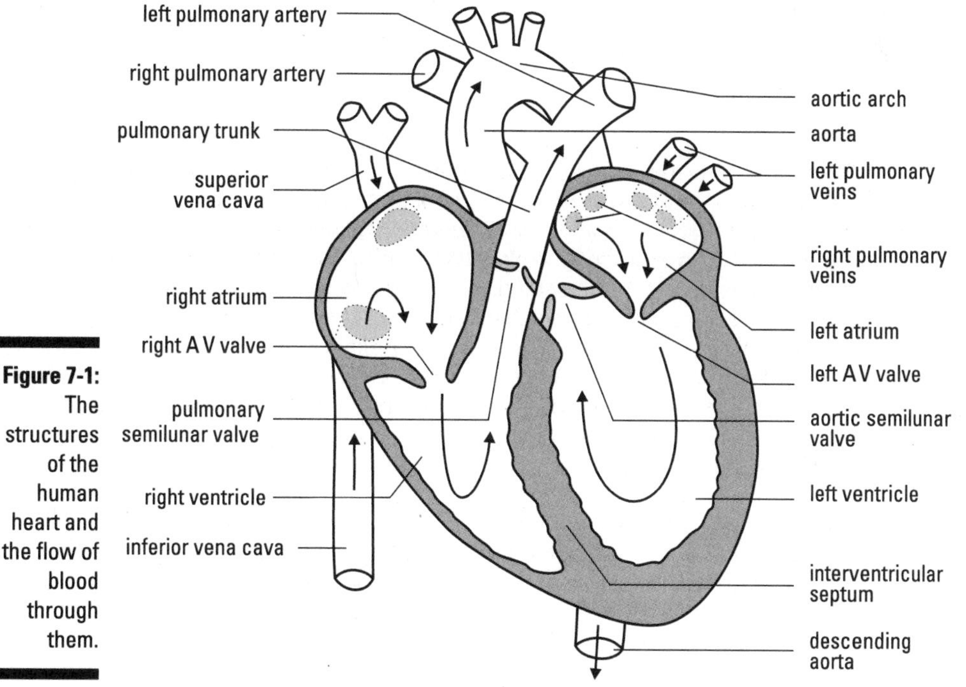

Figure 7-1: The structures of the human heart and the flow of blood through them.

The 8/10th of a second that a heart beats is called the *cardiac cycle*. During that 0.8-second period, the heart forces blood into the blood vessels plus it takes a quick nap. Here's what happens in those 0.8 seconds:

- The left and right atria contract.
- The left and right ventricles contract.
- The atria and ventricles rest.

When the atria and ventricles are resting, the muscle fibers within them are not contracting, or squeezing. Therefore, the relaxed atria allow the blood within them to drain into the ventricles beneath them. This period of relaxation in the heart muscle is called *diastole*.

With most of the blood from the atria now in the ventricles, the atria contract to squeeze any remaining blood down into the ventricles. Then, the ventricles immediately contract to force blood into the blood vessels. This period of contraction in the heart muscle is called *systole*.

If the terms systole and diastole sound familiar, it is probably because you have heard the terms systolic blood pressure and diastolic blood pressure. In a blood pressure reading, such as the normal value of 120/80 mm Hg, 120 is the systolic blood pressure, or the pressure at which blood is forced from the ventricles into the arteries when the ventricles contract; 80 is the diastolic blood pressure, the pressure in the blood vessels when the muscle fibers are relaxed. The "mm Hg" stands for millimeters of mercury (Hg is the chemical symbol for mercury).

Anyway, if your blood pressure is 140/90 mm Hg, which is the borderline value between normal and high, that means your heart is working harder to pump blood through your body (140 versus 120), and it is not relaxing as well between pumps (90 versus 80). Those readings indicate that something is causing your heart to have to work at a much higher level all the time to keep blood flowing through your body, which stresses the heart. The "something" that may be the culprit could be a hormonal imbalance, a dietary problem such as too much sodium or caffeine, a mechanical problem in the heart, a side effect of medication, or blockages in the blood vessels. The high pressure in the "pipes" also may lead to damage. Physical damage from high blood pressure is part of a hypothesis of how fibrous plaques are formed in coronary arteries.

Path of blood through the body

Once the heart contracts and forces blood into the blood vessels, there is a certain path that the blood follows. The blood moves through pulmonary circulation and then continues on through systemic circulation. Pulmonary and systemic are the two circuits in the two-circuit system of higher animals with closed circulatory systems (Figure 7-2).

Sphygmo ma what?

A sphygmomanometer (sfig-mo-ma-nom-e-ter) is the device used to measure blood pressure. Just like temperature is measured on a thermometer filled with mercury, blood pressure readings use a mercury system for measuring pressure.

The cardiac cycle, which describes the rhythmic contraction and relaxation of the heart muscle, coincides with what I am about to explain to you. As each atrium and ventricle contract, blood is pumped into certain major blood vessels, and from there, continues through the circulatory system.

Blood that is lacking oxygen is said to be *deoxygenated*. This blood has just exchanged oxygen for carbon dioxide across cell membranes, and now contains mostly carbon dioxide.

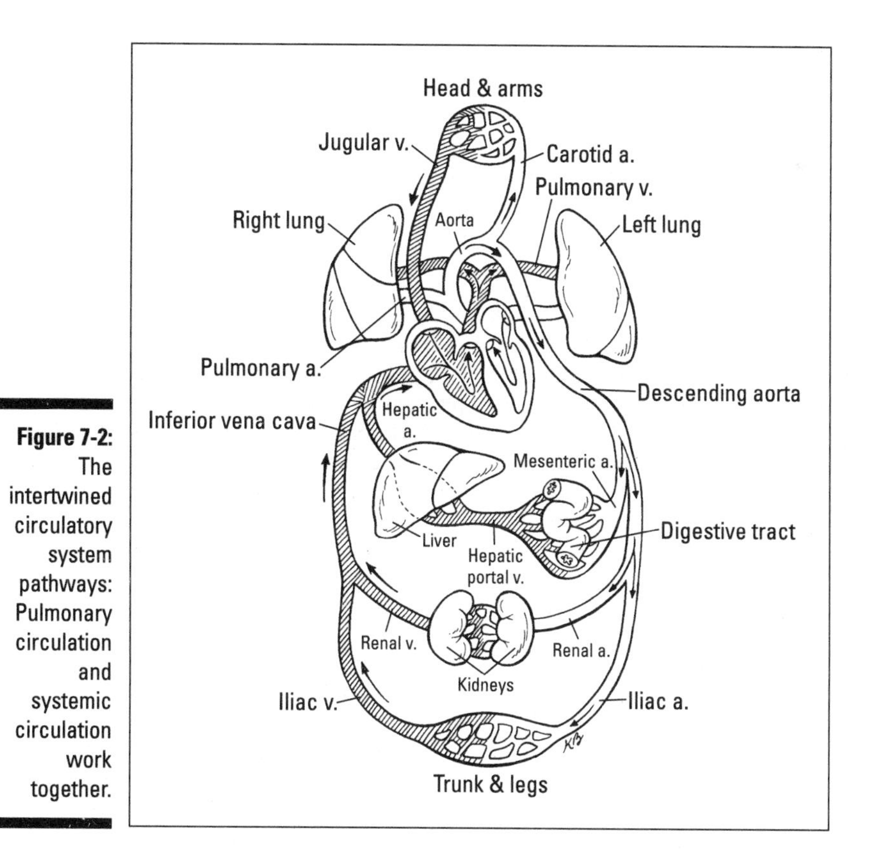

Figure 7-2: The intertwined circulatory system pathways: Pulmonary circulation and systemic circulation work together.

Figure out the heart with art

In color drawings, the arteries are depicted as being red, and the veins are depicted as being blue. There is a physiological reason for this differentiation.

The blood in the arteries is *oxygenated* blood — that is, it is full of oxygen. When oxygen combines with hemoglobin, the molecule that carries oxygen in the red blood cells, it produces the bright red color that you see oozing from a cut. It's kind of the same effect as when something made out of iron — an old car or patio furniture — sits outside for a long period of time. Iron reacts with air and forms rust, which has a reddish color. Hemoglobin contains iron, and the iron is what "holds" the oxygen in the red blood cells.

The blood in the veins is *deoxygenated* blood — that is, it is low in oxygen and full of carbon dioxide. Therefore, it is not bright red, but has more of a purplish hue. To reinforce the path of blood through the heart, I suggest that you find crayons, colored pencils, or markers and color the arteries red and the veins blue. "Vena" means veins, so in Figure 7-1, the superior and inferior vena cava are veins. The two exceptions are that the pulmonary artery should be colored blue because it contains deoxygenated blood, and the pulmonary veins should be colored red because they contain oxygenated blood.

Pulmonary circulation

Deoxygenated blood enters the *right atrium* through the superior vena cava and the inferior vena cava. Superior means higher, and inferior means lower, so the superior vena cava is at the top of the right atrium, and the inferior vena cava enters the bottom of the right atrium.

Note that in Figure 7-1, earlier in this chapter, it looks as if the right atrium is on the left side. Pretend that you are looking at someone's heart through their chest. Their right atrium is on the right side of their body, so that is the direction used, but it is on your left side.

From the right atrium, the deoxygenated blood drains into the *right ventricle* through the right *atrioventricular (AV) valve*, which is so named because it is between the atrium and the ventricle. This valve is also referred to as the *tricuspid valve* because it has three flaps in its structure. When the ventricles contract, the AV valve closes off the opening between the ventricle and the atrium so that blood does not flow back up into the atrium.

As the right ventricle contracts, it forces the deoxygenated blood through the *pulmonary semilunar valve* and into the *pulmonary artery*. *Semilunar* means half-moon and refers to the shape of the valve. Note that this is the only artery in the body that contains deoxygenated blood; all other arteries contain oxygenated blood. The semilunar valve keeps blood from flowing back into the right ventricle once it is in the pulmonary artery.

The pulmonary artery carries the blood that is very low in oxygen to the *lungs*, where it becomes oxygenated.

Systemic circulation

Freshly *oxygenated blood* returns to the heart via the *pulmonary veins*. Note that these are the only veins in the body that contain oxygenated blood; all other veins contain deoxygenated blood.

The pulmonary veins enter the *left atrium*. When the left atrium relaxes, the oxygenated blood drains into the *left ventricle* through the left AV valve. This valve is also called the *bicuspid valve* because it has only two flaps in its structure.

Now the heart really squeezes. As the left ventricle contracts, the oxygenated blood is pumped into the main artery of the body — the *aorta*. To get to the aorta, blood passes through the *aortic semilunar valve*, which serves to keep blood flowing from the aorta back into the left ventricle.

The aorta branches into other arteries, which then branch into smaller arterioles. The arterioles meet up with capillaries, which are the blood vessels where oxygen is exchanged for carbon dioxide.

Exchanging the good and the bad: Capillary exchange

Capillaries bridge the smallest of the arteries and the smallest of the veins. Near the arterial end, the capillaries allow materials essential for maintaining the health of cells to diffuse out (water, glucose, oxygen, and amino acids). However, to maintain health of cells, it is also necessary for the capillaries to transport wastes and carbon dioxide to places in the body that can dispose of them. The waste products enter near the venous end of the capillary. Water diffuses in and out of capillaries to maintain blood volume, which adjusts to achieve homeostasis.

Capillaries are the teeniest of blood vessels. They are only as thick as one cell, so the contents within the cells of the capillaries can easily pass out of the capillary by diffusing through the capillary membrane. And, because the capillary membrane abuts the membrane of other cells all over the body, the capillary's contents can easily continue through the abutting cell's membrane and get inside the adjoining cell.

The process of capillary exchange is how oxygen leaves red blood cells in the bloodstream and gets into all the other cells of the body. Capillary exchange also allows nutrients to diffuse out of the bloodstream and into other cells. At the same time, the other cells expel waste products that then enter the capillaries, and carbon dioxide diffuses out of the body's cells and into the capillaries. An example is shown in Figure 7-3.

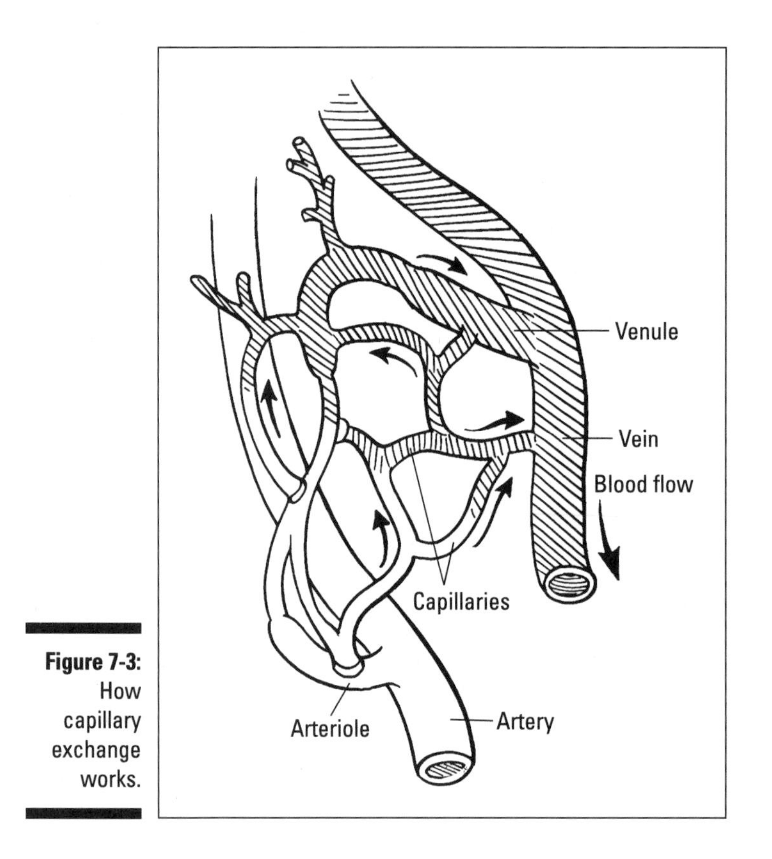

Figure 7-3: How capillary exchange works.

After the capillaries "pick up" the garbage from other cells, the capillaries carry the wastes and carbon dioxide through the deoxygenated blood to the smallest of the veins, which are called *venules*. The venules branch into bigger vessels called *veins*. The veins then carry the deoxygenated blood toward the main vein, which is the *vena cava*. The two branches of the vena cava enter the right atrium, which is where pulmonary circulation begins.

The pressure created when the ventricles contract is what forces blood through the arteries. However, this pressure declines as the blood gets further away from the heart and into the capillaries. Blood pressure does not force blood through the veins as it does through the arteries. What makes blood travel through the veins is contractions of skeletal muscles. As your limbs and trunk move, deoxygenated blood is pushed farther along the venules and veins, eventually returning to the heart. Movement of the skeletal muscles is required to force blood through the veins, which is why disabled people or people in comas need to be turned and moved. Without movement, blood pools in the veins, creating poor circulation.

What makes a ticker tick: Heartbeat generation

Did you ever wonder what made the heart actually beat? What is it that keeps it going and going and going like that little pink drumming bunny? Well, I'm here to satisfy your curiosity.

The heart contains *nodes*, which are special areas of tissue infused with nerves. The most important node is the *sinoatrial (SA) node*, which is the *pacemaker*. Yes, your heart already contains a pacemaker. People who have pacemakers "installed" through surgery have it done because their original natural pacemaker stops working correctly.

The SA node spontaneously contracts the left and right atria and then sends an impulse through the nerves in the node that causes the next node to take action. This next node is the *atrioventricular (AV) node*.

The AV node is located in the lower part of the right atrium, and it generates an impulse that stimulates an area of tissue called the *bundle of His* (pronounced "hiss"). The bundle of His lies between the right and left ventricles and connects with specialized fibers called *Purkinje fibers*. When the impulse reaches the Purkinje fibers, it causes the ventricles to contract. And, the heartbeat is complete.

The sound that the heart makes — described as lub-dub, lub-dub — is attributed to closing of the heart's valves. The first heart sound — the lub — is caused by AV valves closing to prevent backflow from the ventricles into the atria. The second heart sound — the dub — occurs when the semilunar valves close to keep blood in the aorta from flowing back into the left ventricle.

Hearts Can Be Broken: Heart Disease

Hearts *can* be broken. Not just by lovers who cheat, friends who betray trust, relatives who die, or cherished pets that are lost. Hearts and their accompanying circulatory systems can be broken — meaning injured or damaged — not only by trauma but also by genetic predisposition, viruses, and lifestyle choices.

Heart disease is the No. 1 cause of death in the United States. Maybe that's because there are so many facets of heart disease, as opposed to deaths from plane crashes, which obviously are caused by planes crashing. Deaths from heart disease usually are attributed to heart attacks, but heart attacks can be caused by many factors: *atherosclerosis* (blockages in the arteries), *ischemia* (lack of oxygen) *thromboembolism* (blood clots that travel through the bloodstream and block blood vessels elsewhere in the body), or *hypertension* (high blood pressure).

Atherosclerosis occurs when fats, especially cholesterol, accumulate in the lining of the arteries. Cholesterol is part of the fat transport system through the bloodstream, and it is needed as the precursor of steroid synthesis. But, when too much cholesterol is in the body — which can be caused by diet or by genetic factors — it starts to stick to the vessels, rather than pass through them. The fatty deposits are called *plaques*. As the plaques increase in size, they fill more and more of the artery, eventually affecting blood flow.

If arteries are partially blocked, *ischemic heart disease* may occur. People with ischemic heart disease have difficulty breathing during exercise or times of stress because the blocked arteries slow the blood flow, which prevents enough oxygen from being delivered to the heart muscle tissues. The lack of oxygen can cause a pain in the chest that radiates to the left arm. This pain is called *angina pectoris* (*pectoris* refers to the chest, as in the pectoral muscles).

If an artery is blocked by a plaque, blood cells can stick to the plaque, eventually forming a blood clot. A blood clot stuck in a blood vessel is called a *thrombus*. If the thrombus breaks free and moves around the bloodstream, it is called an *embolism*. A *thromboembolism* is a blood clot that breaks free from where it was formed, travels through the bloodstream, and blocks another spot in a blood vessel.

Although the heart fills with blood to pump to the rest of the body, it also must have nutrients and oxygen supplied to the tissues that form it. Because your heart is made of living cells and tissues, it needs to have blood vessels running through it so that nutrients and oxygen can be delivered to the cells within the heart. Arteries that bring blood to the tissues and cells of the heart are called *coronary arteries*.

If a clot or plaque blocks a coronary artery, oxygen cannot be delivered to the muscle tissue, and a *heart attack* occurs. The tissue in the area of the heart where the attack occurs often dies because of the lack of oxygen. The technical term for a heart attack is a *myocardial infarction*. *Myo-* means muscle, *cardi-* means heart, and an *infarct* refers to dead tissue.

What's Blood Got to Do with It?

Blood is the fluid that sustains life. Some blood cells carry oxygen, which is necessary for metabolic reactions, some blood cells fight off invading substances that could destroy your cells, and other blood cells help to form clots, which keep your body from losing too much of this precious fluid and help in wound healing. The fluid portion of the blood carries nutrients needed to fuel each cell in the body. It also shuttles wastes that need to be transported to the excretory system to be passed out of the body and carbon dioxide that needs to be transported to the lungs to be exhaled. So, although the heart and blood vessels form the circulatory system, they would be totally unnecessary without blood.

The elements of formed elements

The formed elements of the blood consist of the "solid" parts. I use the word solid only to differentiate from the liquid portion of the blood. The cells definitely are not hard and solid, or they wouldn't be able to squeeze through capillaries.

Red blood cells

The red blood cells, which are also called erythrocytes (*erythro-* means red; *cytes-* = cells) have the important responsibility of carrying the oxygen throughout the body. Hemoglobin, the iron-containing molecule that harnesses oxygen, exists in the red blood cells. Hemoglobin not only binds oxygen and transports it to capillaries, it also helps to transport carbon dioxide from the capillaries back to the lungs to be exhaled.

By transporting oxygen and hemoglobin, blood is an extremely important part of *homeostasis*. Homeostasis is how your body tries to constantly achieve and maintain balance. Homeostasis is a conglomerate of processes that allows your body to adjust to changes in external temperatures and hormone levels. But, many of the processes that occur to help your body adjust to changes could not happen without the blood transporting certain hormones, nutrients, oxygen, or electrolytes.

If a person has too few red blood cells, as determined by a red blood cell count, or if there is not enough hemoglobin in the red blood cell, he or she is diagnosed with *anemia*. Because hemoglobin carries oxygen, anemia often causes people to feel fatigued. Anemia can be caused by dietary deficiencies, metabolic disorders, hereditary conditions, or damaged bone marrow. Table 7-1 compares the different types of anemia.

Red blood cells are created in the red bone marrow. They live about 120 days shuttling oxygen and carbon dioxide, and then certain white blood cells destroy them in the liver and spleen. As the red blood cells are destroyed, the iron they contain is recycled back to the red bone marrow to be used in new cells. The rest of the material in the old red blood cells is degraded and transported to the digestive system, where much of it ends up in fecal matter.

Table 7-1 Causes and Effects of Different Types of Anemia

Type of Anemia	*Cause*	*Effect*
Aplastic anemia	Bone marrow damaged by radiation or chemicals	Bone marrow cannot produce enough red blood cells (RBCs)
Hemolytic anemia	RBCs prematurely destroyed	Low number of functional RBCs in bloodstream
Iron-deficiency anemia	Dietary deficiency	Fatigue
Pernicious anemia	Digestive system cannot absorb vitamin B_{12}, which is needed to properly form RBCs	Too many immature RBCs in bloodstream, not enough properly functioning mature RBCs

White blood cells

The white blood cells, which are also called leukocytes (*leuko-* = white), are involved in functions controlled by the *immune system*. The immune system is responsible for fighting infections. If a person has a low white blood cell count, it means that the immune system is not functioning properly. If a white blood cell count is too high, it indicates that the person has some type of infection.

There are five important types of white blood cells:

- *Basophils* release histamines. Histamines are those annoying little chemical molecules that cause you to swell up with hives, itch like crazy, sneeze, wheeze, and get teary-eyed when you are around something to which you are allergic. You may wonder exactly what the benefit of something like histamine is. It is this: All of those reactions cause inflammation, which enlists the help of other stronger white blood cells. Plus, sneezing and getting watery eyes are physiologic reactions to help flush the offending allergen from your mucous membranes.
- *Eosinophils* "eat" other cells. The technical term for the eating of a cell is phagocytosis, so eosinophils are said to phagocytize complexes formed between antigens (the invading offender) and antibodies (a "home team" defender).
- *Lymphocytes* kill cells that contain viruses. They are the McAfee's and Nortons of computer fame that scan the body looking for the viruses. There are two types of lymphocytes: B cells and T cells. T cells are the type of virus hunters measured in a person with acquired immunodeficiency syndrome (AIDS). If the T-cell count decreases, it indicates that the human immunodeficiency virus that causes AIDS is winning the fight in that infected person's body.

- *Monocytes* are precursors to macrophages, meaning "big eater." Macrophages digest bacteria and viruses.
- *Neutrophils* are the most abundant white blood cells in the body. These cells phagocytize bacteria, and in doing so keep your system from being overrun by every germ with which it comes into contact.

Platelets

Platelets are pieces of cells that work to form blood clots. The process of clotting is explained in the section called, "Clotting to prevent leaking."

Plasma puts the "stream" in bloodstream

If the red blood cells, white blood cells, and platelets are the "solid" portion of the blood, the plasma is the "liquid" portion. When blood is put into a test tube and spun in a *centrifuge* — a machine that separates the blood cells from the plasma — the formed elements gravitate to the bottom of the tube (and look red because of the red blood cells), and the plasma is a clear layer on top. Inside the blood vessels, the formed elements flow in the plasma. Think of the plasma as a river and the blood cells and platelets as leaves floating in it. The plasma is the "stream" in bloodstream.

The plasma contains many important proteins, without which you would die. Two major proteins the plasma contains are gamma globulin (also called immunoglobulin) and fibrinogen. Gamma globulin is a broad term for a class of proteins that make up the different types of antibodies. The production of antibodies, which help to fight infections, is controlled by the immune system. Fibrinogen is a protein involved in blood clotting which is described in the section called "Clotting to prevent leaking."

Draining, fighting, and absorbing: The lymphatic system

The lymphatic system runs along with the circulatory system. When tissues contain too much fluid, homeostasis can be disrupted. To remove excess fluid from the body, the fluid must be transported to the excretory system. To get to the excretory system, the fluid must get into the circulatory system so that the blood can transport the fluid to the excretory organs. Lymphatic capillaries are intertwined with capillaries of the circulatory system. Fluid in the lymphatic vessels, called *lymph,* diffuses through the membranes of both types of capillaries

Think of your blood as a train that travels the system of "rails" (blood vessels) through your body. Then, the lymph is the taxi that travels the roadways (the lymphatic vessels) to get to the train station — the lymphatic ducts that drain into the subclavian veins (*sub-* = beneath; *clavian* refers to the clavicle, which is also called the collar bone).

The lymphatic system also helps to fight infections by serving as a transport system for cells that fight invasions by bacteria or viruses. The spleen is an organ involved in the production of blood cells and immune cells; however, its structure is very similar to that of a lymph node. The spleen kind of bridges the gap between the circulatory and lymphatic systems. Because the spleen contains many lymphocytes — a white blood cell that fights infection — it "cleans" the blood by filtering blood, killing any invading organisms, and sending the waste back into the bloodstream to be removed from the body.

A third important function of the lymphatic system is that lymphatic capillaries absorb fats in the intestines. The lymphatic vessels transport the fats to the bloodstream. Some of those fat molecules traveling the bloodstream end up being deposited in adipose tissue, creating layers of fat on the body. This function really is necessary, but may seem unfortunate to some people.

Clotting to prevent leaking

When you cut your finger chopping carrots or on a piece of broken glass, your body goes through a certain series of events to make sure that you do not bleed to death.

First, after shrieking in pain or muttering a few choice words, the injured blood vessel constricts. The constriction reduces blood flow to the injured blood vessel, which helps to limit blood loss. Tourniquets help to squeeze off blood flow in much the same way when major blood vessels are injured.

With the injured blood vessel constricted, the platelets present in the blood that is passing through that vessel start to stick to the collagen fibers that are part of the blood vessel wall. Eventually, a platelet plug forms, and it fills small tears in the blood vessel.

Once the platelet plug is formed, a chain of events is initiated to form a clot. The reactions that occur are catalyzed by enzymes called *clotting factors*. There are 12 clotting factors, and the mechanism that occurs is very complex. So, I give you just the highlights of what happens.

- After a platelet plug forms, the coagulation phase begins, which involves a cascade of enzyme activations that lead to the conversion of prothrombin to thrombin. Calcium is required for this reaction to occur.
- Thrombin itself acts as an enzyme and causes fibrinogen — one of the two major plasma proteins — to form long fibrin threads.

- Fibrin threads entwine the platelet plug forming a mesh-like framework for a clot.
- The framework traps red blood cells that flow toward it, forming a clot. Because red blood cells are tangled in the meshwork, clots appear to be red. As the red blood cells trapped on the outside dry out, the color turns a brownish red, and a scab forms.

Transporting Materials Through Plants

Plants may not cut themselves and need to form clots, but they do need to get materials from the ground up through their stems to their parts that are above ground level. Why? Plants absorb nutrients and water through their roots, but photosynthesis — the process by which plants create their fuel — does not take place in the roots. Photosynthesis occurs in the leaves, which are not underground, and photosynthesis requires light in order to begin.

Just as animals contain vascular tissues — in animals, blood vessels — plants also contain vascular tissues. The vascular tissues of a plant are the *xylem*, which transports water and minerals up from the roots to the leaves, and *phloem*, which transports sugar molecules, amino acids, and hormones both up and down through the plant.

The leaves of plants also contain veins, through which nutrients and hormones travel to reach the cells throughout the leaf. Veins are easy to see in the leaves from maple trees; in some plants, the veins are hard to see, but they're in there.

Stemming from the seeds

A plant's stem is vital to the plant's survival. In some plants, such as those that survive just one or two growing seasons — that is, *annuals or biennials* — the stems are soft. These stems are in contrast to the *woody stems* of plants that live year after year — *perennial* plants. Because the stems in annual and biennial plants are not woody, they are called *herbaceous stems* (*herb-* likens them to soft, green plants like grasses and herbs).

The structure of a plant stem depends on the structure of the seed from which the plant sprang. If you ever looked at a bean sprout, you may have noticed the kidney-shaped tissue from which the bean sprouted. Those kidney-shaped tissues are part of the seed and are called *cotyledons*. Flowering plants with one cotyledon are called *monocots*; flowering plants with two cotyledons are called *dicots*. Monocots and dicots have different types of stems, as well as some other major structural differences. Table 7-2 outlines these differences.

Table 7-2 Features of Monocots and Dicots

Feature	*Monocots*	*Dicots*
Cotyledons in seeds	One	Two
Vascular bundles	Scattered throughout	Definite ring pattern
Xylem and phloem	Stem	In stem
Leaf veins	Run parallel	Form a net pattern
Flower parts	In threes and multiples of threes	In fours and fives and multiples of fours and fives

In the stems of herbaceous dicots, the very center of the stem consists of *pith*, which has many thin-walled cells called *parenchymal cells*. The thin walls allow the diffusion of nutrients and water between the cells. Surrounding the pith is the ring of vascular bundles, which contains the xylem and phloem, with a thin layer of *vascular cambium* between the xylem and phloem. Outside the vascular bundle ring is the *cortex* of the stem. The cortex contains a layer of endodermis, some more parenchymal cells, and mechanical tissue, which supports the weight of the plant and holds the stem upright. On the surface of the stem is the epidermis and the cuticle.

Woody stems are much different than herbaceous stems. Woody stems — and this includes branches — develop buds to add new growth. Trees do not get bigger because the trunk grows; they get taller because the branches elongate.

In woody plants, there are two areas of xylem and phloem: the primary and secondary areas of growth, with a ring of cambium in between. As woody plants grow, new tissue is added to the vascular cambium year after year. As new xylem tissue grows every year, it creates *rings* inside the woody stem. These are the rings of annual growth that you can count inside a tree to tell how old it is. As these rings of xylem accumulate year after year, the circumference of the woody stem increases. The newly created xylem cells transport water and minerals up through the stem. This part of the woody stem is called *sapwood*. The older xylem tissue gets filled with material such as gums and resins. This part of the woody stem is called *heartwood*.

The phloem of woody plants gets pushed farther and farther outward as the xylem tissue increases in size year after year. Eventually, the phloem is compacted against the bark of the woody stem. However, in woody plants, the phloem really is active only during the first year of the plant's life. The only phloem that serves to transport materials through the woody plant is the phloem that is newly formed during the most recent growing season.

Moving fluids and minerals through plants

iove through the xylem.
There are several different "modes
phloem; their main function is to
urished.

It's not dew

er concentration of minerals than
which makes sense because the
reas the soil is a much larger area.
d *root pressure*, which forces water
nore water and minerals are "pulled"
root cells via osmosis. This force
n of tiny droplets on the ends of
reason the droplets are seen in the
loss of water from leaves — doesn't
ntil morning. Those droplets are not
roplets are proof that water and min-
nsported through the entire plant.

s, through which water and minerals
t tall plants? Gravity works against the
o more active processes are involved.

spiration

ater is pulled out of cells on the surface
iration. However, there is a theory that
s, they stick together and become one
vaporation of the "top" of the big, huge,
on, a tension is created that passes
lecule — all the way down to the roots.
the roots into the xylem tissue. Air
anspiration pull. Water moves around
alls of xylem cells to keep the long
pter 9 for information on how stomata in
l transpiration.

Transport through the phloem: Translocation

Movement through the phloem is a bit trickier than it is through the xylem. Phloem contains sieve tubes, and the sieve tube elements that make up the sieve tubes are arranged in a vertical line through the entire length of the phloem. Each sieve tube element has holes in their cell walls through which material can pass. These holes look like little sieves, and they appear to filter the material passing through. Cytoplasm runs through these little holes and serves as a connecting fluid between each sieve tube element.

Aphids suck

Aphids, those tiny little insects that can destroy your houseplants when you're not looking, live on the sap flowing through the phloem of a plant. Aphids have long, pointy attachments called *stylets* — think straw — that they insert into the phloem of a plant. Picture a hummingbird sucking nectar out of a flower. It's not the insertion of this stylet that harms the plant. In fact, aphids can go right into a sieve tube without the plant "feeling" a thing. Because of this, an aphid can stay attached to a plant for hours, just sucking out the sap. It's the loss of sap — and the cumulative effect of many, many aphids on a plant — that causes the damage. The aphids fill up, while the plant is starved of its water and mineral mixture.

When sieve tube elements become sieve tube elements, they lose their nuclei. Without nuclei, they are unable to function as "regular" cells and have genetically controlled processes. Instead, each sieve tube element has a companion cell (how cute). The companion cell contains a nucleus and other organelles required for normal cell function. It is thought that the companion cell exerts some control over how the sieve tube functions.

Phloem is responsible not only for moving minerals up through the plants, but also for moving carbohydrates made in the certain plant cells to the rest of the plant's cells. The xylem carries water upward only; the phloem carries sugars both upward and downward. The *mass flow hypothesis* is, so far, the best description of how translocation occurs in the phloem. This theory, proposed way back in 1927, works on the premise that substances move from a *source* to a *sink*, and it goes like this:

- Carbohydrates made in one tissue are actively transported to sieve tube elements. The concentration of solutes — the just-produced carbohydrates — is highest at the *source* of production and lowest at a "sink," which is an area of cells in the phloem that has a low concentration of carbohydrates.
- Water moves from the area of higher concentration to lower concentration — the movement of water follows the movement of solutes. The carbohydrates are actively transported to the sieve tube elements, but water passively diffuses into the sieve tube elements.
- Pressure builds at the source — that is, the sieve tube elements — because cell walls are unyielding and do not expand to accommodate the increased water content. This increase in pressure causes movement of water and carbohydrates to sieve tube elements at a sink. (Think of how when you fill an ice-cube tray with water, when one square is full, the water overflows — or sinks — into the next square. This is the premise behind the source and sink model.) Movement occurs through the tiny holes in the sieve tube elements and continues through the sieve tube: source to sink, source to sink.

- As a sink receives water and carbohydrate, pressure builds at the sink. But before it turns into a source, carbohydrates in a sink are actively transported out of the sink and into needy plant cells. As the carbohydrates are removed, the water then follows the solutes and diffuses out of the cell, relieving the pressure.

Starch — a complex carbohydrate — is insoluble in water, so it acts as a carbohydrate storage molecule. So, whenever a plant needs fuel — such as at night or in the winter when photosynthesis does not occur as well — starches can be broken down into simple carbohydrates. This allows a tissue that normally would be a sink to become a source. And, any cell in the plant is a potential sink. When a cell takes in those simple carbohydrates from the cytoplasm that passes through the sieve tubes, it is receiving energy. Because cells can act both as sink and source, and phloem transport goes both upward and downward, plants are pretty good at "spreading the wealth" of carbohydrates and fluid to where they are needed. As long as a plant has a continuous incoming source of minerals, water, and light, they can fend for themselves.

Chapter 8

Take a Deep Breath: Gas Exchange

In This Chapter

- Finding out what happens in cells after animals take in oxygen and give off carbon dioxide
- Understanding the difference between respiration and breathing
- Figuring out the chemiosmotic theory

This chapter is short and sweet. Because respiration is such a key function to survival, it is mentioned in several chapters. The process of respiration is key to how plants create their fuel and how animals get oxygen into their systems, so I explained it in detail in Chapter 5.

Also, respiration contains some of the most important metabolic reactions that occur in organisms, so some of these cycles — especially the Krebs cycle — have been presented several times in the book.

But, I want to tie the processes together for you. Living things need energy; that's what Part II of this book is about. One way that energy is acquired and converted is through respiration. And, the respiratory reactions are the same processes that occur during the breakdown of nutrients. So, these cycles were covered in Chapter 6. But here in Part III, you are looking at how living things metabolize. And gas exchange is a key component of the processes.

Life's a Gas

From the moment the primordial soup started bubbling, gases were filling the atmosphere. Inorganic molecules were struck by ultraviolet light rays from the sun, as well as by lightning, and, bam! complex organic molecules formed. Once early cells started using these organic molecules as food, they gave off oxygen, just like plants do now. As the oxygen built up in the atmosphere, other gases evaporated from the pot.

Gases and elements have been part of this earth far longer than plants or animals. Life on earth began with the elements and gases. As a result, life on earth is impossible without them. Plants and animals formed from cells using those materials when the earth was forming. The elements and gases are part of every living organism on earth. You and your fellow organisms can survive without food or water far longer than you can survive without oxygen.

So, how do living organisms deal with gases? Well, basically it depends on the organism. A few factors need to be taken into consideration before nature decides what mechanisms of respiration the organism should use.

First, what is the organism's environment like? Does the organism live on land (and have oxygen available in air), or does it live in water? The size and shape of the organism also matter, as does its metabolic rate (how fast it respires) and whether it has a way of transporting gases inside itself.

Oxygen in air versus oxygen in water

If you swim, and especially if you don't, you know that much less oxygen is available in the water than in the air. After all, the difference between the two is what makes you have to pop out of the surface regularly while you're doing the breaststroke.

Air has more than three times the oxygen that water does, and it is much less dense. There are no pockets of air that are lacking oxygen — except maybe deep within a cave — but there can be pockets of water that have little or no oxygen in them. As a result, it is easier to get oxygen from air then from water, if you are designed for the task.

A fish expends 25 percent of its total energy just moving water over its gills. Because water is more dense, and because more water needs to pass over the gills to get a good amount of oxygen out, it takes more energy. Humans, however, expend only 1 to 2 percent of their total energy breathing — which leaves humans energy to expend on other things, like jobs and studying science.

One aspect of water that is more advantageous than air, though, is that oxygen is carried better in water. Oxygen and water form a solution. In land animals, oxygen must "go into solution" before it can diffuse into cells. So, surface membranes where gas exchange takes place — that is, where oxygen is taken into the body and carbon dioxide is released from the body — must remain moist. And, that, my friends, is why your mouth, nostrils, and lungs always are moist.

Gases are exchanged at every cell in your body, too, not just in your respiratory system. If you took off your skin, your body would be extremely moist. It wouldn't stay that way for long, though, because water evaporates from the body very rapidly over moist surfaces. Your skin, and the skin on other animals, is a protective barrier. Skin prevents gas exchange from occurring too rapidly. It keeps land-living organisms from drying out.

Why body size and shape matter

Oxygen goes into solution with water, but once in the water, it diffuses very slowly. So, oxygen must be moved through most organisms. Only the flattest, smallest organisms have no oxygen transport mechanisms. Flatworms are one example. These worms are so small and flat (hence their name) that all of their cells are near a gas exchange surface. Oxygen can come in, and carbon dioxide can go out as they please. This method of exchange also limits the size of the organism.

Most of the larger, rounder animals living here on earth have circulatory systems that shuttle oxygen around bodies. In humans, oxygen is carried by the red blood cells through blood vessels, and then it diffuses into regular cells through capillary exchange. This process is described in more detail in Chapter 7.

Chapter 7 also gives you information on the four types of gas exchange systems. Integumentary exchange occurs through the skin. Gills are used to exchange gases in aquatic environments. Tracheal systems are found in insects, and lungs are used by land animals.

What does the metabolic rate have to do with it?

If you added up the results of everything you took into your body — all the food, all the oxygen, all the water — and you summed up every reaction that occurred in your body, you would get your metabolic rate. Because most of the reactions that occur in the body are aerobic (they require oxygen), one way to determine the metabolic rate is to measure how much oxygen is taken in by the organism over a unit of time (for example, 1 liter of oxygen per hour).

Several factors are taken into consideration when determining the metabolic rate of an organism. The size and weight of the organism have an effect, as do the species of the organism and the environment in which it lives. The metabolic rate usually is measured when the organism is in its normal state (resting but not sleeping) just going through the normal metabolic processes that sustain life; it is called the *standard metabolism*.

Temperature plays a role in the metabolism of heterothermic animals, however. *Heterothermic animals*, such as fish and insects, experience changes in their body temperature when the temperature of the environment changes. Standard metabolic rates rise when the temperature rises, and they decrease when it is cold. They have very little insulation on their bodies. When their metabolic rate is very low — that is, when temperatures are very low — they are extremely inactive. This characteristic supplies one reason why you don't see many insects outside during the winter or early in the morning. Heterotherms have partial regulation of body temperature; poikilotherms (cold-blooded animals), such as snakes, cannot regulate their body temperature or metabolism.

Homeothermic animals, such as birds and mammals, maintain a stable body temperature, although it is high. Homeothermic mammals, such as you, have higher metabolic rates because metabolic reactions are occurring more rapidly, and their bodies produce more energy. The energy is released in the form of heat. But, to maintain the balance between the body heat they lose, they take in more oxygen than heterothermic animals.

Homeothermic animals such as yourself also have two ways of protecting against the loss of too much heat. These animals have insulation in the form of feathers or hair and body fat to keep them warm. (Wouldn't you sometimes rather have feathers than fat?) The second protective measure is homeostasis.

Homeostasis is a special process that maintains balance within the body of things like pH level, glucose level, body temperature, etc. The brains of these animals sense alterations from the normal body temperature and stimulate the appropriate response to bring the body temperature back into the normal range: shivering or sweating.

Cycling Through the Cycles

Respiration is different than breathing. *Breathing* is the physical act of inhaling and exhaling. Breathing is the mechanism that land (terrestrial) animals use to bring oxygen into the bodies and to remove carbon dioxide from their bodies.

Respiration, however, is one big term that encompasses several metabolic processes. Respiration is the overall process for producing energy in the form of adenosine triphosphate (ATP) from the fuel and oxygen that an organism acquires.

Aerobic respiration occurs when oxygen is available. In humans, this is the first choice for metabolism of nutrients and production of ATP.

Anaerobic respiration — metabolism without oxygen — is used as a back-up system during times that there is not enough oxygen available. The metabolic pathways of anaerobic respiration are entered only to keep cells from dying. Because, without aerobic respiration producing molecules that accept electrons, all respiratory cycles come to a grinding halt. When that happens, ATP is not produced; if ATP is not produced, a cell soon dies. Every cell needs a steady supply of ATP to keep cycling through its respiratory cycles. The anaerobic pathways that occur in plants — alcoholic fermentation — and animals — lactic acid fermentation — are described in detail in Chapter 6.

Aerobic respiration

Aerobic respiration has three steps: *glycolysis*, the *Krebs cycle*, and *oxidative phosphorylation*.

- **Glycolysis** is the process that breaks down glucose and converts it to pyruvate (see Chapter 6).
- **The Krebs cycle** takes the pyruvate and puts it through conversions that result in the production of the coenzymes NADH and $FADH_2$, as well as a molecule of ATP and some carbon dioxide (see Chapter 6).
- **Oxidative phosphorylation** is the process that takes the NADH and $FADH_2$ and passes them through an electron transport chain until ATP is produced (see Chapter 6).

So, at the end of the entire process of respiration — that is, after all three steps have been completed — the result is 36 molecules of ATP for you to expend any way you wish. The more you use, the more you make, though, so don't be stingy with your energy.

Where it all happens

The Krebs cycle occurs in the fluid part (the *matrix*) of mitochondria. Mitochondria are the "powerhouses" of the cell. They are organelles with a history and a very important function.

Mitochondria are thought to have been primitive cells that became incorporated into bigger primitive cells that then went on to evolve into animals. The same is thought to be true of chloroplasts, except that the mitochondria's equivalent organelles were engulfed by primitive cells that evolved into plants. What is amazing is that mitochondria and chloroplasts look and function very much alike.

Mitochondria have a convoluted membrane inside them — called *cristae* — that divides the mitochondria into two compartments: inner and outer. The inner compartments of the mitochondria contain the matrix. The cristae house the electron transport chain proteins.

Cracking the Krebs cycle

Dr. Hans Krebs, along with other noted scientists, did experiments with muscle cells and found that they quickly metabolize glucose along with a compound called *pyruvate.* Pyruvate is a three-carbon compound that becomes a two-carbon acetyl group attached to a compound called coenzyme A at the beginning of the Krebs cycle. The scientists found that if they added several different organic acids to the muscle cells with which they were experimenting, the metabolism of carbohydrates increased. They eventually determined that the process occurred in a cycle, so that it happened nearly continuously (as opposed to a series of steps that start then finish completely).

Adding coenzyme A to make acetyl coenzyme A

The first step in the Krebs cycle is the addition of coenzyme A (CoA) to pyruvate to create acetyl coenzyme A. Sound confusing? Remember that this is a cycle. Therefore, the products of the reactions are also used in the reaction. When coenzyme A is added to the three-carbon pyruvate molecule, carbon dioxide (CO_2) is given off. There is a carbon atom in carbon dioxide, so the loss of that carbon atom accounts for why the resulting compound at this step is a two-carbon molecule, acetyl coenzyme A (acetyl CoA).

Acetyl CoA) then enters the cycle. Coenzyme A is given off, and the resulting compound is citrate. Citrate is converted into isocitrate (an isomer of citrate). At this point in the cycle, an electron carrier molecule enters. This molecule is called nicotinamide adenine dinucleotide (NAD), and it has a positive charge (NAD^+). NAD^+ enters the Krebs cycle at three points. The addition of NAD^+ to isocitrate results in the loss of a carbon dioxide molecule as well as the loss of reduced NAD (which is written as NADH because it took a hydrogen ion, and thus an electron, from NAD^+). Don't worry. NADH gives the electron back. If it didn't, there wouldn't be NAD^+ available to continue the cycle, and no energy would be produced, thereby shutting down the cell.

The busiest point of the Krebs cycle

The compound that results from the isocitrate reaction is α-ketoglutarate, which is a five-carbon molecule. (Remember that the cycle starts with a six-carbon molecule, but a carbon was lost when carbon dioxide was given off in the isocitrate reaction.) At this step in the cycle, three compounds are added, and three compounds are given off. NAD^+ is added, resulting in NADH being given off. Water is added, and carbon dioxide (loss of another carbon!) is given off. Guanosine diphosphate (GDP) and an inorganic phosphate (P_i) also are added at this step. The combination of a two-phosphate molecule — guanosine diphosphate — (*di-* = two) plus the inorganic phosphate equals a three-carbon molecule: guanosine triphosphate (GTP). *Tri-* means three, and GTP is given offin this step. At the end of this step, the α-ketoglutarate molecule (five carbons) is converted to succinate, a four-carbon molecule.

From succinate on through the three remaining steps in the Krebs cycle, the compounds all have four carbons. After succinate is produced, fumarate, malate, and oxaloacetate are produced. Following the production of oxaloacetate, acetyl CoA is added, and the cycle begins again. Ultimately, one molecule of glucose provides 36 molecules of adenosine triphosphate (ATP; the unit of measuring energy in a living organism). If you need a different look at the Krebs cycle for all this to sink in, head to Chapter 6.

Chloroplasts also have a membrane inside of them that creates inner and outer compartments. The inner membrane of a chloroplast is called the thylakoid membrane, and it is where the electron transport proteins are located.

The chemiosmotic theory

The last step of respiration is oxidative phosphorylation. In this step, the coenzymes are converted to ATP, because ATP is the much more efficient form of energy.

What happens during oxidative phosphorylation is that the coenzymes NADH and $FADH_2$ are passed from electron carrier to electron carrier through an electron transport chain. Think of a bucket brigade here that works by one fireman dumping a bucket full of water into the next fireman's bucket. The buckets are the electron "carriers," and the water inside the buckets represents the coenzymes.

At the end of the electron transport chain, the coenzymes have given up energy so that phosphate molecules could be added to adenosine diphosphate (ADP) to create ATP. The *chemiosmotic theory* describes the entire process, which is illustrated in Figure 8-1.

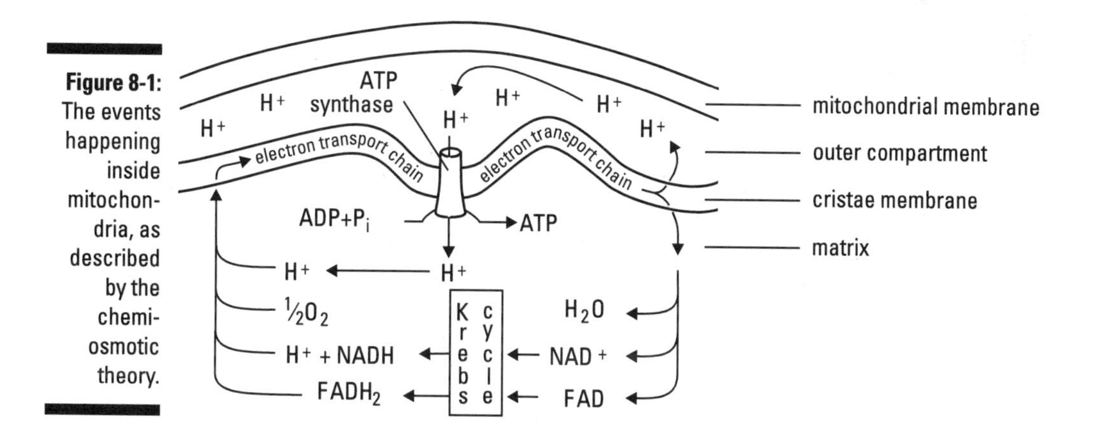

Figure 8-1: The events happening inside mitochondria, as described by the chemiosmotic theory.

The chemiosmotic theory starts with the Krebs cycle occurring in the matrix of the mitochondria. During the Krebs cycle, NADH, $FADH_2$, and hydrogen ions (H^+) are being produced and are then shuttled through the electron transport chain, along with oxygen.

The Krebs cycle is an aerobic process; that's why oxygen is necessary for it to take place.

When NADH and $FADH_2$ are moving through the electron transport chain, the hydrogen ions are pumped out of the matrix, across the cristae, and into the outer compartment. The hydrogen ions gather in the outer compartment of the mitochondria — that is, outside the cristae — which creates a *proton gradient* and an *electron gradient*. A *gradient* is a stash of potential energy. So, the protons, which are the hydrogen ions, are held in the outer compartment until they are needed. Likewise, the energy from the electron transport chain builds up until it is needed.

There are proteins in the cristae that allow the hydrogen ions to flow back into the matrix. The H^+ flow through the cristae via tiny openings called channels, so the proteins are called *channel proteins*. In reality, they are *ATP synthases*. A *synthase* is an enzyme that initiates the production (synthesis) of a substance — in this case, ATP.

As the hydrogen ions move back and forth through the channels, they create energy. This energy is enough to kick-start the production of ATP.

Once the ATP is made, it is quickly consumed by cells so that the entire process can continue. Life truly is a cycle.

Chapter 9

Throwing Out the Trash: Eliminating Waste to Maintain Homeostasis

In This Chapter

- Understanding the concept of homeostasis
- Clarifying the digestive process
- Defining nitrogenous wastes and body fluids
- Figuring out why and how organisms and plants eliminate waste products

You know this chapter deals with how living organisms get rid of waste. How does homeostasis fit in? Homeostasis is a key concept in biology. It is the description for when the internal conditions of living organisms (for example, temperature and levels of things like glucose, calcium, or potassium) remain stable (within a normal range), regardless of what is going on in the external environment. So, if the garbage created within each cell accumulated, the body would become a dump. Toxins would build up and could alter the levels of things like your electrolytes (for example, sodium, potassium, and calcium ions). Homeostasis attempts to maintain your system in a normal range; if toxins would build up, homeostasis would be disrupted, and you would become very sick. So, in this chapter, I describe what constitutes cellular garbage and explain homeostasis further. Then I help you figure out what organisms do with that trash by exploring the digestive and excretory systems.

What's in the Trash?

Each cell in your body is actively involved in metabolism, which basically is the process of using nutrients from food to provide fuel for cellular processes. Metabolism is like a wood-burning stove that heats a home (your body). Food is like the logs that are thrown on the fire. When logs burn, ash is created. Ash is the waste created from the using of energy. If the ash is not removed from the fireplace, eventually the fire can no longer burn. When food is broken down, as much of the nutrients as possible are used to fuel the body. After as much energy as possible is extracted from food through digestion and metabolism, the remainder is excreted, or removed.

Homeostasis

Suppose that it's really, really cold outside — snowing even — and you run out to your mailbox in a short-sleeve shirt. While you're out there, a neighbor stops by to chat. Your body wants to maintain its body temperature around 98.6°F. Your skin senses the cold conditions outside, and nerve impulses are sent from receptors in your skin to your brain that say, "Hey! It's really cold out here!!"

In an attempt to stay around 98.6°F, your body makes adjustments automatically. Goose pimples form, which actually are the hair follicles on your body tightening to make your body hair stand up higher to help insulate your body. If that doesn't help to maintain the normal temperature, you start to shiver. Shivering is an attempt by your body to create heat through movement.

If your chatty neighbor is still rambling on, and shivering doesn't help keep you warm, your body's "thermostat" will begin to drop (if it goes too far, hypothermia begins), and your "furnace" will kick on to create heat internally so that homeostasis — maintaining relatively normal values — occurs.

Disease occurs when homeostasis can't be achieved. Suppose that your chatty neighbor spread the flu virus to you when he sneezed during your conversation. Your body needs to fight off the invading virus, which likes living at your normal body temperature. At 98.6°F, the virus can happily reproduce, making you sicker and sicker. Your body wants to be in homeostasis, but if it maintained normal body temperature, the virus would take over your entire body. In defense against the virus, your body temperature rises above the normal range (fever), which makes your body a very uncomfortable place for the virus to live. The virus starts to slow down in the hotter temperature, which allows the cells of your immune system to attack it. But, the fact that your body temperature goes above normal (fever) means that homeostasis is disrupted, and that indicates disease (the flu). Once your body has effectively suppressed the viral attack, the fever "breaks," and your body temperature returns to normal. The disease state is over, and homeostasis returns.

Remember that fever is a natural, healthy process. If you develop a fever, let it do its job of making your body inhospitable to a virus or bacteria. Don't quell the fever with aspirin, ibuprofen, or acetaminophen right away. If the fever continues for days, though, you should see the doctor. A fever caused by a virus usually breaks on its own as your body fights the virus. However, an infection caused by bacteria requires an antibiotic to help your body fight. The fever associated with a bacterial infection usually is higher and lasts longer than a fever caused by a virus.

Chew on This: Workings of the Digestive System

Imagine biting into a big, juicy cheeseburger. Did you salivate at the thought of it? If so, your body is prepared for you to take an actual bite. The enzyme in your saliva — salivary amylase — is there to start digesting the carbohydrates, most likely those in the bun. Go ahead and take a real bite. Chew slowly so that your teeth can break down some of that food. Swallow and pay attention to the feeling of the bits of food being squeezed down your esophagus into your stomach. That action is called peristalsis, and it occurs throughout your entire digestive tract (see Figure 9-1).

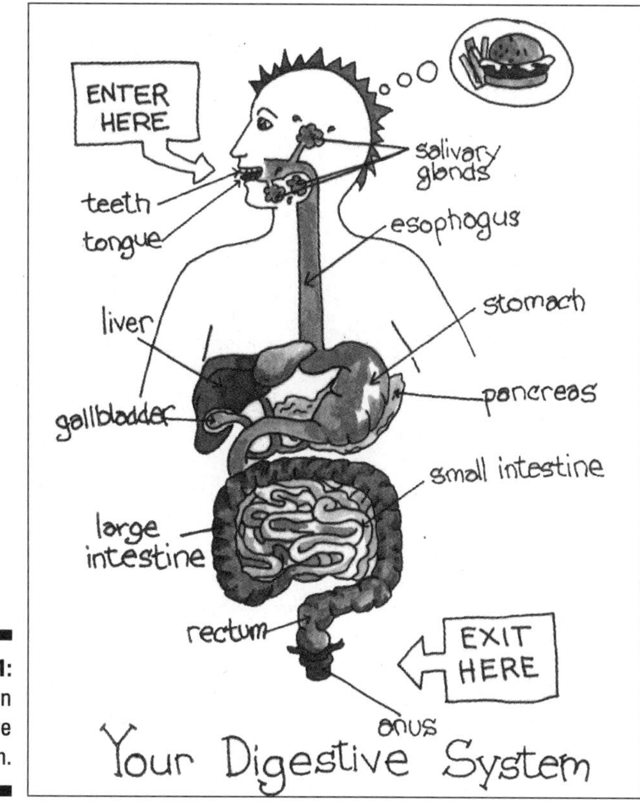

Figure 9-1: The human digestive system.

Once the cheeseburger bits are in your stomach, they are referred to as a *bolus*. The bolus is drowned in gastric juice, which is made up of the enzyme pepsin and hydrochloric acid (HCl).

If you eat too much, your stomach produces more acid, and the contents of your overly full stomach can be forced back up into the esophagus, which runs in front of the heart, giving you heartburn.

The enzyme and acid act to break down the food and release the nutrients. For example, the hamburger bun contains mostly carbohydrates, very little protein, and a small amount of fat. The meat contains very little carbohydrate, and large amounts of protein and fat. Likewise for the cheese. Those carbohydrates, protein, and fat all are important for proper nutrition (it's excesses of them that get you in trouble!), but they must be in their smallest forms to be used by each cell. This final stage in digestion occurs in the small intestine.

Digested food from the stomach is pumped into the small intestine, which gets flooded with fluid and enzymes from the liver (bile) and the pancreas (pancreatic amylase, trypsin, and lipase).

The suffix *-ase* usually denotes an enzyme. Amylases break down starch. Salivary amylase is found in the saliva; pancreatic amylase is created in the pancreas and secreted into the small intestine. The prefix *lip-* refers to fats (such as lipids, lipolysis). So, lipase is an enzyme that breaks down fats. *Tryp-* is a prefix used with proteins (such as the amino acid tryptophan). But, instead of trypase, the pancreatic enzyme that breaks down proteins is called trypsin.

Those chemicals help to break the molecules from the digesting food into its smallest components. The smallest form of carbohydrate is glucose, which is a sugar molecule. Proteins can be reduced to amino acids; fats can be reduced to fatty acids and glycerol. The smallest forms of the nutrients pass through the walls of the small intestine and are absorbed into the bloodstream.

The scoop on poop

The useable nutrients are absorbed into the bloodstream from the small intestine. The leftover material continues on to the large intestine, where fecal matter (feces, or "poop") is created. The large intestine absorbs water and some electrolytes from the leftover material, and that water is returned to the body to prevent dehydration. If too much water is absorbed, constipation occurs; if too little water is absorbed, diarrhea occurs.

Once the feces are created, they pass to the colon, where they are stored. When the colon is full, a signal is sent to your brain telling you that you need to relax your anal sphincter and release the feces.

Why poop is brown and other smelly info

Ever wonder why poop is brown? Pigments in bile (yellow and green) are part of the leftover fecal material. Also, dead red blood cells that have been replaced with fresh, new, healthy red blood cells end up in the large intestine. Combine the yellow, green, and red, and you get brown.

And since we're on the subject, I bet you're wondering why poop smells like, well . . . you know. A variety of bacteria live in the large intestine. *Escherichia coli* (*E. coli*) is the most common intestinal bacteria, and *E. coli* survives by eating some of the leftover material that forms the feces. When the bacteria digest some of your leftover material, they give off the odor of feces. How does having bacteria in your intestines benefit you? As the bacteria further degrade your leftovers, they produce vitamin K, some B vitamins, and amino acids. All those products are absorbed through the lining of the large intestine into your bloodstream, which transports them around the body for beneficial uses. For example, vitamin K is necessary for proper blood clotting.

Back to the bloodstream

Important, useful molecules pass through the walls of the small intestine into the bloodstream. The bloodstream carries those molecules throughout the entire body. Every nook and cranny are supplied by blood capillaries, so every nook and cranny receive nutrients from the food you digested.

The walls of capillaries are very, very thin. Just outside of the capillary walls is a fluid called *interstitial fluid.* This fluid fills every space between every cell in the body, cushioning and hydrating the cells, and serving as part of the "matrix" through which nutrients and wastes are passed. The nutrients gained from digested food diffuse through the capillary walls, across the interstitial fluid, and are absorbed by the cells. At the same time, waste produced by the cell's metabolic processes diffuses out of the cell, across the interstitial fluid, and into the capillary, where it can be carried to the kidney for excretion.

Sixty percent of a human's body weight is from fluid. Of that 60 percent, 20 percent is extracellular fluid, or fluid that exists outside the cells. This extracellular fluid is made up mostly of interstitial fluid (16 percent) and (blood) plasma (4 percent). Interstitial fluid is the fluid that exists between and around every cell of the body. It provides the medium for diffusion of nutrients and wastes and also helps to cushion the cells. Plasma is the fluid that blood cells flow in. Of the original 60 percent, 40 percent comes from intracellular fluid, or fluid that exists inside the cells of the body.

Nitrogenous Wastes

Nitrogenous wastes are unnecessary, excess materials that contain nitrogen, and too much nitrogen in the body is not healthy. Compounds with nitrogen in them are byproducts of metabolism. These include urea, uric acid, creatinine, and ammonium.

Proteins are made up of amino acids. When proteins are digested, they are broken apart into amino acid chains. When the excess amino acids are degraded, urea forms in the liver. Uric acid forms when nucleotides are broken down (nucleotides make up DNA and RNA). When you use your muscles, you use up some of the energy-containing compounds that exist in the muscle tissue. The muscle tissue is degraded and rebuilt. When it is degraded, creatinine is released, and it must be excreted because it contains nitrogen. Ammonia forms in the body as a result of cells breaking down nutrients and then creating energy-storage molecules. Most cells in your body go through the Krebs cycle (see Chapter 8), and ammonia is a waste product of those reactions. Ammonia is converted to urea.

Kidney structure and function

In humans, the kidneys are the organs held responsible for the production of urine. There are two kidneys, one on each side of your back, just below the ribs. Like most organs in the body, the function of the kidneys is closely tied to its structure (see Figure 9-2). Each kidney has three distinct areas:

- The renal cortex, which is the outer layer
- The renal medulla, which is the middle layer
- The renal pelvis, which becomes a ureter

Each kidney contains more than 1 million nephrons, which are microscopic tubules that make urine. Each nephron contributes to a collecting duct, which carries the urine into the renal pelvis. From there, the urine flows down the ureter, which is the tube that connects the kidney to the bladder.

Each of the million tiny nephrons in the kidney is a mass of even tinier tubules. The main part of the nephron consists of the proximal (near) and distal (far) convoluted tubules, which become the nephron's collecting duct. At the beginning of the proximal convoluted tubule is a ball-like structure made up of the glomerulus, which is the site where the nephron's tubule intermingles with a capillary, and the glomerular capsule (also called Bowman's capsule). In the glomerulus, the transfer of waste products from the bloodstream takes place through the capillary wall into the tip of the proximal convoluted tubule. Also at this site, any materials that are filtered by the nephron and are to be returned to the bloodstream are reabsorbed from the glomerulus through the capillary wall so that they can be recirculated. Venules (smallest veins) join the capillaries (smallest arteries), and together, they join the renal vein, which carries blood away from the kidney.

Urine is spurted from the ureter into the top of the bladder continuously. The bladder holds a maximum of about 1 pint of urine, but you begin to feel the need to urinate when it is only one-third full. (That way, you have time to find a bathroom, if need be, before you are completely full. Or, you can finish reading a chapter before you absolutely must get up and go.) When the bladder is two-thirds full, you start to feel really uncomfortable.

Urine leaves the body through the urethra, which is a tube at the bottom of the bladder that opens to the outside of the body. It is held closed by a sphincter muscle. When you want to start urinating, that sphincter muscle relaxes, opening the urethra and letting the urine flow out.

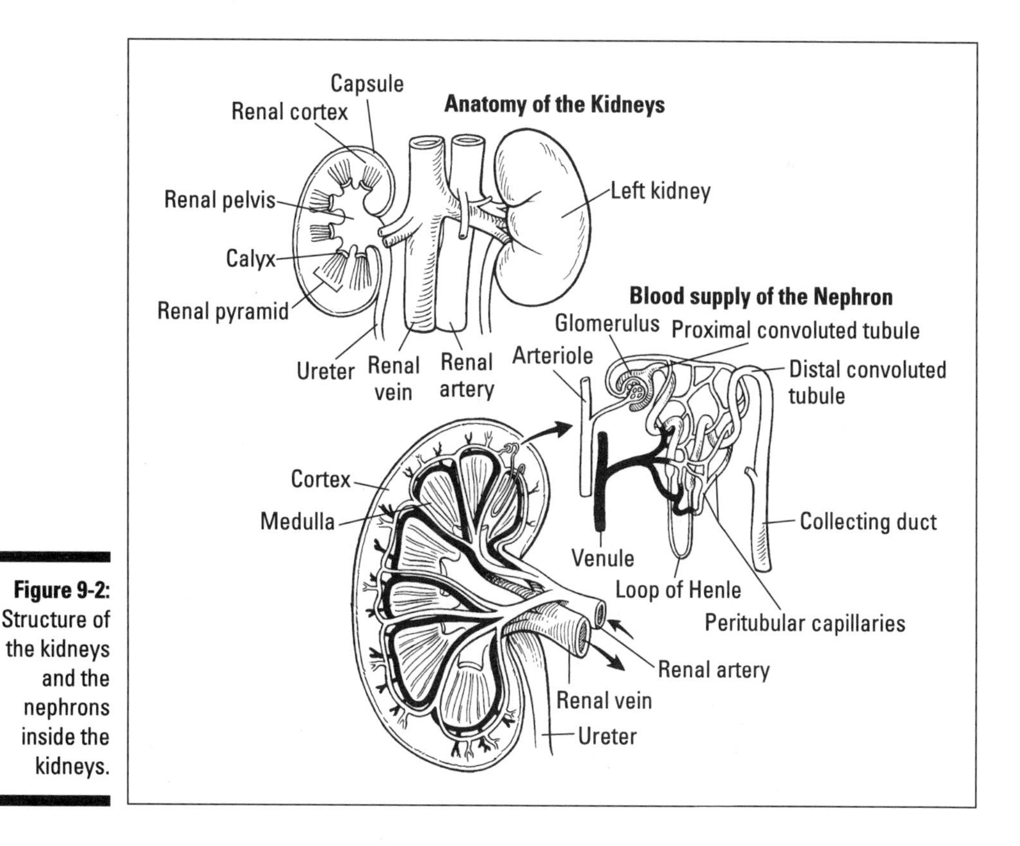

Figure 9-2: Structure of the kidneys and the nephrons inside the kidneys.

Diffusion and osmosis

Imagine that you dress up as "the church lady" from *Saturday Night Live*. You don a dress with big polka dots and big fake pearls. (It's okay, guys; this is just pretend.) If you stand in the corner of a church and put on an extra heavy dose of perfume, as some church ladies are wont to do, someone across the room would eventually be able to smell it. The molecules of perfume diffuse (spread) through the air. The diffusion of the perfume required no energy, so it is called passive transport.

Osmosis, on the other hand, is also a form a passive transport, but it describes the movement of water across a membrane. Imagine a 20-gallon saltwater fish tank. The concentration of salt is pretty high. In contrast, a 30-gallon tank of fresh water has a very low concentration of salt. If the two tanks were separated with a membrane (visualize a thin layer of skin between the tanks), this is what would happen: Water from the bigger, less concentrated tank would move to the smaller tank, so that the concentrations in both tanks would become equal. Things in nature want to be balanced (remember homeostasis?). Balanced concentrations are isotonic. Solutions with low concentrations of solute are hypotonic (*hypo-* means low), and solutions with high concentrations of solute are hypertonic (*hyper-* means high).

How animals other than humans excrete their wastes

Like humans, other animals also have a urinary system that is basically a system of filtering tubes. For example, worms have a tube that runs down the center of their bodies (see Figure 9-3). If you look at a worm up close, you'll notice that it has segments. On each segment are two openings called *nephridia.* Fluid enters the nephridia, which filter the fluid, secreting necessary materials into the fluid filling the cavity of the worm's body (the coelom) and passing the remainder through the collecting tubule. Capillaries surround the collecting tubule, and they reabsorb materials that are secreted into the fluid surrounding the collecting tubule. The waste material becomes concentrated and is excreted at the end of the worm through an excretory pore.

In fish (see Figure 9-3), urination depends on their environment. Fish that live in the ocean are less salty than the saltwater they call home. Therefore, they lose water through osmosis, and they take in water constantly but rarely urinate. Salts that accumulate in their system are secreted out through their gills. Freshwater fish that live in creeks, lakes, and rivers, are saltier than the water in which they hang their hats. So, water constantly diffuses into the fish. Therefore, they rarely "drink," but they urinate constantly and absorb salts through their gills to remain isotonic (and maintain homeostasis).

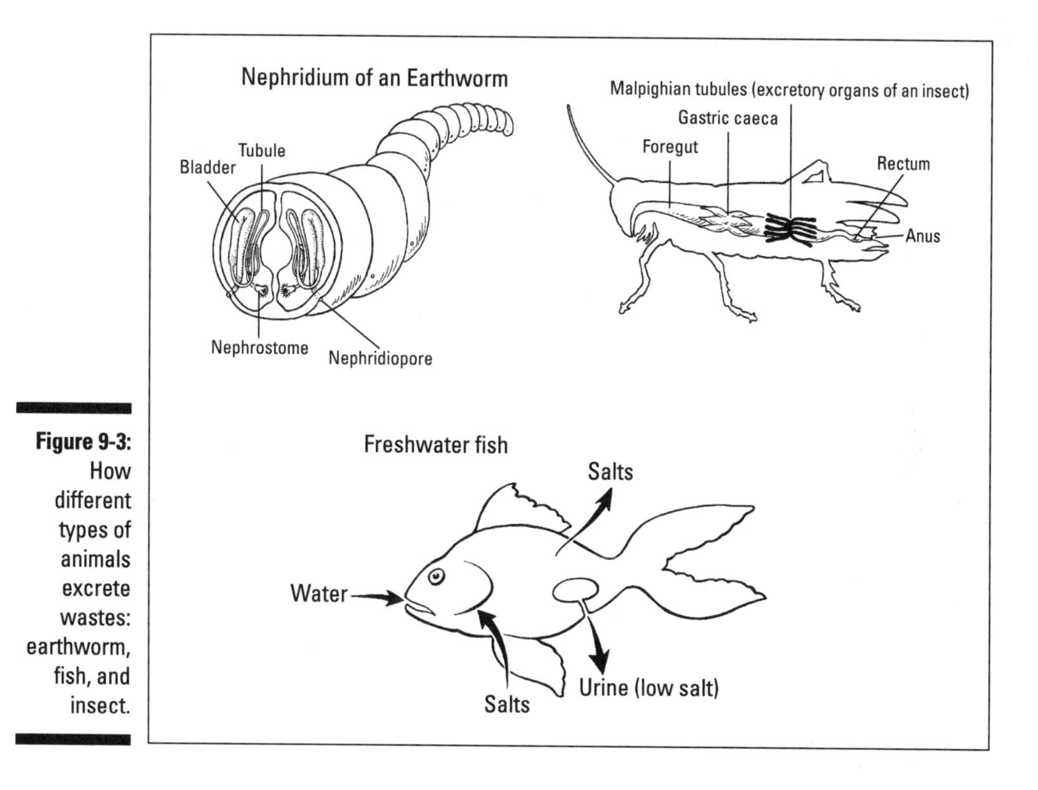

Figure 9-3: How different types of animals excrete wastes: earthworm, fish, and insect.

Insects, like humans, have a system of tubes that remove waste (see Figure 9-3). The midgut of insects (the middle of their digestive tract) has tubes that collect fluids from the mixture of blood and lymph that surround their cells. The fluid contains both the wastes to be excreted and the salts and water that are to be remain inside the insect. The salts and water to be retained pass through the walls at the bottom of the insect's digestive tract. Wastes become concentrated and continue down the tube; at the end, they are excreted through an anus.

Figuring out how plants excrete wastes

On each leaf, there are openings that you cannot see with your naked eye. Each opening is called a *stoma;* a bunch of stomas are called *stomata* (Figure 9-4). The purpose of these stomata is to allow water and carbon dioxide into the plant. When the stomata are open, water and carbon dioxide can enter the plant, and light can combine with both molecules to allow photosynthesis to occur. At the same time, the waste product of plants — oxygen — can escape through the stomata. However, if the stomata are open too long, the leaves dry out. (Dying plants have dried out leaves because their stomata remained open in an attempt to get water flowing through the plant.) To prevent this unfortunate event from happening, each stoma has two guard cells surrounding it.

Urinary problems

I'm not talking long lines for the women's room and no lines for the men's room here, yet. I'm talking about true medical problems. Later in life, *incontinence* — the inability to hold urine in the bladder — often becomes an embarrassing problem for men and women. The need to urinate becomes more urgent and frequent, and sometimes urine leaks out of the bladder uncontrolled. Incontinence is thought to be more prevalent in women, however, because of the stress put on the sphincter muscles associated with the bladder during pregnancy and childbirth.

In men, *enlargement of the prostate gland* starts at about age 50. The prostate gland sits right under the bladder and surrounds the urethra. Its function is to add fluid to semen as the semen passes through the urethra in the penis. As the prostate gland enlarges, it presses on the urethra, squeezing urine back up into the bladder. It also makes urinating painful. If the condition progresses far enough, urine can back up into the kidneys, which can lead to kidney disease. As people age, the number of nephrons in a kidney declines, and the overall size of the kidney decreases, which can contribute to reduced renal function and lead to serious problems in older people.

So, based on these aging-associated problems, incontinence would make older women need to use the bathroom more frequently, but an enlarged prostate would make an older man need to use the bathroom less often. Those problems alone certainly could account for the long line at the women's room and the short line at the men's room. Then, if renal function decreases with age, it would seem that there would be more young people using bathrooms than older people. Sounds about right. I've also noticed that women are the ones that usually take kids (both boys and girls) to the bathroom and do the diaper changes, so that could add to the longer line. Then you have teenage girls who need to find a mirror every chance they can get. . . . Well, you get the picture: busy, busy ladies room and usually empty men's room. I say make all bathrooms unisex (and make more changing tables available to men!!).

Photosynthesis is the process that plants use to create energy. The chlorophyll-containing cells allow light to combine with carbon dioxide and the hydrogen molecules from water to produce carbohydrates. Oxygen is given off as a waste product.

The guard cells control the opening and closing of the stoma (see Figure 9-4). Think of one of those old-fashioned (okay, "retro") plastic coin purses that you squeeze at each end to create an opening. When you apply pressure, the coin purse opens. In plants, that "pressure" is applied by water flowing into the guard cells. The guard cells expand, and the stoma opens. When water leaves the guard cells (due to a variety of reasons), the guard cells shrink, and the stoma closes. The opening and closing of the stomata keep transpiration (the flowing in of carbon dioxide and flowing out of oxygen) at a minimum, while photosynthesis is allowed to occur as often as possible.

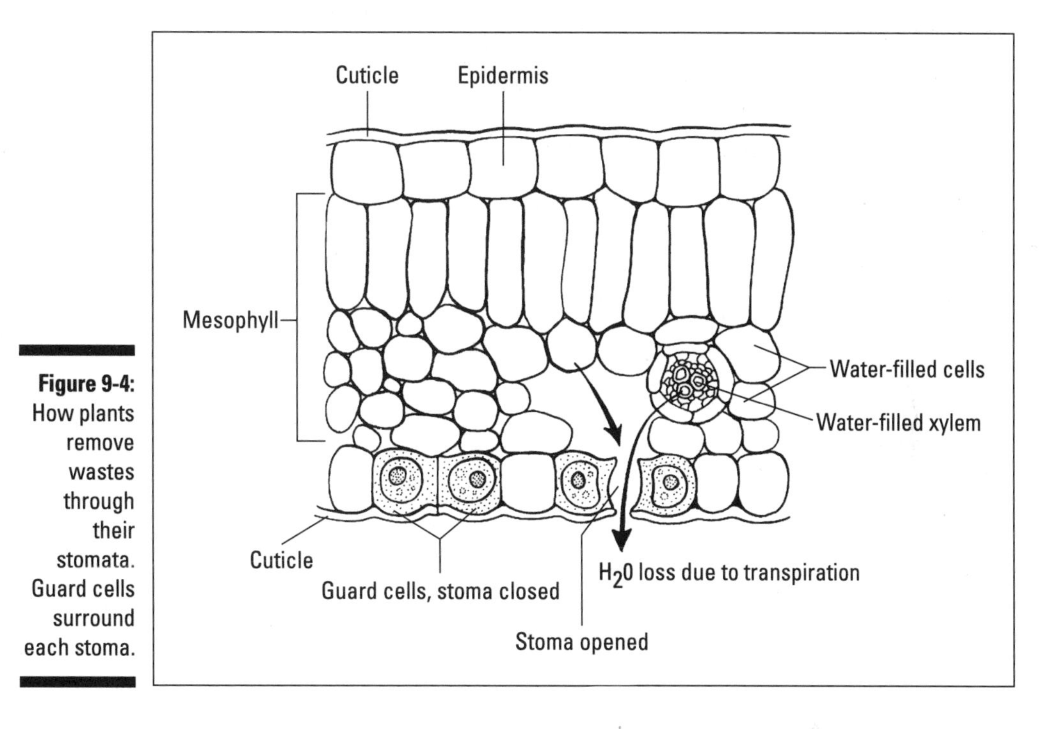

Figure 9-4: How plants remove wastes through their stomata. Guard cells surround each stoma.

Chapter 10

Better Living Through Biology

In This Chapter

- Discovering how enzymes work
- Regulating reactions and responses by hormones in animals and plants
- Understanding the structure of nerves and how nerve impulses are generated and carried
- Figuring out how the brain and senses work
- Finding out how muscles contract and how other animals and plants move

With all the metabolic processes and reactions going on in animals and plants, organisms need to exert some control to avoid chaos. Enzymes help to avoid chaos by controlling metabolic reactions. Some metabolic reactions produce hormones, which then regulate other cycles in the animal or plant. Animals with nervous systems have a brain and nerves that control the release of hormones throughout the body. Or, the nerves can relay information picked up by the sense organs and send it to the brain. Sometimes the information relayed by the nerves results in the movement of muscles in the body. However, plants, as well as some animals, do not have muscle tissue; these organisms have other forms of locomotion and movement.

Revving Things Up with Enzymes

Enzymes are proteins that allow certain chemical reactions to take place much quicker than the reactions would occur on their own. Enzymes function as *catalysts*, which means that they speed up the rate at which reactions occur. Usually, the reactions are part of a cycle or pathway, with separate reactions at each step. Each step of a pathway or cycle usually requires a specific enzyme. Without the specific enzyme to catalyze a reaction, the cycle or pathway cannot be completed. The result of an uncompleted cycle or pathway is the lack of a product of that cycle or pathway. And, without a needed product, a function cannot be performed, which negatively affects the organism.

Getting things going with catalysts and activation energy

Reactions are not impossible without enzymes, but they do happen much more quickly. Enzymes do not change during reactions, nor do they change the other contents of the reaction. They just speed up the rate at which all parts of the reaction react.

A + B <–> C + D

(reactants) <–> (products)

In a chemical reaction, the reaction is said to be completed when equilibrium is reached. *Equilibrium* is the state when the amounts on each side of the reaction have stabilized. Chemical reactions have forward directions and backward directions, and reactions tend to move in both directions until no more products are created from the reactants, and products are no longer converted back into reactants. That is the point of equilibrium. The *equilibrium constant* is written as K_{eq}.

Reactions will occur with the *free energy* available in the system (system is referring to the area where the reaction is occurring). There is always some energy in the system before a reaction begins, and this free energy is called *G*. The amount of change in the free energy of a reaction is labeled ΔG (the Greek letter delta, Δ, is used to represent change).

Exergonic reactions give off energy, so they represent a negative change in free energy ($-\Delta G$) — that is, the free energy is given off, so there is a "loss" of free energy. In actuality, the energy is just transferred. *Endergonic reactions* absorb energy into the system, so the free energy in the system increases ($+\Delta G$). This increase appears to be a "gain" in energy, when really it is just another energy transfer. Exergonic reactions will continue until equilibrium is reached, because they yield energy; endergonic reactions kind of quit while they are ahead. Because endergonic reactions take in energy, the reactions peter out so that less energy is taken in. They usually do not reach equilibrium.

There are two theories as to how reactions occur: collision theory and transition state theory.

- In the *collision theory*, it is thought that reactions occur because molecules collide; the faster they collide, the faster the reaction occurs. The energy level that must be reached for the molecules to collide is called the *activation energy* (E_a). The activation energy is affected by heat, because a higher temperature increases the energy of each molecule.

- In the *transition state theory*, reactants are thought to form bonds and then break bonds until they form products. As this forming and breaking happens, free energy increases until it reaches a transition state (also called activated complex), which is viewed as the midpoint between reactants and products. Reactions proceed faster if there is a higher concentration of activated complex. However, if the *free energy of activation* is high (ΔG_t), the transition state is low, and the reaction is slow. The reaction rate is proportional to the concentration of the activated complex. If the activation energy is lower, the reaction occurs faster because more activated complexes can form.

In living organisms, the reactions that need to occur have high activation energies. So, to get reactions to occur, either the temperature must be increased, or the activation energy must be decreased. But the internal temperature of a living thing cannot be raised too high like chemicals in a laboratory can be. Instead, living organisms rely on enzymes to lower the activation energies so that reactions can occur quickly.

And occur quickly they must! Without enzymes, toxic chemicals could build up in the body to dangerous levels, or the energy-producing Krebs cycle would not be able to produce ATP, which is the main fuel of the body produced from food that is eaten and digested. For example, the first enzymatic reaction discovered was the one that breaks down urea into products that can be excreted from the body. The enzyme urease catalyzes the reaction between the reactants urea and water, yielding the products carbon dioxide and ammonia, which can easily be excreted.

urease

Urea + water <–> carbon dioxide + ammonia

Enzymes are needed at each step of the Krebs cycle, which operates under aerobic conditions to produce energy. One of the steps in the Krebs cycle is the conversion of succinate to fumarate. The reaction requires the presence of the enzyme *succinate dehydrogenase*. The names of enzymes usually end in *-ase*; this name also explains what happens in the reaction: Succinate has a hydrogen atom removed. If an organism could not produce the enzyme succinate dehydrogenase, then the Krebs cycle would be halted at this point, and the organism would not be able to create a useable form of energy. Death would be certain.

Cofactors and coenzymes: Coexisting with enzymes

Enzymes are made mostly of proteins, but they also have some nonprotein components. When these nonprotein components must be included in order for the enzyme to act as a catalyst, then the nonprotein component is called a *cofactor*. Examples of cofactors are potassium, magnesium, or zinc ions.

How cyanide kills

The poison cyanide kills quickly because it prevents an enzyme from doing its job. During the Krebs cycle, products are made that must travel through the electron transport chain so that they can be converted to ATP. This requires the transfer of electrons to oxygen, because the Krebs cycle is an aerobic process. One of the enzymes that is required in this process is *cytochrome oxidase*. Cytochrome oxidase catalyzes the transfer of electrons through the electron transport chain to oxygen. Cyanide prevents cytochrome oxidase from catalyzing the reaction, and ATP production stops immediately. And, because ATP is used quickly — there usually are no extra stores of ATP — death happens quickly, within minutes, because there is no ATP to fuel the brain, heart, or lungs, not to mention all the metabolic reactions that take place.

A *coenzyme* is a type of cofactor. Coenzymes are small molecules that can separate from the protein component of the enzyme and react directly in the catalytic reaction. An important function of coenzymes is that they transfer electrons, atoms, or molecules from one enzyme to another.

Vitamins are closely connected to coenzymes. The function of vitamins is that they help to make coenzymes. Niacin, which is one of the B vitamins, helps to make nicotinamide adenine dinucleotide (NAD), which is one of the coenzymes that carries electrons from Krebs cycle through the electron transport chain to produce ATP. Without NAD, very little ATP would be produced, and the organism would be low in energy.

Controlling the enzymes that control you: Allosteric control and feedback inhibition

Enzymes may control the rate at which reactions take place, and their absence certainly puts a stop to many cycles and pathways. However, in nature's system of checks and balances, enzymes are under control, too. This system helps to regulate the reactions that require enzymes.

In the laboratory, the effectiveness of an enzyme, for example, can be affected by changing the temperature or the pH level. But this cannot be done in a living organism without disrupting homeostasis — that overriding process of keeping the internal environment of the organism stable in response to changes in the external environment. Because enzymes are part of the internal environment, changes in temperature or pH can denature or inhibit the enzymes. Remember that enzymes are made mostly of proteins, and proteins become denatured — separated and destroyed — at high temperatures.

Enzymes work on a substrate. The substrate is the chemical that they affect. But, to work on the substrate, the catalytic site of an enzyme must combine with the substrate. However, because enzymes are specifically made for certain reactions, the correct enzyme must "fit" with the substrate just as jigsaw pieces fit together or keys fit into locks. In fact, the way that enzyme-substrate complexes are formed is often described by the "lock and key model." If an enzyme and substrate do not fit together properly, no reaction takes place. Once an enzyme and substrate are combined properly, the reaction quickly takes place.

Allosteric enzymes have sites in addition to their catalytic sites that allow regulatory molecules to bind to them. So, when the enzyme is needed for a reaction, the substrate binds to the enzyme at the catalytic site to activate it. But, when the need for the reaction has ended, a regulatory molecule binds the enzyme at the allosteric site, effectively shutting it down. *Allosteric control* is one method of regulating enzyme activity. *Feedback inhibition* is another.

During feedback inhibition, a pathway proceeds normally until the final product is produced at too high of a level. Then, some of the final product serves to shut down the pathway by inhibiting the activity of the initial enzyme. The process of feedback inhibition prevents cells not only from having to use energy creating excess products but also from having to make room to store the excess products. It's like keeping yourself from spending money on huge quantities of food that you invariably never eat and just end up storing until the food rots.

All Hormones Are Not Raging

Hormones are specialized substances that coordinate the activities of specific cells in certain areas of the body. In living organisms, cells form tissues, and tissues form organs. Different activities occur at each level, and these activities must be coordinated so that the right product is created and transported to the right place in the body.

Hormones are produced by cells in glands, and they are secreted by the gland into the bloodstream. The bloodstream then transports the hormone to certain tissues, where the hormone has its effect.

The *endocrine system* is the system of hormone production and secretion within an organism. The endocrine system often is compared to the nervous system, which is composed of the brain, spinal cord, and nerves. Both nerve impulses and hormones carry information throughout the body. Both the endocrine and nervous systems coordinate internal activities.

Sometimes the nervous system can work without the brain, as in a *reflex arc*. A reflex arc gives sensory nerves direct access to motor nerves so that information can be transmitted immediately, such as "Ow! That's hot!" when you touch a turned-on iron. The sensory nerves detect the excessive heat and instantly fire off a message to the motor nerves that says, "Pull your hand away, brainchild!" The motor nerves get the proper muscles into action to actually move your hand away before you even "think" about it. Your brain comes into play after the fact when you sense the pain of the burn or think to yourself, "That was stupid. I guess I should run cold water on this burn."

The endocrine system keeps a check on cellular processes and components of the bloodstream and can make adjustments. However, the reaction time is not immediate. Hormones are chemical messengers created in a gland in one area of the body and carried in the bloodstream to a target tissue elsewhere in the body, where the hormone must be absorbed into the tissue before it takes effect. These actions do not happen in a split second, as do nerve impulses. Hormones certainly aren't "raging" — flowing steadily through the bloodstream as needed, but not raging.

The word *endocrine* stems from a Greek word meaning *within.* Endocrine glands, which produce hormones, secrete their products into the bloodstream, which remains within the body. On the other hand, *exocrine glands* secrete products to the outside of the body. Examples of exocrine gland secretions are sweat and saliva.

General functions of hormones

Whether a hormone is from a plant, invertebrate animal, or vertebrate animal, they regulate several important functions. The functions listed here are just a few examples of the many functions that hormones serve.

- **Assuring that growth occurs properly.** In humans, growth hormones must be secreted at normal levels by the pituitary gland throughout childhood and adolescence. The extremes of too much or too little growth hormones are obvious: giants or midgets, respectively. In invertebrate animals, such as insects, growth hormone is responsible for molting, which is the shedding of the outer layer — the exoskeleton. In plants, several plant hormones control proper growth of roots, leaves, and flowers.
- **Ensuring that development and maturation occur properly and on time.** In insects, *metamorphosis* — the process of changing body forms during developmental stages — is controlled by a substance called juvenile hormone. Metamorphosis is the process that changes a larva or caterpillar into a pupa and then into a moth or a butterfly. In plants, indoleacetic acid is one hormone that affects aspects of development such as root growth, secondary growth in stems, leaves separating from the stem, and promoting the development of buds.

- **Making sure that reproduction occurs at the best possible time.** For humans, who have steady supplies of food year-round and sheltered environments in which to live, reproduction can occur whenever the urge hits. But for other animals and plants, reproduction needs to occur during certain seasons of the year when the climate and food supplies are optimal. This is not to say that humans do not have reproductive hormones! Human reproductive hormones, such as testosterone and estrogen, create the desire to mate and affect the ability to reproduce (see Chapter 13).

Table 10-1 lists some plant hormones, as well as some of the many hormones found in mammals. Several of these hormones are mentioned throughout this chapter.

Table 10-1 Some Important Plant and Mammal Hormones

Plant Hormones	*Mammal Hormones*
Abscissic acid	Antidiuretic hormone
Cytokinins	Adrenaline (epinephrine)
Ethylene	Aldosterone
Gibberellins	Amylase
Indoleacetic acid	Estrogen
	Gastrin
	Growth hormone
	Insulin
	Oxytocin
	Progesterone
	Secretin
	Testosterone
	Thyroid-stimulating hormone

How hormones work

Proteins are needed everywhere in the body. Cell membranes, tissues, enzymes, and hormones all are proteins. The nucleus of each cell contains genetic material that controls the production of proteins and RNA. The genes present on strands of deoxyribonucleic acid (DNA) are switched "on" when a certain protein needs to be produced. Then, they are turned "off" when the level of the protein is high enough in the body. But before the proteins can be created, the specific hormones that regulate whether the genes are turned on or off must get inside the nucleus of the cell to reach the DNA. Of course, hormones do not control the expression of every gene; only a relatively small number of genes are directly hormonally regulated.

Steroid hormones, which are made of cholesterol, diffuse from the bloodstream across the plasma membrane of the cell they are trying to get into. Once inside the cell, some of the hormones pass through the cytoplasm of the cell and diffuse into the nucleus. Inside the nucleus, steroid hormones bind to receptor proteins. Once the receptor protein and hormone form a complex, the genes that direct the production of the needed substance are "turned on" to start creating the substance. Chapter 14 explains in detail the way that genes control the production of proteins.

Peptide hormones, which are made of proteins, bind to receptor proteins on the plasma membrane of the target cell. Then, the receptor protein causes a second messenger to be produced, and the second messenger causes changes in the cell. The following compounds are a few examples of second messengers:

- *Cyclic adenosine monophosphate (cAMP)* is made from ATP. When a hormone stimulates the production of cAMP, the cAMP activates a protein kinase enzyme, which starts a cascade of other protein activations. Ultimately, it is the enzyme that causes the necessary effect in the cell.
- *Inositol triphosphate (IP_3)* is created from phospholipids on the membrane of the target cell. When IP_3 is created, it causes a calcium ion to be released from the endoplasmic reticulum inside the cell. The calcium ion activates enzymes, and the enzymes cause the necessary effect in the cell.

Ouch! Nerve Impulses

The nervous system, like the endocrine system, transports messages from nerve endings to brain and from brain to cells, tissues, and organs. In fact, in some respects, the nervous system and endocrine system overlap. Sometimes cells of the nervous system secrete chemical messengers instead of neurotransmitters. These specialized nervous system cells are called *neurosecretory cells*, and they produce *neurosecretions*.

Neurosecretions, which are classified as hormones because they carry information from sensor cells to target cells, can be released directly into the bloodstream or transported to storage cells, from which they are later released into the bloodstream.

Being treated for a hormone disorder? Thank some castrated chickens

As many people with high hopes were heading to California to pan for gold, a scientist with different high hopes was re-attaching testicles to a group of castrated chickens. In 1849, one A. A. Berthold found that when he transplanted testes into previously neutered chickens, the chickens developed the characteristics of normal roosters: They grew combs, they began to crow, and they started to "exhibit the generally pugnacious behavior expected of roosters." Couldn't have said that better myself! Even if Berthold reconnected the testes somewhere in the body other than the normal location, the effects were seen.

Berthold surmised, correctly, that the testes contained a substance that caused the development of male characteristics and that the nervous system had nothing to do with it. At the beginning of the 20th century, the term hormone was born to describe these newly discovered substances. Around 1909, Dr. Thomas Addison discovered the hormone disorder that bears his name — Addison's disease. And, by 1922, the hormone insulin, which is in short supply in people with diabetes mellitus, was discovered. Even today, hormones are still being discovered. It is now known that there are hormones that not only cause reactions to occur but also inhibit reactions to stop them from occurring.

One purpose of neurosecretions is to carry information to target cells that are not near the nerve cells that produce them. The hypothalamus, which is deep within the brain, detects conditions in the external environment of an organism as well as the internal environment of the organism. In attempts to maintain homeostasis, the hypothalamus produces neurosecretions that are released into capillaries in the hypothalamus. Blood vessels then carry the secretions to the pituitary gland, which lies at the base of the hypothalamus, and the pituitary gland controls the secretion of many important hormones. The way that the hypothalamus and pituitary glands work together shows how the nervous system and endocrine system are connected.

Just passing through: Neuron structure

The nervous system contains two types of cells: neurons and neuroglial cells. *Neurons* are the cells that receive and transmit signals. The *neuroglial cells* are the support systems for the neurons — the neuroglial cells protect and nourish the neurons.

Each neuron contains a *nerve cell body* with a nucleus and organelles such as mitochondria, endoplasmic reticulum, and Golgi apparatus. Branching off the nerve cell body are the *dendrites*, which act like tiny antennae picking up signals from other cells. At the opposite end of the nerve cell body is the *axon*, which is a long, thin fiber with branches at the end that sends signals. The axon is insulated by a myelin sheath made up of segments called Schwann cells (see Figure 10-1). Nerve impulses are received by the dendrites, travel down the branches of the dendrites to the nerve cell body, and are carried along the axon. When the impulse reaches the branches at the end of the axon, it is transmitted to the next neuron. Impulses continue to be carried in this way until they reach their final destination. The final destination depends on what type of neurons they are.

Getting off here? Last stops for impulses

There are three types of neurons, each with different functions. The function of the neuron determines where those neurons transmit their impulses.

- **Sensory neurons:** These neurons are also called *afferent* neurons. (Think of them as being affected by a sight, sound, smell, touch, or taste.) Their function is to receive initial stimuli from sense organs — eyes, ears, tongue, skin, and nose — as well as by impulses generated within the body in response to adjustments that are necessary to maintain homeostasis. For example, if your internal body temperature is rising because of high heat outside, sensory organs will transmit an impulse carrying the message that action needs to be taken to cool down the body. Or, if you touch the tip of a knife, the sensory neurons in your finger will transmit impulses to other sensory neurons until the impulse reaches an interneuron.
- **Interneurons:** These types of neurons are also called *connector neurons* or *association neurons*. What they do is "read" impulses received from sensory neurons. Interneurons are found in the spinal cord or brain. When an interneuron receives an impulse from a sensory neuron, the interneuron determines what response should be generated. If a response is required, the interneuron passes the impulse on to motor neurons.
- **Motor neurons:** These neurons are also called *efferent* neurons, and their function is to stimulate *effector* cells. When the motor neurons receive a signal from the interneurons, the motor neurons work to stimulate an effect. When the effector cells are stimulated, they generate reactions. For example, motor neurons may carry impulses to the muscles in your hand to stimulate the movement of muscles to pull your hand away from the sharp knife. Or, in an effort to maintain homeostasis when your body temperature is rising, the motor neurons may stimulate the sweat glands to produce sweat in an attempt to release some heat to the outside, thereby decreasing your internal temperature.

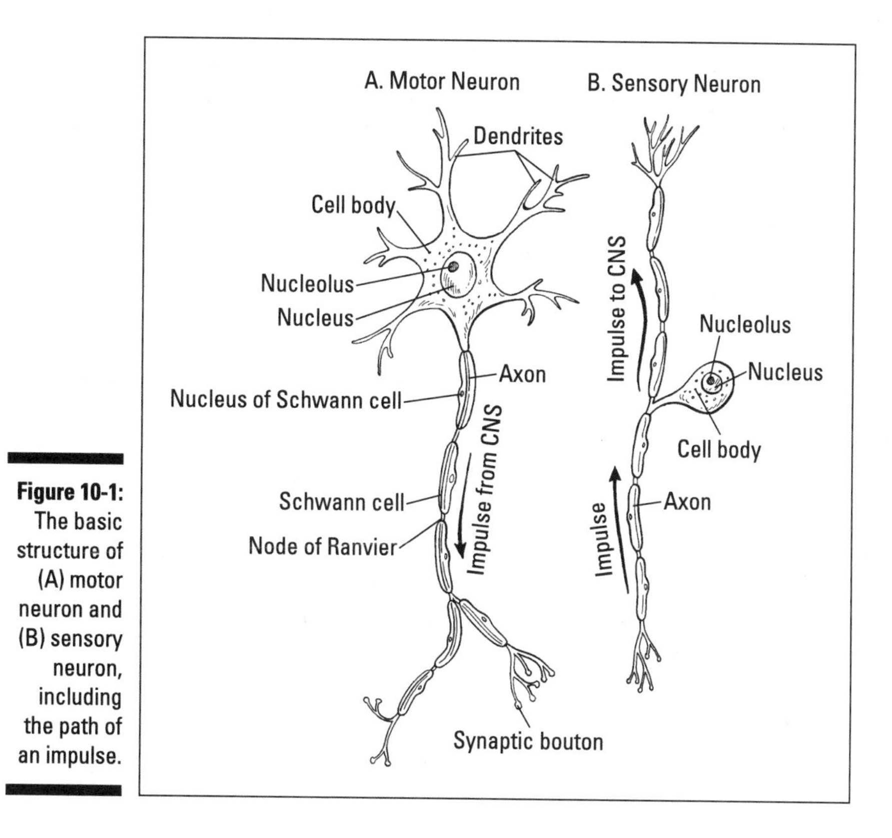

Figure 10-1: The basic structure of (A) motor neuron and (B) sensory neuron, including the path of an impulse.

Creating and carrying impulses

Although it happens quicker than a lightning strike, nerve impulses pass through a neuron in about *7 milliseconds* (see Figure 10-2). The impulse is shuttled through the membrane of the neuron by electrical changes on the membrane of the neuron. The impulse is passed to the next neuron by a chain of chemical events.

Here's what happens in six easy steps to get an impulse from one end of a neuron to the other end. Remember, all of what you are about to read happens faster than you can blink your eye. Isn't the body just amazing?

1. **Polarization of the neuron's membrane occurs: sodium on outside, potassium inside.**

   Neurons — the cells of the nervous system — have nerve cell bodies that are surrounded by cell membranes just as any other cell in the body has a membrane. When a neuron is not stimulated — it's just sitting there inside your body with no impulse to carry or transmit — its membrane is *polarized*. Being polarized means that the electrical charge on the outside of the membrane is positive, whereas the electrical charge on the inside of the membrane is negative. The outside of the cell contains excess sodium ions (Na^+); the inside of the cell contains excess potassium ions (K^+). I know what you're thinking: How can the charge inside the cell be negative if there are positive ions in there? Good question. The answer is that in addition to the K^+, there are negatively charged protein and nucleic acid molecules inside the cell, which makes the inside negative compared to the outside.

An even better question for you to come up with while reading this section is "If cell membranes allow ions to cross, how does the Na^+ stay outside and the K^+ stay inside?" If this thought crossed your mind, you deserve a huge gold star. The answer to this one is that the Na^+ and K^+ do, in fact, move back and forth across the membrane. However, Mother Nature thought of everything. There are *Na^+/K^+ pumps* on the membrane that pump the Na^+ back outside and the K^+ back inside. The charge of an ion inhibits membrane permeability.

2. **Resting potential gives the neuron a break.**

   When the neuron is inactive and polarized, it is said to be at its resting potential. It remains this way until a stimulus comes along.

3. **Action potential: Sodium ions move inside the membrane.**

   When a stimulus reaches a resting neuron, the *gated ion channels* on the resting neuron's membrane open suddenly, which allows the Na^+ that was on the outside of the membrane to go rushing into the cell. As this happens, the neuron goes from being polarized to being *depolarized.* Remember that when the neuron was polarized, the outside of the membrane was positive, and the inside of the membrane was negative. Well, once more positive ions go charging to the inside of the membrane, the inside becomes positive, as well, which removes the polarization.

   Each neuron has a *threshold level.* The threshold is the point at which there is no holding back. Once the stimulus goes above the threshold level, more gated ion channels open, allowing more Na^+ inside the cell. This causes complete depolarization of the neuron, creating an *action potential.* In this state, the neuron continues to open Na^+ channels all along the membrane. When this occurs, it is called an *all-or-none phenomenon.* "All-or-none" means that if a stimulus does not exceed the threshold level and cause all of the gates to open, no action potential results; however, once the threshold is crossed, there is no turning back: Complete depolarization occurs and the stimulus will be transmitted.

   When an impulse travels down an axon covered by a myelin sheath (refer to Figure 10-1), the impulse must move between the uninsulated gaps that exist between each Schwann cell. The gaps are called the *nodes of Ranvier.* During the action potential, the impulse undergoes *saltatory conduction* (think "salt," as in the sodium ions that allow this to happen) and jumps from one node of Ranvier to the next node of Ranvier, which increases the speed at which the impulse can travel.

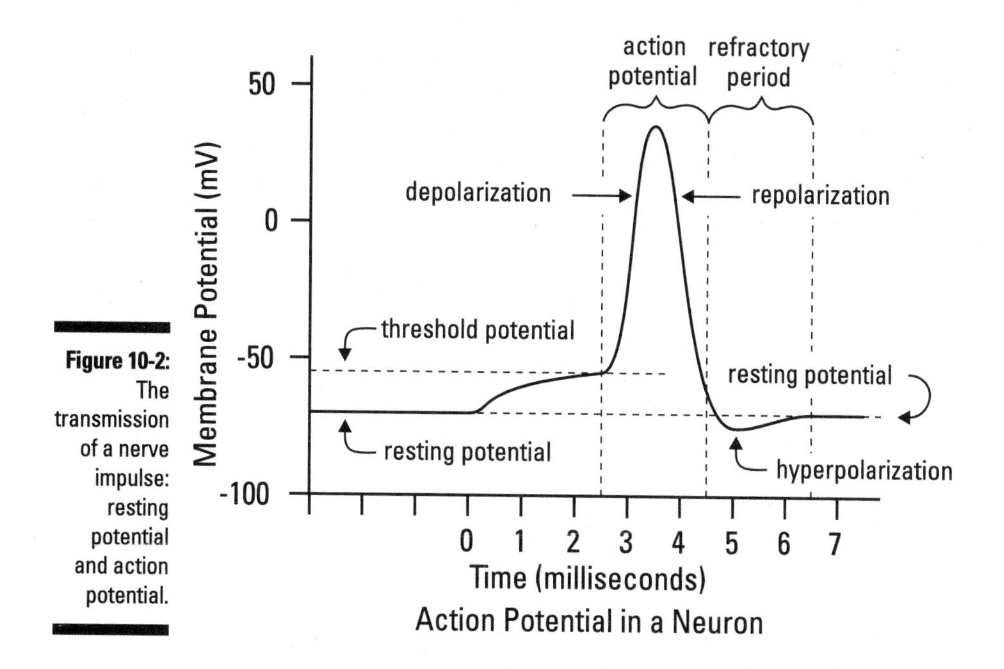

Figure 10-2: The transmission of a nerve impulse: resting potential and action potential.

4. **Repolarization: Potassium ions move outside, sodium ions stay inside the membrane.**

   Once the inside of the cell becomes flooded with Na^+, the gated ion channels on the inside of the membrane open to allow the K^+ to move to the outside of the membrane. With K^+ moving to the outside, the balance is restored by repolarizing the membrane, although it is opposite of the initial polarized membrane that had Na^+ on the outside and K^+ on the inside. Just after the K^+ gates open, the Na^+ gates close; otherwise, the membrane would not be able to repolarize.

5. **Hyperpolarization: More potassium ions are on the outside than there is sodium inside.**

   When the K^+ gates finally close, the neuron has slightly more K^+ on the outside than it has Na^+ on the inside. This causes the membrane potential to drop slightly lower than the resting potential, and the membrane is said to be hyperpolarized. This period doesn't last long, though (well, none of these steps take long!); once the impulse has traveled through the neuron, the action potential is over, and the cell membrane returns to normal.

6. **Refractory period puts everything back to normal: Potassium returns inside, sodium returns outside.**

   The refractory period is when the Na^+ and K^+ are returned to their original sides: Na^+ on the outside; K^+ on the inside. While the neuron is busy returning everything to normal, it does not respond to any incoming stimuli. Once the Na^+/K^+ pumps return the ions to their rightful side of the neuron's cell membrane, the neuron is back to its normal polarized state and stays in the resting potential until another impulse comes along.

Multiple sclerosis (MS)

People who suffer with multiple sclerosis (MS) are affected with a gradually worsening condition. MS patients develop lesions on the myelin sheaths covering the axons of their nerves. The lesions are inflamed, irritated areas that form hardened scars, which are called scleroses. These hardened areas interfere with proper conduction of impulses through the neuron. And, as more and more of these lesions and scleroses develop throughout the body, movement of the body becomes increasingly difficult, then nearly impossible.

Just as there are gaps between the Schwann cells on an insulated axon, there are gaps between the axon of one neuron and the dendrites of the next neuron. Neurons do not touch. The impulse must traverse that gap to continue on its path through the nervous system. The gap between two neurons is called a *synaptic cleft* or a *synapse*. The way that an impulse is carried across the synapse may involve electrical conduction (in many invertebrate animals and in the brain of vertebrate animals; however, most often, the following chemical changes occur to allow the impulse to continue on its way:

1. **Calcium gates open.**

   At the end of the axon from which the impulse is coming (the *presynaptic cell,* referring to the fact that the axon is preceding the synapse), the membrane depolarizes, which allows gated ion channels to open. This time, the ion that is allowed into the cell is calcium (Ca^{2+}).

2. **Synaptic vesicles release a neurotransmitter.**

   When the Ca^{2+} rushes into the end of the presynaptic axon, *synaptic vesicles* connect to the presynaptic membrane. The synaptic vesicles then release a chemical called a *neurotransmitter* into the synapse. Table 10-2 lists some common neurotransmitters and their functions.

3. **The neurotransmitter binds with receptors on the postsynaptic neuron.**

   The chemical that serves as the neurotransmitter diffuses across the synapse and binds to proteins on the membrane of the neuron that is about to receive the impulse. The proteins serve as the receptors, and different proteins serve as receptors for different neurotransmitters — that is, neurotransmitters have specific receptors.

4. **Excitation or inhibition of the postsynaptic membrane occurs.**

   Whether excitation or inhibition occurs depends on what chemical served as the neurotransmitter and the result that it had. For example, if the neurotransmitter causes the Na^+ channels to open, the membrane of the postsynaptic neuron becomes depolarized, and the impulse is carried through that neuron. If the K^+ channels open, the membrane of the postsynaptic neuron becomes hyperpolarized, and inhibition occurs. The impulse is stopped dead if an action potential cannot be generated.

If you're wondering what happens to the neurotransmitter after it binds to the receptor, you are really getting good at this biology stuff. Here's the story: After the neurotransmitter produces its effect, whether it is excitation or inhibition, it is released by the receptor and goes back into the synapse. In the synapse, enzymes degrade the chemical neurotransmitter into smaller molecules. Then, the presynaptic cell "recycles" the degraded neurotransmitter. The chemicals go back into the presynaptic membrane so that during the next impulse, when the synaptic vesicles bind to the presynaptic membrane, the complete neurotransmitter can again be released.

Table 10-2 Characteristics of Common Neurotransmitters

Neurotransmitter Function	*Source*	*Effect*
Acetylcholine	Secreted at gaps between the neurons	Stimulates or inhibits contraction of muscles, depending on receptor. and muscle cells
Dopamine	Created from amino acids	Lack of dopamine in brain causes muscular rigidity and tremor seen in patients with Parkinson's disease; schizophrenia is treated with drugs that block the dopamine receptors.
Epinephrine	Created from amino acids	Can be stimulatory or inhibitory; responsible for "fight-or-flight" response.
Norepinephrine	Released by postganglionic axons	Increases blood pressure; is released in response to low blood pressure and stress.
Serotonin	Produced through enzymatic reaction involving tryptophan; produced by animals, bacteria, and many plants	Associated with mood, tension, learning, and memory.

What a Sensation! The Brain and the Five Senses

In humans, the nervous system is divided into the central nervous system (CNS), which consists of the brain and spinal cord, and the peripheral nervous system (PNS), which contains all the nerves that run everywhere in the body. The spinal cord, which is attached to the brain, runs down the center of your body, so equate that with the CNS. All the nerves that branch off the spinal cord, including the cranial nerves and spinal nerves, and reach to the periphery of your body make up the PNS.

Within the PNS, there is the *somatic nervous system*, which directs motor nerve fibers to skeletal muscles, and the *autonomic nervous system*, which directs motor nerve fibers in smooth muscle, cardiac muscle, and glands. The smooth muscle and cardiac muscle do not come under the same control as skeletal muscles because they make up things such as your heart, which needs to work automatically, without any conscious or subconscious intention.

The autonomic nervous system has two divisions, both of which work opposite each other to maintain homeostasis:

- The *sympathetic nervous system* automatically stimulates the body when action is required. This is the part of the nervous system responsible for the "fight-or-flight" response, which stimulates a surge of adrenaline (epinephrine) to give the body quick energy so that danger can be escaped. The sympathetic nervous system also quickens the heart rate to move blood through the blood vessels faster and releases sugar from glycogen stores in the liver into the blood so that fuel can is readily available to the cells.
- The *parasympathetic nervous system* stimulates more routine functions, such as the secretion of digestive enzymes or saliva. In contrast to the sympathetic nervous system, the parasympathetic nervous system slows down the heart rate after the "fight-or-flight" response is no longer needed.

The brain

The brain is the master organ of the body. It takes in all information relating to the body's internal and external environments, and it produces the appropriate responses.

Inside the skull, the *meninges* cover the cerebrum (the large, gray, bumpy part of the brain). The meninges are strong membranes that cover the brain and spinal cord. *Cerebrovascular fluid* flows between the membranes. An infection here is called *meningitis* for inflammation of the meninges.

TECHNICAL STUFF

Cerebral palsy

The cerebrum of the brain controls thought processes, the ability to understand, and movements of the face and limbs. A brief lack of oxygen during birth can cause damage in the motor areas of the cerebrum. The result is *cerebral palsy*, in which children show the characteristic spastic weakness of the arms and legs. Cerebral refers to the cerebrum, and palsy indicates a slightly paralyzed position. Children affected with cerebral palsy do not have genetic defects; the cause of their affliction is largely attributed to the lack of oxygen or injury during birth.

The *cerebrum* is the largest part of the brain and is the part responsible for consciousness. The cerebrum is divided into left and right halves, which are called *cerebral hemispheres*. Each cerebral hemisphere has four *lobes* named for the bones of the skull that cover them: frontal, parietal, temporal, and occipital. Specific areas of the lobes are responsible for certain functions, such as concentration, understanding speech, recognizing objects, memory, and so on.

At the center of the brain are the *thalamus* and *hypothalamus*, which form the structure called the *diencephalon*. The hypothalamus generates many neurosecretions, which are carried to the pituitary gland at the base of the hypothalamus. The hypothalamus controls homeostasis by regulating hunger, thirst, sleep, body temperature, water balance, and blood pressure. The *pituitary gland* is called the master gland because, along with the hypothalamus, it helps to maintain homeostasis by secreting many important hormones.

At the base of the brain are the cerebellum and the brain stem. The *cerebellum* coordinates muscle functions such as maintaining normal muscle tone and maintaining posture. The *brain stem* is formed by three structures: the midbrain, the pons, and the medulla oblongata. The *spinal cord* is a continuation of the brain stem that runs down through the vertebrae of the spine.

Reflex arcs are connections between sensory neurons, the spinal cord, and motor neurons. They are good examples of how the nervous system protects you by making you get out of danger almost before you realize you are in danger. Here's an example: You are cooking a fabulous gourmet dinner for your beloved, and you grab the lid of a pot without using a hot pad; you know, without thinking. You just want to check on the vegetables. Your nervous system has other ideas.

When you grab that extremely hot lid, the endings of the sensory nerves in your skin detect the heat and send an impulse up through the axon of a sensory neuron to the nerve cell body of the sensory neuron. The impulse continues through sensory neurons until it reaches an interneuron in the spinal cord. The interneuron determines the appropriate response — which, in this case, would be stimulating the muscles to pull your hand away. The excitatory impulse is transferred to the cell body of a motor neuron and travels down the axon of the motor neuron until it reaches muscle tissue. The muscle responds by contracting to pull your hand away from the hot lid. With all these words describing what happens, it makes it seem like this process takes quite a while. But think about when you've touched something hot by mistake. You pulled your hand away immediately thanks to a quick-reacting reflex arc. Without the reflex arc protecting you, you might just unknowingly hold that hot lid in your hand until real damage is done!

The five senses

The sense organs — eyes, ears, tongue, skin, and nose — help to protect the body. The sense organs are filled with receptors that relay information through sensory neurons to the appropriate places within the nervous system. Each sense organ contains different receptors. *General receptors* are found throughout the body because they are present in skin, visceral organs (visceral meaning in the abdominal cavity), muscles, and joints. *Special receptors* include chemoreceptors (chemical receptors) found in the mouth and nose, photoreceptors (light receptors) found in the eyes, and mechanoreceptors found in the ears. Table 10-3 compares the various types of receptors found in the nervous system.

Table 10-3 Types of Receptors and Their Functions

Receptor	*Site*	*Function*
Chemoreceptors	Taste buds, cilia in nasal cavity	Detect chemicals in food and air
Mechanoreceptors	Cilia in ear	Detect movement of ear drum and ossicles (ear bones)
Osmoreceptors	Hypothalamus	Detects concentration of solutes in the bloodstream
Photoreceptors	Retina of eye	Detect light
Proprioceptors	Muscles	Detect positioning and movement of limbs
Stretch receptors	Lungs, tendons, ligaments	Detect expansion or elongation of muscle tissue

Oooh, that smell: Olfaction

If you walk in the door of your home and smell an apple pie baking and peppers and onions sautéing, how do you distinguish the smells of both foods? How do you know that apple pie is apple pie and that pepper and onions are in fact peppers and onions and not eggplant and zucchini? The key is the olfactory cells.

Olfactory cells (remember the mnemonic device "smells like an old factory?") line the top of your nasal cavity. On one end, olfactory cells have cilia — hair-like attachments — that project into the nasal cavity. On the other end of the cell, are olfactory nerve fibers, which pass through the ethmoid bone and into the olfactory bulb. The olfactory bulb is directly attached to the cerebral cortex of your brain.

As you breathe, anything that is in the air that you take in enters your nasal cavity: hydrogen, oxygen, nitrogen, dust, pollen, chemicals. You don't "smell" air or dust or pollen, but you can smell chemicals. The olfactory cells are chemoreceptors, which means the olfactory cells have protein receptors that can detect subtle differences in chemicals. As you breathe upon walking into your kitchen, the chemicals from the apple pie and peppers and onions waft into your nasal cavity. There, the chemicals bind to the cilia, which generate a nerve impulse that is carried through the olfactory cell, into the olfactory nerve fiber, up to the olfactory bulb and directly to your brain. Your brain then determines what you are smelling. If the scent is something you have smelled before and are familiar with, your brain recalls the information that has been stored in your memory. If you are sniffing something that you haven't experienced before, you need to use another sense, such as taste or sight, to make an imprint in your brain's memory. You'll also spout out that infamous saying, "What's that smell?"

Mmm, mmm, good: Taste

Did you ever notice that when you smell something that you really like that you start to salivate? Or, did you every notice that when you have a cold you can't taste things very well? Both of these circumstances are because the senses of smell and taste work closely together. If you cannot smell something, you cannot taste it, either. The connection between olfaction and taste allows this old scout trick to work: Hold someone's nose shut while they have their eyes closed and tell them to take a bite of an apple. Instead, have them bite into an onion. They will think they bit into an apple because their brain was telling them "apple," and they had no sensory information coming in to tell them otherwise. If they were able to smell the onion, however, the "apple" message would quickly be changed to "onion."

Taste buds on your tongue contain chemoreceptors that work in a similar fashion to the chemoreceptors in the nasal cavity. However, the chemoreceptors in the nose will detect any kind of smell, whereas there are four different types of taste buds, and each detects different types of tastes: sweet, sour, bitter, and salty.

Eat carrots for better night vision

You may think that eating carrots for better eyesight is an old wives' tale. Well, take it from a youngish wife, eating carrots *can* help your night vision. Here's how. The rods of the retina detect dim light and movement because of a chemical they contain. The chemical, called *rhodopsin*, contains the protein *opsin* and the pigment *retinal*. When light strikes the rods of the retina, rhodopsin breaks apart into opsin and retinal, which generates the nerve impulse that travels to the brain. When your eyes are adjusting to darkness, it is hard to see initially because the rods in your retina are busy forming rhodopsin so that the small amount of light available can split rhodopsin and generate nerve impulses. The more rhodopsin there is in your eye, the more sensitive your eyes will be in darkness. Retinal, one of the components of rhodopsin, is derived from vitamin A. And what contains an abundance of vitamin A? Carrots.

A common misconception is that the little bumps on your tongue are the taste buds. As with all misconceptions, this idea is wrong, too. The little bumps on your tongue are called papillae, and the taste buds actually lie down in the grooves between each papilla.

Foods contain chemicals, and when you put something into your mouth, the taste buds in your tongue can detect what chemicals you are ingesting. Each taste bud has a pore at one end with microvilli sticking out of the pore, and sensory nerve fibers attached to the other end. Chemicals from food bind to the microvilli, generating a nerve impulse that is carried through the sensory nerve fibers and eventually to the brain.

The sense of taste allows you to enjoy food, which you must ingest to live. But it also serves a higher function. You could live without tasting your food; that is not so extremely important. What is important is that when the taste buds detect chemicals, the signal that is sent to the brain sets in motion the production and release of the proper digestive enzymes needed to break down the food you are ingesting. This function allows your digestive system to work optimally to retrieve as many nutrients as possible from food.

Now hear this: Sound

The ear not only is the organ of hearing, but it also is responsible for maintaining equilibrium — or balance. To maintain equilibrium, the ear must detect movement. To hear, the ear must respond to mechanical stimulation by sound waves.

The outer ear is the external opening to the ear canal. Sound waves are shuttled through the ear canal to the middle ear. The eardrum sets the mechanics in motion. When a sound wave hits the eardrum, the eardrum moves tiny bones — the malleus, incus, and stapes — which subsequently move. This movement is picked up by the mechanoreceptors in the inner ear, which exist on hair cells containing cilia between the end of the semicircular canals and the vestibule. When the cilia move, the cells create an impulse that is sent through the cochlea to the eighth cranial nerve, which carries the impulse to the brain. The brain then interprets the information as a specific sound.

But, as I said, the ear also allows you to stand up straight and keep from falling over. The fluid within the semicircular canals of the inner ear moves, and that movement is ultimately detected by the cilia. When the fluid doesn't stop moving, you can develop motion sickness. The cilia transmit impulses to the brain about angular and rotational movement, as well as movement through vertical and horizontal planes, which helps your body to keep its balance.

Seeing is believing: Sight

When you look at an eye, the *iris* is the colored part. The iris actually is a pigmented muscle that controls the size of the *pupil*, which dilates to allow more light into the eye or contracts to allow less light into the eye. The iris and pupil are covered by the *cornea*. Behind the pupil (the "black hole" in the very center of the iris) is an anterior chamber. Behind the anterior chamber is the *lens* of the eye. The ciliary body contains a small muscle that connects to the lens and the iris. The *ciliary muscle* changes the shape of the lens to adjust for far or near vision. The lens flattens to see farther away, and it becomes rounded for near vision. The process of changing the shape of the lens is called *accommodation*. People lose the ability of accommodation as they grow older, prompting the need for glasses.

Behind the lens of the eye is the *vitreous body*, which is filled with a gelatinous material called vitreous humor. This substance gives shape to the eyeball and also transmits light to the very back of the eyeball, where the *retina* lies. The retina contains *photoreceptors*, which detect light. Two types of sensors detect light: *rods*, which detect motion, and the *cones*, which detect fine detail and color. The rods work harder in low light, and the cones work best in bright light. There are three types of cones: one that detects blue, one that detects red, and one that detects green. Color blindness occurs when one type of cone is lacking. For example, if the red cones are lacking, the person cannot see red, but will see green instead. Sounds dangerous at traffic lights, doesn't it?

Everybody has a blind spot

There is one area in the retina where there are no rods or cones. This area is called the *optic disk*, and it is where the optic nerve passes through the retina. Casually, this area is known as the *blind spot*, because vision here is impossible. Everybody is blind in that spot.

Macular degeneration

The *macula lutea* is a small area in the retina where there is a large concentration of cones. Vision here normally is very sharp, and detection of colors is acute. However, with age, this area of the retina loses its acuity — that is, it degenerates. Therefore, *macular degeneration* describes a condition in which vision becomes distorted. Vision may be blurry; what should look straight (like a phone pole) looks wavy; things may look larger or smaller than they really are; and colors may look faded.

There are several causes of macular degeneration, which, with the aging of the population in the United States, has become the No. 1 cause of blindness. Macular degeneration can be caused by overgrowth of new blood vessels around the macula lutea. These blood vessels leak, however, and the oozing blood destroys the macula lutea. Macular degeneration also can be an inherited condition; 15 percent of people with relatives who have the condition develop the condition themselves after they turn 60. People with light blue or light green eyes develop macular degeneration more often than those with dark eyes; excessive sun exposure, which affects those with lighter pigmentation more severely, may be a contributing factor. Smoking and high blood pressure also can contribute to this condition. Treatment includes vitamin and mineral supplements; zinc may prevent the progression of the disease. Laser treatment may help stop the overgrowth of the leaky blood vessels.

When light strikes the rods and cones, nerve impulses are generated. The impulse travels to two types of neurons: first to *bipolar cells* and then to *ganglionic cells*. The axons of ganglionic cells form the *optic nerve*. The optic nerve carries the impulse directly to the brain. Approximately 150 million rods are in a retina, but only 1 million ganglionic cells and nerve fibers are there, which means that many more rods can be stimulated than there are cells and nerve fibers to carry the impulses. Your eye must combine "messages" before the impulses are sent to the brain. So, isn't it amazing that you can see in real time?

A touchy-feely subject: Touch

The eyes, ears, nose, and tongue all contain special receptors: photoreceptors, chemoreceptors, and mechanoreceptors. However, the skin contains general receptors. These simple receptors can detect touch, pain, pressure, and temperature. Throughout your skin, you have all four of these receptors interspersed. If a metal rod is passed over your skin, at some places you will feel only pressure, whereas other areas will detect the coolness of the metal. The general receptors work the way that reflex arcs work. Skin receptors generate an impulse when they are activated, which is carried to the spinal cord and then to the brain.

The skin is not the only tissue in the body to have receptors, however. Your organs, which are made of tissues, also have receptors. The lungs contain *stretch receptors* that detect when the lungs have expanded. The hypothalamus in the brain contains *osmoreceptors*, which detect levels of solutes (particles such as ions) in the blood. These receptors allow the hypothalamus to generate effects, such as the release of a hormone, to maintain homeostasis. Organs also have pain receptors, but pain in the organs often is "referred" to the skin. For example, people having a heart attack often say they feel the pain in their left arm. The left arm is the site of *referred pain* from the heart.

Joints, ligaments, and tendons contain *proprioceptors*, which detect the position and movement of the limbs. Information from these receptors travels to the cerebellum of the brain, which controls the positioning of body parts and posture. Muscle fibers also contain stretch receptors; however, the stretch receptors in muscle detect when the muscle fibers are elongated, whereas the stretch receptors in the lungs detect expansion of the lung cavity. When muscle fibers stretch, an impulse is generated and transmitted to the spinal cord, which then causes contraction of the muscle. The muscles are constantly sending information through the nervous system so that the brain can maintain normal muscle tone.

Shake Your Groove Thing: Moving Those Muscles

I'm sure you know that muscles are extremely important to your body. Toned muscles certainly make your body look better and be stronger, but muscles have even more important functions than improving your looks and strength.

- *Muscles allow you to stand upright.* The force of gravity is strong. Without muscles contracting, the force of gravity would keep you down. Muscle contraction requires that the force of gravity be opposed, and muscle contraction allows your body to assume different positions.
- *Muscles allow you to move.* I don't mean just walking or running, but every little movement that your body performs, right down to the blink of an eye, the dilation of your pupils, and your facial expressions.
- *Muscles allow you to digest.* Muscles all along the digestive system keep food moving downward and outward. Peristalsis, the squeezing of food down through the esophagus, stomach, and intestines, is created by the contraction of muscles.
- *Muscles affect the rate of blood flow:* Blood vessels contain muscle tissue that enables them to dilate to allow blood to flow faster through the vessel or contract to slow down the flow of blood. Muscle contraction also is responsible for movement of blood through the veins.

- *Muscles help to maintain normal body temperature.* When muscles contract, they give off heat. That heat is used to maintain body temperature, because some heat is continually lost through your skin. In an effort to maintain homeostasis, the body shivers when cold in an attempt to generate heat.
- *Muscles hold your skeleton together.* The ligaments and tendons at the ends of muscles wrap around joints, holding together the joints — and therefore the bones of your skeleton.

Muscle tissue and physiology

There are three types of *muscle tissue* in your body: cardiac, smooth, and skeletal. Muscle tissues are made up of *muscle fibers*, and muscle fibers contain many, many *myofibrils*, which are the parts of the muscle fiber that contract (Figure 10-3). Myofibrils are perfectly aligned and make the muscle look striated. The repeating unit of these striations — light and dark bands — is called a *sarcomere.*

Cardiac muscle is found in the heart. Fibers of cardiac muscle have one nucleus (uninucleated), are striated (have light and dark bands), are cylindrical in shape and branched. The fibers interlock so that contractions can spread quickly through the heart. Between contractions, cardiac fibers completely relax so that the muscle does not get fatigued. Cardiac muscle contraction is totally involuntary and occurs without nervous stimulation, and does not require conscious control.

Smooth muscle is found in the walls of internal organs that are hollow, such as the stomach, bladder, intestines, or lungs. Fibers of smooth muscle tissue are uninucleated, shaped like spindles, and are arranged in parallel lines. Smooth muscle fibers form sheets of muscle tissue. Contraction of smooth muscle is involuntary. It contracts more slowly than skeletal muscle, which means it can stay contracted longer than skeletal muscle and not fatigue as easily.

Skeletal muscle is probably what you think of when you picture a "muscle." Fibers of skeletal muscle have many nuclei (multinucleated), are striated, and are cylindrical. The skeletal muscle fibers run the length of the muscle (think of the "grain" in meat). Because some skeletal muscles are long, such as the hamstrings, skeletal muscle fibers can be long. Skeletal muscle is controlled by the nervous system; read the above sections on the nervous system for information about motor neurons and reflex arcs, which control muscles. Movement and contraction of skeletal muscle can be stimulated consciously — you consciously decide that you are going to stand up and walk across the room, which requires muscle. Therefore, skeletal muscle contraction is said to be voluntary. Figure 10-3 shows you how skeletal muscle is connected to the nervous system, as well as how the skeletal muscle contracts.

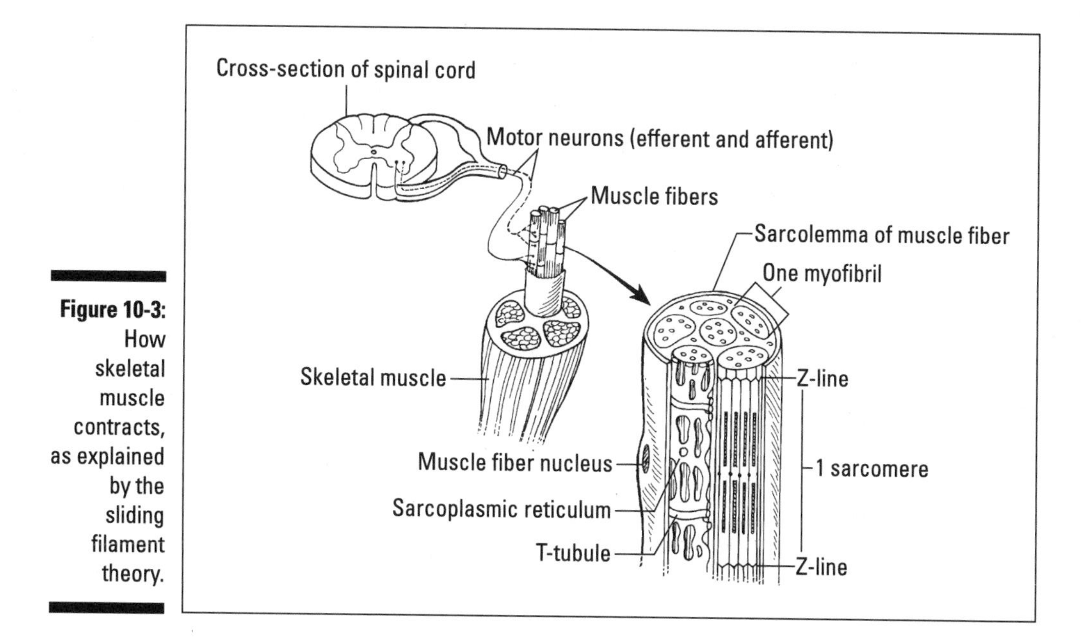

Figure 10-3: How skeletal muscle contracts, as explained by the sliding filament theory.

Looking at a muscle contraction

Contraction of muscles is described by the *sliding filament theory* (see Figure 10-3). The filaments in the sliding filament theory are the *thin filaments* and *thick filaments* that make up myofibrils.

Thin filaments are made up of two strands of *actin*, which is a protein that is wound in a double helix (like DNA is). Thin filaments have molecules of *troponin* and *tropomyosin* at binding sites along the actin double helix.

Thick filaments contain groups of *myosin*, which also is a protein. Myosin strands are bulbous at one end; however, multiple strands of myosin are mixed together in opposite directions, so it appears that both ends of thick filaments have bulbous ends.

Myofibrils have actin and myosin filaments that are lined up parallel to one another in an alternating pattern. The actin filaments are attached to a Z line; myosin filaments lie between the actin filaments, unattached to Z lines. From Z line to Z line is one sarcomere, which is a unit of contraction.

ATP is required for muscle contraction. If no ATP is available (such as if the body's oxygen stores are depleted), the metabolism switches to using lactic acid to create ATP. Lactic acid is produced during anaerobic respiration, but lactic acid cannot be directly used by muscles for contraction. A muscle fiber contains only enough ATP to sustain a contraction for about 1 second. Energy is stored in phosphocreatine molecules (ATP + creatine), which are formed during periods of no contraction. Creatine is a substance that some jocks take to try to bulk up. Phosphocreatine is quickly broken down to release more ATP as the low amounts of ATP in a muscle cell are used up. Humans also take every two molecules of adenosine diphosphate (ADP; two phosphates each for a total of four) formed per contraction and make a new ATP (three phosphates)and a molecule of adenosine monophosphate (AMP; one phosphate). As AMP levels rise, glycolysis is stimulated to synthesize more ATP.

ATP binds to the bulbous end of a myosin filament, and when it does, the ATP splits into ADP plus a molecule of inorganic phosphate (P_i). The ADP and P_i stay attached to the myosin.

A calcium ion also is required for a muscle to contract. The thin filaments of a myofibril contain troponin and tropomyosin. Calcium binds to the troponin, which causes the tropomyosin to move out of the way so that the binding sites on actin are open.

After the actin's attachment sites are exposed, myosin binds to the actin. Cross-bridges are formed to connect actin and myosin, which causes myosin to release the ADP and P_i.

When myosin releases the ADP and P_i so that it can link to actin, the shape of the bulbous end of myosin changes. As this change occurs, the actin filament slides toward the middle of the sarcomere, pulling the Z lines at the end of the sarcomere closer together. This shortens, or contracts, the muscle fiber.

The cross bridges linking the actin and myosin are released when another ATP molecule attaches to the bulbous end of myosin.

Rigor mortis

When an animal dies, it no longer has the ability to perform metabolic functions. It no longer breathes, so respiration cannot occur, and the blood cannot be filled with oxygen for transport. The heart no longer pumps, anyway, so the blood cannot be circulated through the body. The brain no longer functions to send the signals for what hormones need to be produced to maintain homeostasis or what movement needs to be performed. But, most of all, the cells no longer perform their metabolic cycles. The blood is not exchanging nutrients for wastes. The cells do not receive glucose from which they can make ATP. And, without ATP made continually available to muscle fibers, movement stops completely. Without ATP flooding the myofibrils, the process of contraction cannot take place, and that includes the very last step of muscle contraction. In order for a myofibril to relax, more ATP must cause the cross bridges between actin and myosin to break. At the point when there is no more ATP available to the myofibrils to generate a subsequent contraction, the last contraction becomes permanent, and the corpse stiffens. This condition is called rigor mortis, which means rigidity of death.

Movement in other organisms

Animals are not the only organisms that move. Some bacteria must move to decompose matter and to get into cells to cause infection. And, I'm sure you've seen a plant move toward the sunlight. How does it do that? The combination of hormones and the ability to move. Here's the story.

Motility, or the ability to move (motor), is achieved through a variety of specialized structures.

Flagella and cilia are hair-like structures that project from a cell membrane. Their movement is kind of a waggle, and when they move, they displace the fluid inside the cell — the cytoplasm. This action is similar to how boats move through water. Consider this: You are a cell sitting in a rowboat. The oars you are holding are your "flagella." As you move the oars, water is displaced. When the water moves back into place, the water helps to push the boat along, just as flagella help certain cells move through the bloodstream.

Cilia wave back and forth to help trap particles. Animals such as clams have cilia to help them trap food. However, cilia also are present in your ears and nose to help trap dirt particles and prevent them from getting into your inner ear or nasal cavity. Cilia also can contain receptors, such as the mechanoreceptors in your ear or chemoreceptors in your nasal cavity.

Microtubules, intermediate filaments, and microfilaments are protein fibers that exist in muscle cells and some unicellular organisms, such as amoeba. Table 10-4 provides information about these protein fibers.

Table 10-4 Functions of Microtubules, Intermediate Filaments, and Microfilaments

Microtubules	*Intermediate Filaments*	*Microfilaments*
Support cellular activities	Help to maintain the shape of the cell	Contain actin, which is involved in contraction
Provide motility		Help unicellular organisms like amoeba change their shape

Centrioles and *basal bodies* organize microtubules. Centrioles that exist in the cytoplasm outside of the nuclear envelope produce microtubules. The microtubules form spindle tubules, which help cells to divide during cell division. Basal bodies help to form flagella and cilia.

Plants cells do not contain centrioles. Instead of having cells that provide motility, plants have hormones that cause them to move. But, really, they can't move far — they are rooted in the ground!

Plants undergo *tropism* when they want to "move." Movement in a plant is really a change in growth pattern. Tropism is the change in growth pattern in response to a stimulus from the environment, such as a change in light, temperature, or nutrient availability. When something in the plant's environment changes, the plant produces a hormone to help it adjust. Here are a few examples:

The hormone *auxin* helps plants achieve *phototropism*, which is a change in response to an increase or decrease of light. Auxin is produced in the stem and travels by active transport to stimulate elongation of the plant — that is, it makes the plant grow higher. If the plant receives equal light on all sides, the stem will grow straight. If light is uneven, the auxin moves toward the darker side of the plant. This might seem backward, but when the shady side of the stem grows, the stem, in its crookedness, will actually bend toward the light. This action keeps the leaves toward the light, so that photosynthesis can continue.

Auxin combines with other hormones called *gibberellins* to work together to affect plant growth in response to gravity. This response is called *gravitropism*. The hormones work to keep the stem and leaves growing upward and the roots growing downward, even when the stem or roots may be horizontal at times.

Hormones are made from proteins and fats, but there is one protein that works with a light-absorbing molecule to help plants adjust to changes in the light and dark pattern. This response is called *photoperiodism*, and the substance that helps the plants adjust is called *phytochrome*. Just as people are naturally awake during the day and sleep at night, plants also have a *circadian rhythm* that tells them when there is less light available. Phytochrome is made in the leaves of a plant; it absorbs red light, and apparently keeps the plant on the right schedule of light and dark hours. They phytochrome in essence "stores light" so that energy from light is available through dark hours to allow photosynthesis to continue. Pretty cool, eh?

Part IV

Let's Talk About Sex, Baby

"OOO-KAY, LET'S SEE. IF WE CAN ALL REMAIN CALM AND STOP ACTING CRAZY, I'M SURE I'LL EVENTUALLY REMEMBER WHAT NAME I FILED THE ANTIDOTE UNDER."

In this part . . .

Living things are not considered living unless they can reproduce. Reproduction is a necessity of life in more ways than the one you are thinking of right now. Without it, there would have been one generation of life billions of years ago, and then that would have been it. You certainly wouldn't be sitting here reading this book right now, and there would be no study of biology because there would be no life.

In this part, you figure out how cells divide, because if that doesn't happen, new organisms can't be made. You find out how more plants and more animals come to populate the earth. Then, you examine the basics of genetics, including heredity, how genes carry the instructions for a new organism, and how DNA is copied and "read" to create the proteins that make up the organism.

Chapter 11

Dividing to Conquer: Cell Division

In This Chapter

- Finding out why mitosis results in the exact replication of a parent cell
- Understanding the cell cycle and the various stages of mitosis
- Seeing how cytokinesis helps each new daughter build her own cytoplasmic home
- Finding out why sexually reproducing organisms form cells with the haploid number of chromosomes
- Discovering why the mechanics of meiosis are critical to genetic variety and how mutations can occur

Hopefully, this chapter can give you an understanding of just how important the processes of *mitosis* — cell division — and *meiosis* — sex cell formation — are to life. You discover that all living things are capable of reproducing themselves in order to survive — and in some cases better enable their offspring to survive in a changing environment. The chapter looks at the cell cycle and various stages of mitosis and demonstrates why this method results in the exact replication of a parent cell. Then the focus is on the more specialized form of cell division, meiosis, which results in the formation of sex cells. After studying the mechanics of meiosis, you find out how genetic variety can occur in sex cells and how these differences result in variations ranging from mathematical ability to straight hair.

Keep On Keepin' On

Biology is, of course, all about life. And, when you think about it, life is really all about continuation — living things keep on keepin' on from one generation to the next, passing on critical genetic information and — in some cases — adapt to better cope with their environments. Certainly, this is one of the core differences between living organisms and inanimate objects. Think about it: Have you ever seen a chair or table replicate itself? And wouldn't it be nice if your car would develop a water-repellant skin as an adaptation to wet weather? Forget it! Only living things have the ability to pass on genetic information, replicate, differentiate, and adapt to environmental changes through reproduction.

To understand the process of reproduction, you need to know a few key terms and processes. You must have a general understanding of how individual cells — whether they are single-celled organisms or one of the several trillion cells in your body replicate — through the process of mitosis. To understand how cells are able to pass on genetic information, you need a basic understanding of the cell cycle and the various stages of mitosis. You also need to have an idea of how the process of cell division can run amuck, leading to such nasty outcomes as mutations and cancer. Finally, you also need to grasp the mechanics of meiosis, in order to see how genetic variety occurs.

Understanding the basics of reproduction goes a long way toward helping you figure out both the process and variety of life. It is basic to your understanding the differences in individual cells. You've probably wondered how hair and toenails can be so different and still exist on the same being. Well, maybe not. But you may have wondered what makes your hair brown and your cousin's a bright auburn (especially when you're paying for that expensive hair coloring). A feel for the basics of meiosis helps you see how these genetic differences occur. Of course, you always knew you were truly unique. Now you can discover how you got that way.

Reproduction and Life

The study of reproduction is basic to the study of life and living things. Sad but true, all living things die. Without reproduction, all life would cease to exist in just one generation. That's why the study of reproduction is a matter of life and death in biology. And, at the heart of any study of reproduction, is an understanding of the process of cell division. And, at the heart of the process of cell division is an understanding of the contents of the cell's nucleus, and at the heart of an understanding of the contents of the cell's nucleus is. . . . Well you get the picture. By focusing on smaller and smaller components of a cell, biologists have begun to gain a deep knowledge of the mechanisms of cell division and reproduction.

To grasp the process of reproduction and cell division, you have to get personally involved with the inhabitants of the *eukaryotic* cell's nucleus. Eukaryotic cells are those with true nuclei, such as in plants, fungi, and animals. (Refer back to Chapter 2 for a more detailed look at cells.)

The nucleus is the high rent district — the place where the hereditary materials of the cell are assembled into chromosomes, those tightly coiled molecular repositories of deoxyribonucleic acid (DNA). DNA is the cell's blueprint, without which all construction on new cellular neighborhoods would come to a screeching halt. DNA molecules are extremely long, but in chromosomes, they are squished together and packed with a special kind of protein called histone. In fact — as unappetizing as it sounds — the long DNA molecules actually wind around globules of protein to form structures called chromatids. As the structures coil even more and thicken, they become the stars of the show, chromosomes.

You need to know about two other nucleic structures to understand the basics of cell division:

- The *nucleolus* is the site of *ribosome* manufacture. Ribosomes are little duplexes made up of protein and ribonucleic subunits. The ribosomes are where amino acids are assembled into proteins.
- *Nucleoplasm* is another component of the nucleus. This appetizing mixture of water and the molecules is used in the construction of ribosomes, nucleic acids, and other nuclear material. (Best served with a DNA chaser.)

What Is Cell Division?

Cell division is the process by which new cells are formed to replace dead ones, repair damaged tissue, or allow organisms to grow. During cell division, two events occur that together replicate the old cell and pass on critical genetic information. These two events are *mitosis* and *cytokinesis*. Together, these events form the cell cycle (see Figure 11-1). But first, the various nucleic structures vital to life and the continuation of life must form, grow in volume, and get organized so that they can pack up and move into their new homes.

All eukaryotic cells go though a basic life cycle, which, once begun, is continuous (and it's a good thing, or that cut or broken bone would never heal). *Interphase*, the stage between actual cell division, makes up the first three of four stages of the cell cycle, including the first gap phase (G_1), the synthesis stage (S), and the second gap phase (G_2; see the section on interphase).

Finally, *mitosis* — the equal distribution of genetic information from a parent cell to two daughter cells' nuclei — occurs. This fourth part of the cell cycle is not part of interphase. Mitosis is followed by cytokinesis: a division of the cell's *cytoplasm*. Cytoplasm is the fluid material cushioning the cell's nucleus and containing nutrients used to perform these cellular processes. Mitosis and cytokinesis usually happen in sequence. Voila! Two new daughter cells have been created, each with its own deluxe and very complete set of genetic information.

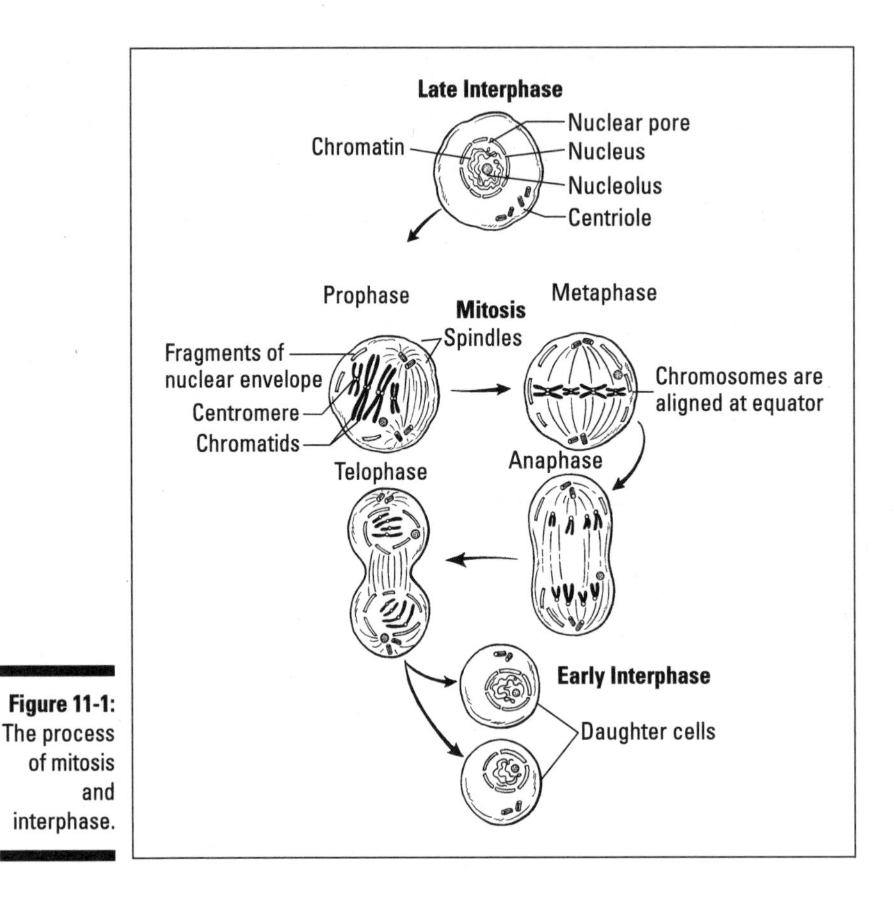

Figure 11-1: The process of mitosis and interphase.

Getting organized: Interphase

Division of genetic material would be impossible unless cell structures were created and organized for orderly cell division. This happens during interphase, which — as you might guess from its name — is the collection of processes that actually takes place *between* occurrences of cell division. The cell is not dividing during this time, but instead, getting its house in order by engaging in the metabolic functions that make it unique. The nuclear membrane is intact, and individual chromosomes are not apparent during this period.

You might say that interphase is the time that the cell asserts its individuality. Activities such as nerve-cell transmission or glandular-cell secretion are the kinds of activities that take place during interphase. The phases that together form interphase are as follows:

- **G_1 Phase:** First, during the G_1 (gap-phase one) stage, the cell grows in volume as it produces various cell components including transfer RNA (tRNA), messenger RNA (mRNA), ribosomes, and enzymes. This is typically the longest phase of the cell cycle. Some cells never leave it; they are said to be terminally differentiated. This phase immediately follows an occurrence of mitosis. It's when the cell is making the proteins and enzymes it needs to perform its unique functions. For example, a pancreas cell might be producing insulin for secretion, or a muscle cell might be learning how to contract so that movement can occur. Think of this phase as the cell's infancy and toddler period. Imagine a newborn's body building the bones and muscles necessary for mobility, until one day the baby is actually able to get up and walk.

Note that during this initial portion of interphase, each chromosome is made up of just a *single* double-stranded piece of DNA (it is always in the form of a double-helix) and its associated protein.

- **S Phase:** During S phase, the cell puts the pedal to the metal for that key process — replication of DNA to prepare for distribution of genes to daughter cells. (Note: If you are male, don't be offended that offspring are referred to as only daughter cells. Consider it one small way of making up for all those brightly hued feathers on male birds.) It's during this phase that the DNA within the cell nucleus replicates (copies exactly — and I mean *exactly).* Then the process is complete (drum roll, please), and there are 92 chromatids arranged as twin threads, two per chromosome, in most human cells, where once there were just 46 chromosomes. Each of these chromatids becomes a new chromosome — identical to its parent — and each contains its own set of genes, which is different from the sets on other chromosomes. (Each chromosome has thousands of different genes, so all the genetic messages that make an organism unique are clear. You wouldn't want the message that gives you blue eyes to mix with the one that creates your red hair. Think of the implications!)

  This replication process is how the parent cell ensures equal gene distribution between the two daughter cells. After all, it's always important to avoid sibling rivalry and jealously, even at a cellular level.

- **G_2 Phase:** During this phase, the cell is packing its bags and getting ready to hit the road for mitosis by synthesizing the spindle-fiber proteins needed to aid in the movement of chromosomes. Now the cell is all packed up and ready to hand off equal amounts of genetic material to its offspring. But there has to be a way to move the chromosomes from mother to daughters and then position them correctly in the new cells. So, in "little organs" (cutely dubbed *organelles*) outside the nucleus, the cell organizes proteins into a series of fibers or *microtubules*, which — in the first phase of mitosis — form *spindles.* Think of these spindles as teeny people movers (just substitute chromosome for "people," and you've got it). In the next phase of the process, the spindles are arranged from pole to pole of the cell, helping to move chromosomes into place. Following the entire mitosis process — poof — the spindles will simply disappear, with no demolition experts necessary.

So, you want to be a millionaire?

Here's a quick pop quiz to help prepare you for your next trivia game (or time in the hot seat).

True or False: All human cells boast a full complement of 46 chromosomes during the G_1 phase.

False: Most human cells have 46 chromosomes, with two exceptions: sex cells with 23 and the poor red blood cells that lack nuclei. Because red blood cells have no nuclei, they don't get those stylish chromosomal accessories.

Note: As you explore the next stage of cell division — mitosis — it's important to understand that, despite the fact that certain distinct processes take place during each stage, the boundaries between them are fluid and always moving. You might think of the stages, in fact, as seasons. It's critical that certain process occur — buds sprouting in spring, for example — but one flows like a gently moving stream into the other. As much as it might be nice if this were the case, no gun goes off signaling the end of winter and the beginning of spring. In the same way, when all the pieces are in place, a cell flows into the realm of mitosis.

Mitosis: One for you, and one for you

After interphase is over, the cell has done all the growing and replicating it's going to do, and now it's time to prepare to vacate its cramped quarters and move its genetic riches into two daughter cells.

During mitosis, the cell is making final preparations for the impending split. Processes during mitosis ensure that genetic material is distributed equally (again, eukaryotic cells are model parents intent on avoiding bickering between their daughter cells).

These processes take place in four stages. Remember, the cell cycle is a continuous process, with one stage flowing into another virtually seamlessly. But biologists have divided mitosis into the following stages to aid in their study:

- Prophase
- Metaphase
- Anaphase
- Telophase

Prophase

It's during this first stage of mitosis that the chromosomes become visible (well, not to the naked eye; you do need a pretty powerful microscope). Soon the thin, tangled threads of the chromatids coil and thicken to become chromosomes. Remember, each of these chromosomes carries identical genes on to their new residences.

At the same time the chromosomes are fattening up, the spindle fibers formed in interphase are doing what all good spindle fibers do — turning into actual spindles to help move the mobility challenged chromosomes to their final destinations. By this time, the centrioles (they're hanging on to the spindle fibers) have duplicated and moved to the poles of the cell, throwing off spindles like clotheslines running from one end of the cell to the other.

The other important event occurring is the disintegration of the nuclear membrane, giving the chromosomes free rein over the entire cell. (Well, sort of free rein. They're free to go anywhere they want, as long as they end up lined up exactly along the equatorial plane by the middle of the next stage. It's something like the "free" rein parents give teens.)

Metaphase

By metaphase, the nucleus of the cell has completely disappeared, making it possible for the chromosomes to get to their pre-assigned positions. (There is no free will here. That's restricted to higher life forms.) At first, they attach themselves at their beltlines to any old spindle around. But, pretty soon — like the good little soldiers they are — they move toward the center of the cell, until they form a perfect row along the equatorial plane. At this stage, there are still 46 of these guys and a total of 92 chromatids.

Anaphase

Now the chromatids part company, almost ready to start life on their own. But like a lot of young adults, they're not ready to go too far — yet. The chromosomes separate at their centers, and the chromatids move along their little spindle/people movers to the opposite poles. This event is sort of the "coming out" party for a chromosome. After this separation and movement, they are called *daughter chromosomes.* But they're not quite ready to move out of the family estate — just yet.

Telophase

As sort of a graduation gift, each identical set of chromosomes gets its own nuclear membrane and set of nuclear bodies (nucleoli). Those microtubules that made up the spindles dissolve, meaning that the spindles disappear — just as if someone pulled in the clotheslines. Now the two cells are ready to move into their own digs.

Cytokinesis

The last order of business is to give the new *daughter nuclei* suitable homes through a process called *cytokinesis.*

In animal cells, the process begins with a mere indentation or furrow in the center of the cell. The furrow squeezes the cell membrane into the cytoplasm until two separate cells are formed (imagine squeezing a ball of silly putty at the center until it becomes two balls of silly putty and you've got the idea.) This process is referred to as *cell cleavage.*

In plant cells, the process is slightly different because a rigid cell wall is involved, preventing the "silly putty" process of cell cleavage. Instead, a new cell wall forms at the center of the cell, with *vesicles* (small bags of necessary intercellular material) that join to form a double membrane called the *cell plate*. The plate eventually moves outward to marry the cell membrane (call it a marriage of convenience). The cell plate then separates the two cells, and they move apart to become two new daughter cells.

Once cytokinesis is complete, the new cells move immediately into the G_1 stage of interphase. No one stops to applaud the great accomplishment of successfully completing the mitosis process. This is too bad, really, because it is a process so key to life and the continuation of life. In complex organisms, it's the root of renewal and regeneration; it's used to heal wounds and regenerate parts of the body. In simple organisms — fungus for example — it is the means of reproduction. (Not much fun, but effective.)

Meiosis: It's all about sex, baby

Most human cells have 46 chromosomes for a reason, whereas sex cells have half that number: 23. Here's why: The most successful plants and animals have developed a method of shuffling and exchanging genetic information, constantly developing new combinations designed to function better in a changing environment. This process usually involves organisms that have two sets of genetic data, one from each parent.

Through *sexual reproduction,* a new individual is formed through the union of *gametes* (sex cells). But before the union of gametes can happen, the two sets of genetic information present in most cells must be reduced to one. Without this process, the *zygote* — the cell resulting from the union of two sex cells — would have four sets of chromosomes. Then, with the next generation, it would have eight, then 16, then 32 — well you get it — it would be a real mess. The resulting organisms would be unable to decode the genetic information, and the final result would be death.

So, gametes (eggs in females and sperm in males) have what's known as the *haploid* number of chromosomes. The word comes from the Greek *haplos*, meaning single. So each gamete contains a single set of chromosomes — 23. When the two gametes unite, they combine their chromosomes to reach the full complement of 46 in a normal *diploid* cell. Logically enough, diploid comes from the Greek *diplos*, meaning double.

The algebraic equation used to illustrate the relationship between haploid and diploid cells is:

n (haploid) + n = $2n$ (diploid)

And that may be the simplest equation you'll ever see in a science book.

Meiosis, then, is the process that cuts the diploid number in half so that orderly reproduction can take place (see Figure 11-2). It has two major parts, each with mitosis-like phases (Table 11-1).

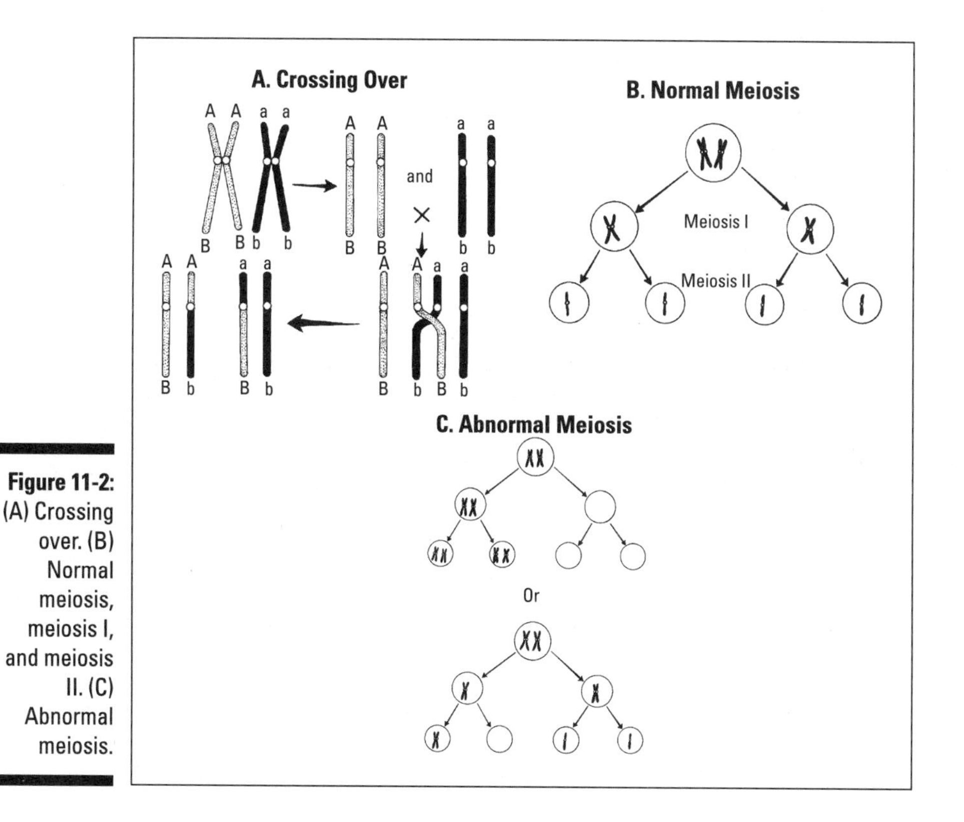

Figure 11-2: (A) Crossing over. (B) Normal meiosis, meiosis I, and meiosis II. (C) Abnormal meiosis.

Table 11-1 Comparison of Mitosis and Meiosis

Mitosis	*Meiosis*
One division is all that's necessary to complete the process.	Meiosis requires two separate divisions for completion of the process.
In the more simple mitosis, chromosomes do not synapse.	Homologous chromosomes must synapse to complete the process, which occurs in prophase I.
Here, homologous chromosomes do not cross over.	Crossing over is an important part of meiosis and one that leads to genetic variation.
Centromeres divide in anaphase.	Centromeres only divide in anaphase II, not anaphase I.
Daughter cells have the same number of chromosomes as their parent cell. (Diploid)	Daughter cells have half the number of chromosomes as their parent cell. (Haploid)
Daughter cells have genetic information that is identical to that of their parent cells.	Daughter cells are genetically different from parent cells.
The function of mitosis is reproduction in some simple organisms, but in most organisms it functions as a means of growth, replacement of dead cells and damage repair.	Meiosis creates sex cells, the first step in the reproductive process for complex organisms, both plants and animals.

You also need to know that, in human cells, the members of the 23 chromosome pairs are *homologous*, which means similar but not identical. One member of the pair, for example, may carry the information for black hair, while the other for blond.

In the human male, meiosis takes place after puberty, when the diploid cells in the *testes* (male sex organs) undergo meiosis to become haploid. In females, the process begins just a little earlier: in the fetal stage. (Okay, a *lot* earlier.) But, while a little girl is dog-paddling around in her mother's womb, a series of diploid cells complete only the first part of the process and then migrate to the ovaries, where they hang out and wait until puberty. With the onset of puberty, the cells take turns entering meiosis II. (Just one per month, no pushing or shoving, please!) Usually, one single egg cell is produced per cycle, although exceptions occur , which, if fertilized, are then called fraternal twins, or triplets, or quadruplets, and so on. The other meiotic cells simply disintegrate. (Well, it's not so simple. Ask any human female.)

When a sperm and egg cell — each with 23 chromosomes — unite in *fertilization*, the diploid condition of the cell is restored. Further divisions by simple mitosis result in a complete human being. Well, that and four years of college and etiquette lessons.

To produce the haploid condition in the gametes, the process of meiosis goes through two divisions, appropriately called meiosis I and meiosis II. Both of these divisions contain stages that are similar to those that occur during mitosis.

Meiosis I

Prior to meiosis, DNA replication occurs — just as in the case of mitosis — so the gamete has the usual, full wardrobe of 46 chromosomes and 92 chromatids. But not for long!

- **Prophase I:** Many of the processes from prophase are common to this phase. The chromatids coil and thicken to form chromosomes, the nucleoli disappear, the spindles form, and the nuclear membrane disintegrates. But, in meiosis, there is an additional event that is absolutely critical to the process: *synapsis.*

  The synapsis process begins when the chromosomes move and lie next to each other. At this point, the associated chromosomes can swap equal amounts of DNA in an event called *crossing-over.* This swapping of materials results in four completely unique chromatids. The new arrangement of four chromatids is called a *tetrad.*

- **Metaphase I:** During metaphase, the synapsed chromosomes make their way to the center of the cell as single units. Once they get to the equator, their arrangement is really a matter of chance. And the deck of genetic cards is shuffled once again.

- **Anaphase I:** It's during this stage that the chromosome number is reduced from diploid to haploid. Haploid is reached only after meiosis I. You start with a diploid (2N) cell, then the DNA divides. You are now at a similar starting point as in mitosis, 4N. You then split up the "2N" chromosomes in meiosis I followed by a splitting of these 2N cells to 1N in meiosis I. Anaphase I begins when the two members of each pair of homologous chromosomes move away from each other and toward the poles of the cell. Thus, 23 chromosomes end up on one side of the cell and 23 on the other. How do they choose with which genetic companions to gather? It's pretty simple, really, it just depends on which side of the equatorial plane they were during metaphase. For obvious reasons, this process is called *segregation.* It's also important to know that there are no bullies in these chromosomal gangs. Each pair of chromosomes makes its own decision about how to segregate, so the process is referred to as *independent assortment.*

- **Telophase:** During telophase, the cell takes a step back (or forward, depending on your perspective) to an interphase-like condition. The once tightly coiled chromosomes straighten out, the nuclear membrane re-forms around them, and nucleoli reappear.

Meiosis I is, of course, a throwback to mitosis — any old eukaryotic cell worth its cytoplasm could do these things. The real trick is reduction (as anyone who has tried to reduce knows only too well). Now, in this last stage of meiosis I, each daughter cell has one member of each homologous chromosome pair. So, here it is, the key fact we've been leading up to all this time. Each cell receives one-half of the total number of chromosomes considered normal for the species, but each cell still has a complete set of genetic material. And that, ladies and gentlemen, is one of the neatest tricks in the cellular arena. Don't try it at home!

Following the exhausting stunts performed during meiosis I, many cells kick back for a while and perform some regular metabolic activities. The one interphase-like activity they avoid at this stage is replication, and for obvious reasons: If the cell were to double its chromosome number at this stage, all the activity of the previous stages would be for naught (see the section on nondisjunction). It would be like dieting for weeks to lose that extra five pounds and then regaining it with one brownie binge — just before your wedding.

Meiosis II

During meiosis II, the two cells continue their dance of division so that — in most cases — four cells are the end result.

- **Prophase II:** As in both mitosis and Prophase I, the nuclear membrane disintegrates, the nucleoli disappear, and the mechanisms for spindle formation pop into view. The big difference between I and II, however, is that the cells are now haploid rather than diploid. And this time around, the cells avoid synapsis, crossing-over, segregation, and independent assortment. These things are hard work. Once per meiosis is enough.
- **Metaphase II:** Nothing too exciting here, folks. Just as in any old metaphase, the chromosomes attach at their centromeres to spindles and then line up at the equatorial plane. But remember that the pairs of chromosomes are no longer together in the same cell, so each member moves separately. (Free, free at last!)
- **Anaphase II:** In anaphase II, as in the mitosis version of anaphase, the centromeres of the chromosomes actually split in two, with the daughter chromosomes now moving to the poles of the cell. In this phase, they function more like their counterparts in mitosis than those in meiosis I. Because there are no paired homologs at this stage, there is none of the segregation or independent assortment you saw in anaphase I.
- **Telophase II:** The nuclear membrane and nucleoli reappear, the chromosomes stretch out for the briefest of rests, and the spindles disappear. It's time for cytokinesis. Now there are four — count them, four — haploid cells, where at the beginning of meiosis there was just one diploid (but the ever-essential round of replication just prior to meiosis makes it effectively start at 4N). Such a deal!

Viva la Difference!

The whole point of understanding meiosis is to see how genetic variation occurs. Five factors, described in the preceding section, influence genetic variation in offspring. The following is a more detailed look at crossing-over, segregation, independent assortment, fertilization, and mutations.

Mutations

Several agents and events can cause damage to the DNA molecule, including x-rays and chemicals like nicotine. When this damage occurs to the DNA of a sex cell, future generations are affected and mutations can occur. Sometimes, a whole strand of DNA is broken, making it impossible for the cell to synthesize protein properly — or sometimes at all. This problem is called *chromosomal mutation,* and it's serious stuff. If the damage is bad enough, the cell can die. And if enough cells, die, it's curtains for the whole organism. Worse still is if the cells become immortal and metastatic (spreading), like in cancer, and then kill us.

Crossing-over

At the very beginning of the meiosis process, as the homologous chromosomes come together for synapsis, crossing-over occurs, resulting in a new gene combination and new chances for variety (which is the spice of life for happy haploid cells). While the chromosomes are positioned close together, they exchange equivalent portions of chromatids, thus swapping genes. This can happen at a number of different points — or *loci* — on the chromosomes, paving the way for a wide range of genetic variations. Crossing-over is one way of helping to explain how you can have red hair from your mother's father and a prominent chin from her mother. After crossing-over, those two genes wound up on a chromosome of your mother's, which she handed down to you — along with her wedding dress and that picture of your grandmother, which is how you know you share her prominent chin.

Segregation

Segregation is the process following crossing-over when the chromosomes separate and move to the poles of the cell. At this point, *alleles* (alternative forms of gene for a specific characteristic) separate, with one daughter cell getting one and the other going to the second daughter cell. For example, an allele on one of the chromosomes might designate five fingers, while the other allele would be for six fingers. These genes now have an equal chance of being transmitted to the next generation. Whether the offspring eventually has five or six fingers will be finally determined by which gene is contributed, along with the allele that comes from the mate during fertilization. So, rather than checking a potential mate's bank account before marriage, you might do well to look into his bank of alleles!

Independent assortment

If just one pair of chromosomes were splitting and moving to the cell's poles during meiosis, you would have two genetically different gametes. The sex cell, of course, contains 23 pairs of chromosomes, all of which split and move independently to the poles. Independent assortment is a chance event, determined strictly by the position of the chromatids on the equatorial plane prior to the chromosomal split. So, how many genetic variations are possible thanks to independent assortment?

2^{23} or 8,388,608

And that's without crossing-overs or mutations! Don't you think it's amazing you look like your sister at all? Perhaps it is until you realize that you do share 50 percent of your genetic material with each sibling.

Fertilization

The female of the species brings the potential for millions of genetically different eggs and the male the potential for millions of genetically different sperm. So, when you put together the male and female gametes, the possibilities for variation virtually are unlimited. And that's why every human being that has ever been born — and ever will be born — is genetically unique. Well, almost. Of course, you have the issue of identical twins, which develop from the same fertilized ovum. They have the same genes and are the same sex. And then there's the new wrinkle of cloning. Will scientists ever be able to create whole human beings from single, borrowed cells, thus building exact genetic copies? Probably. There is the case of the world-famous cloning poster girl, Dolly.

Nondisjunction

Unfortunately, it's not a perfect world. There is no Tooth Fairy or Easter Bunny, your local baseball team probably won't win the World Series, and meiosis doesn't always work perfectly. Every now and then, just when everything seems perfect in the cellular realm (isn't that always the way?), a glitch occurs in the process. Here's what sometimes happens:

Remember that in order for the process to work properly, the number of chromosomes in diploid cells has to be reduced to haploid. One of the important occurrences in this process is the segregation of homologous chromosomes into separate cells at the very first meiotic division. Occasionally, a pair of chromosomes finds it just too hard to segregate (sort of like siblings that can't make it on their own), and they end up in the same gamete. Bad chromosomes!

What happens next is not pretty. Two of the final four cells resulting from the meiotic process are missing a chromosome as well as the genes it carries. This condition usually means the cells are doomed to die. Each of the other two cells has an additional chromosome, with the genetic material it carries. Well, that should be great for these cells, shouldn't it? It should mean they'll have an increased chance for genetic variation, and that's a good thing, right?

Wrong! An extra chromosome is like an extra letter from the IRS. It's not something to hope for. Many times, these over-endowed cells simply die, and that's the end of the story. But sometimes they do survive and go on to become sperm or egg cells. The real tragedy, then, is when an abnormal cell goes on to unite with a normal cell. When that happens, the resulting zygote (offspring) has three of one kind of chromosome, rather than the normal two. The term biologists use for this occurrence is *trisomy*.

And here's the real problem: All the cells that develop by mitosis to create the new individual will be trisomic (have that extra chromosome). Now, cells are conformists by nature; they don't deal well with change. A viable female must have two X chromosomes. Too much of some things, even good things, can kill you. And, two active X chromosomes in a female with three X chromosomes would kill a potential female, so all women inactivate one of them (randomly) in each cell. The extra X chromosome is turned into an unexpressed Barr body.

One possible abnormality occurring from an extra chromosome is *Down syndrome*, a condition that often results in some mental and development impairment and premature aging. Scientists have now pinpointed the chromosome related to Down syndrome; it's chromosome number 21. If an ovum with two number 21 chromosomes is fertilized with a normal sperm cell with just one number 21, the resulting offspring has 47 chromosomes (24 + 23 = 47), and Down syndrome occurs.

Here's an interesting fact about Down syndrome and certain other genetic abnormalities. You probably already know that the mother's age is a factor in such conditions. But did you know why? Meiosis begins in the fetal stage for females. Then, when meiosis I is completed, the cells rest in the ovaries until puberty, when one per month enters meiosis II in preparation for fertilization. So, if a cell has been waiting its turn for 40 or 45 years, it's pretty darned old — in cellular terms at least. Aging gametes is not such an issue for males because their sex cells don't actually enter meiosis until after puberty, and it's a continuous process, with new cells being produced all the time. Just look at all those aged male movie stars with their new offspring. (All right, it's not a pretty picture, but the babies are normal, at least.)

Pink and blue chromosomes

Ever wish you could have been born the opposite sex so that you wouldn't have to spend so much of your budget on makeup or you wouldn't have to shave your face every darned morning of your adult life? Sorry, it never was really your decision to make. Like all other genetic characteristics, sex is determined at a chromosomal level.

In many organisms — including humans and fruit flies, believe it or not — the sex of an individual is determined by specific sex chromosomes, which biologists refer to as the X and Y chromosomes. When scientists are talking about sex and referring to chromosomes that aren't involved in the determination of gender, they call those chromosomes autosomes. Genetic characteristics exist on specific places of specific chromosomes. So the genes that determine you'll be spending a king's ransom on razors are located on the diminutive Y chromosome, which cozies up to the somewhat larger X chromosome — the one whose genes are responsible for that all that money you pay to cosmetic companies — just as if the two were a homologous pair. Males have both an X and Y chromosome and females have two X chromosomes. So, whether or not an X is lucky enough — or unlucky, depending on your perspective — to get tapped by a Y, her prince in shining armor, determines maleness or femaleness. (You can find more on the different forms of reproduction, including the differences between reproduction in plants and animals, in Chapters 12 and 13.)

Chapter 12

Making More Plants

In This Chapter

- Finding out the difference between asexual and sexual reproduction
- Checking out vegetative reproduction
- Figuring out what flowers are for
- Accepting self-pollination as normal behavior for a plant
- Observing the development of plant zygotes
- Understanding the structure of seeds

If living organisms did not reproduce, life would cease to continue, species by species, until all species became extinct. As the older members of a species grow weak and die, new members develop — the old "circle of life."

Plants are in the circle of life, too. Living organisms cannot live forever. If a species is going to continue, the older members must replace themselves with younger members.

Reproduction is the process by which organisms replace themselves. How they do it varies, however. Obviously plants and animals have different reproductive parts. This chapter focuses on how plants reproduce. It's a short story and a short chapter. Don't worry. The next chapter tells you all about how animals do it.

Asexual Reproduction

Even the most simple one-celled organisms — called *unicellular organisms* — reproduce. Unicellular organisms really never die; they just keep creating fresh new versions of themselves to carry on the species. These little buggers don't hook up with a willing partner. They go it alone by means of asexual reproduction. *Asexual reproduction* is reproduction by cell division: Organisms that reproduce in this way split themselves into two new cells.

The two new cells, although identical to the original cell, are fresh, more vigorous, and more fit. And, one law of biology is "survival of the fittest," which can be interpreted as "only the strong survive." Older, weaker cells are less fit to survive; therefore, to benefit the species, they must replace themselves with stronger members.

Some simple many-celled organisms — called *multicellular organisms* — also produce via asexual reproduction. *Algae*, which are multicellular plant organisms that live in water, go through several cell divisions, eventually producing some specialized reproductive cells called *zoospores*. Each zoospore can develop into a new organism.

Other multicellular organisms form from pieces of a "parent" organism. Some plants reproduce in this way. In vegetative reproduction, if you break off a piece of a plant (the technical term is "take a cutting") and stick it in water, new roots and shoots may grow, creating a whole new plant from a piece of the parent plant. In essence, the new plant is a clone of the parent plant: It has all the same genetic information because the cells are identical.

Strawberry plants reproduce in this way. In addition to producing stems, the strawberry plant produces a *stolon* — casually called a runner — that spreads across the ground. Wherever that stolon starts to put roots down is where a new strawberry plant grows. The strawberry plant is an example of vegetative reproduction because plant material rather than a specialized reproductive structure is responsible for continuing the species.

Sexual Reproduction

Plants do have sex, believe it or not. It may not be an earth-shattering experience for them, and they don't have the nervous system to allow them to enjoy the sensations, but it is sexual reproduction nonetheless. Sexual reproduction differs from asexual reproduction in that an organism has specialized tissues used only for reproductive purposes.

Sexual reproduction also gives a species the advantage of dispersal. *Dispersal* refers to the ability of a species to produce numerous offspring and spread out over large distances.

For example, think of how the fuzzy, white parts of dandelion plants blow in the wind. Each of those white fuzzies has seeds attached. The wind disperses the seeds so that dandelion plants can — unfortunately — spread out all over your yard, your neighbor's yard, and so on. Otherwise, a dandelion plant would be limited to one increasingly larger spot in your yard, and they'd be much too easy to eliminate. Dispersal protects the dandelion species.

Another key advantage to sexual reproduction is the ability to mix genetic material. By creating new individuals, rather than just new clones of an older organism, species can escape some harmful effects of some mutant genes because there are two genes for each feature. If one gene for a specific feature has mutated, the other gene for the same feature may override the change, if it is negative. If the mutation is beneficial, however, it may be allowed to stand, and it becomes part of the genetic makeup of the species from then on.

Over time, as species evolve, weak genes eventually are repressed or eliminated. Weaker members of the species are less successful in reproducing, so their genes are eliminated from the gene pool. The *gene pool* is the entire collection of genes of a species; it provides variety among all the possible characteristics of a species.

Life cycles of plants

In both plants and animals, gametes — sex cells — must be produced so that reproduction can occur. In animals, the production of gametes involves two steps: mitosis and meiosis (see Chapters 11 and 13). For now, understand that mitosis produces new cells from parent cells with the same number of chromosomes as the parent cells (diploid), and meiosis produces twice as many cells with half the chromosome number (haploid) so that haploid cells can be combined during sexual reproduction to produce one totally new whole cell with the correct number of chromosomes. In plants, however, the production of gametes does not directly involve meiosis.

In plants, meiosis in a parent plant results in the production of spores. The spores develop into haploid organisms; this step is a fundamental difference between plants and animals. In animals, no development occurs until two haploid cells combine to produce a new organism. Anyway, the haploid plant organisms grow, becoming multicellular haploid organisms that develop specialized tissues. The specialized tissues of the haploid organism produce gametes. The gametes merge (this is sex for a plant) and produce zygotes that contain the same number of chromosomes as the parent plant — that is, a diploid plant. The zygotes develop and grow into "adult" plants. Within the diploid plant, specialized cells create haploid spores, and the life cycle begins again.

Flowering plants

Most plants on earth are the flowering variety, called *angiosperms*. In addition to flowers flowering, trees, shrubs, vines, and fruit and vegetable plants also "flower." So, this discussion about sexual reproduction in plants focuses just on flowering plants.

Flowers (Figure 12-1) are the specialized reproductive tissues of a plant. The spores and gametes are protected within the flower.

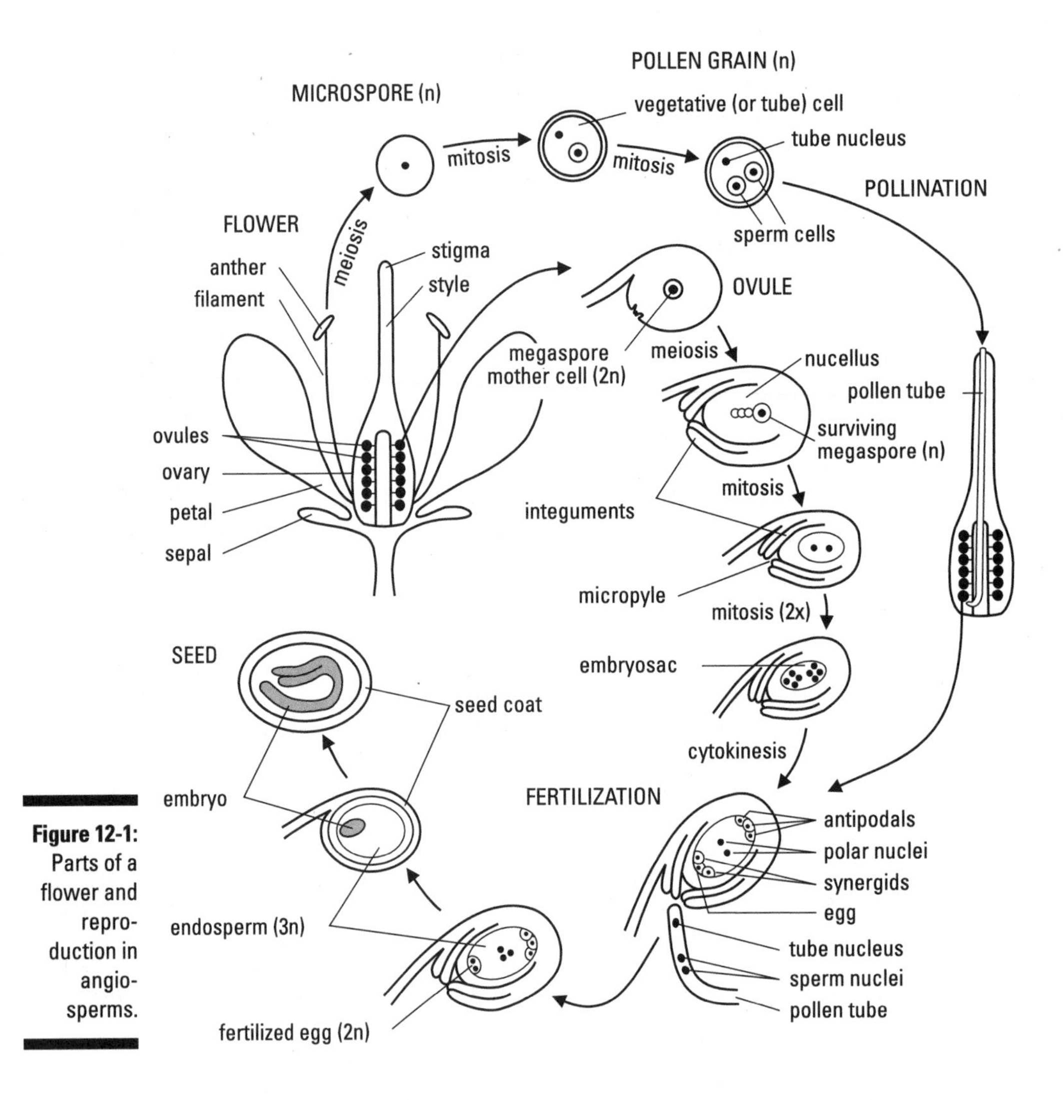

Figure 12-1: Parts of a flower and reproduction in angiosperms.

The stem of a flower is a shoot — a branch of the stem — that has highly specialized leaves. The leaves of a flower are the usually colored, pretty part; their function is to protect the area of the plant where reproduction happens. The tip of the shoot around which the leaves form is called the *receptacle*. The lowest layer of leaves on the flower contains *sepals*; these leaves usually are green. The highest layer contains *petals*; these leaves usually are colored and fragrant. In plants that depend on the spreading of pollen for reproduction, the purpose of sepals and petals is to attract insects or birds, which help to pollinate the plants.

Inside the petals are *stamens*, and the stamens surround the *pistils*. The stamens and pistils are highly specialized and modified, which means that they develop from leaves but look nothing like leaves and have very specific functions. A stamen — the word stamen is derived from the Latin word for thread or fiber — is a filament with an *anther* on top. The anther contains *microsporangia*, which create microspores. *Microspores* then develop into grains of *pollen*. A pistil is made up of a stigma, a style, and an ovary. The pistil is the part of a flower that contains ovules. Inside the ovules, megaspores are produced. Megaspores ultimately develop into female gametophytes.

Plants that flower contain two types of spores: microspores and megaspores. Microspores are created in the anther of a stamen and become pollen grains, which are equivalent to male sperm. Pollen grains are male gametophytes. Megaspores are made in the pistils and become the female gametophytes.

What's a gametophyte, you ask? Well, *gameto-* refers to a sex cell, and *-phyte* means plant. Therefore, a gametophyte is the haploid sex cell in a plant.

At the bottom of a pistil is an ovary, which contains just one ovule or several ovules. The number of ovules depends on the species of plant. The inside of the ovule is where the primary megaspore is produced. The first megaspore — called the megaspore mother cell — produces four megaspores with half the number of chromosomes. Only one of those four megaspores becomes a female gametophyte. The rest wither away. This aspect of plant sexual reproduction is very similar to human sexual reproduction: Of the four haploid cells produced in a woman's ovary, only one becomes an egg, and the other three wither away.

The female gametophyte, which is also called a *megagametophyte*, develops inside the ovule. The cell goes through three divisions, ultimately producing eight nuclei. Three nuclei go to one end of the cell, three nuclei go to the other end of the cell, and two nuclei — the polar nuclei — stay in the middle of the cell. Cell walls form around each of the nuclei at the ends of the cells. Both polar nuclei remain in the middle, with a cell wall forming around them. So now, seven new cells are forming: six from the nuclei at the ends of the gametophyte, and one containing the two polar nuclei (see Figure 12-1). At this stage, the gametophyte, which contains seven forming cells, is called an embryo sac. The embryo sac is a fully developed gametophyte containing seven partially developed haploid cells.

In male gametophytes, the anther at the top of the stamen contains *pollen sacs* where microspores are produced. Microspore mother cells go through meiosis, producing four microspores. Each microspore develops into a pollen grain. Inside the pollen grain, mitosis occurs, and two nuclei are produced. One nucleus is the *tube nucleus*, and the other is the *generative nucleus*. At this point, the grains of pollen can be released into the air. They continue to develop only if they land on the right spot on a pistil so that pollination occurs. If a pollen grain is so lucky as to fall on the right spot — and you thought dating was hard! — its development continues as it germinates.

Pollination and fertilization

As a grain of pollen drifts through the air and eventually falls downward, it must be fortunate enough to land right on top of the stigma. The stigma is the slightly sticky top part of the pistil in a flower (see Figure 12-1). Lilies have great stamen and pistils. If the pollen grain gets stuck on the stigma, the pollen grain starts to grow a pollen tube, which grows down into the style and into the ovary. This process is the act of sexual reproduction for a plant. Yippee! Aren't you glad you're an animal!

A pollen grain contains two nuclei: the tube nucleus and the generative nucleus. As the pollen tube grows down into the style, both nuclei enter the pollen tube. The tube nucleus stays near the tip of the pollen tube; the generative nucleus goes through mitosis to form two sperm nuclei. Each sperm nuclei contains half the number of chromosomes as the parent plant cell. The sperm nuclei enter the ovule and head for the embryo sac, just as human sperm enter the uterus and head for the egg.

Of the two sperm nuclei that enter the ovule, one connects with the cell from the female embryo sac that became the egg cell. This sperm-egg cell fusion becomes the zygote. The other sperm joins with the polar nuclei from the embryo sac, and this union forms the *endosperm nucleus*.

Developing the zygote into an embryo

After a sperm cell and the egg cell fuse, a cell division produces two cells: one large, one small. Several more cell divisions occur after that, producing a line of cells called a *suspensor*. As more of these cell divisions occur, an embryo forms (see Figure 12-1). The embryo continues to form as the seed is being developed.

The embryo forms in such a way that it is properly oriented in space. That is, the parts at the bottom of the embryo grow downward to become roots, and the parts at the top of the embryo grow upward to become shoots.

The embryo contains the *hypocotyl*, which is attached to the bottom end of the suspensor. The hypocotyl becomes the lower part of the stem and the roots. The other end of the embryo is where the cotyledons — seed leaves — develop. Seed leaves are temporary structures; they serve as nutrient storage sites for the developing plant. After the plant is growing above ground and can start producing nutrients on its own through photosynthesis, the cotyledons shrink away. Dicotyledonous plants have two cotyledons to help them get started; monocotyledonous plants have just one.

Seed production

The embryo sac that is formed through cell divisions in the female gametophyte contains a large cell in the middle with two polar nuclei. This cell is involved in the production of the *endosperm*. The endosperm provides nutrient material to the developing embryo inside of it. The endosperm also contains tissues from the ovule that harden to become the *seed coat*. In some types of plants, what is left of the ovary forms around the seed as fruit.

At this point in development, seeds usually dry out, and their water content drops very low. The low water level keeps their metabolism at a minimal level. So, the seeds that you buy in a little packet are very much alive, but they are in a state similar to that of a hibernating bear. Sometimes seeds drop to such a low metabolic level that they need a stimulus, such as cold temperature, to start germinating.

Germination begins when dry seeds take up water and swell. Of course, the seeds cannot just be in water; they need to be in good soil as well. When the seed swells, metabolism speeds up and growth occurs. Seedlings grow out of the seed and develop into a diploid (full chromosome set) plant.

Chapter 13

Making More Animals

In This Chapter

- Differentiating asexual and sexual reproduction in animals
- Understanding how gametes are made
- Figuring out how a mass of cells turns into a newborn
- Seeing how the birds and bees do it

This chapter is where you get to read all about sex. Aren't you excited? Don't get too excited. This chapter contains no centerfold or erotica. It's just the nuts and bolts of how animals reproduce. You find out about what happens inside the body to prepare for reproduction, how different types of animals — yes, including humans — actually mate, and how offspring develop prior to birth.

Asexual Reproduction: This Budding's for You

If you were an organism that reproduced by asexual reproduction, you would be a clone of your "parent," which was a clone of its "parent," and your "children" would be clones of you. Asexual organisms essentially are just fresher, younger versions of the original organism. Asexual organisms don't really die, they just bud off into new versions of themselves and continue on. The only way for an asexual organism to change is for it to be affected by something that causes a mutation. A *mutation* is an alteration in a gene or genes that changes the protein(s) that is produced by the gene(s). With a different protein being created, a characteristic of the organism changes. If the mutation "takes," then the organism produces offspring that are different from itself. Otherwise, it's just live, wear out, and split — live, wear out, and split — live, wear out, and split. You get the picture. It's not too exciting.

For organisms to adapt to their environment, they must be able to change. The best way for organisms to produce new organisms that have a different genetic makeup is to mix genetic material. "Mixing" requires another organism. Variety is the spice of life, as "they" say. It also helps to protect against harmful mutations. The organisms with the strongest genes survive and reproduce. The organisms harboring mutations harmful to the species have a weaker genetic make up and a harder time reproducing. Keeping bad genes out of the gene pool is beneficial to the species and way down the road, it helps with evolution of the species.

Sexual Reproduction — The Ins and Outs

Whereas asexual reproduction involves the entire organism (one whole organism splits into other whole organisms), sexual reproduction starts at the cellular level, involves two parents, and new organisms develop and grow over time.

In animals, sexual reproduction begins with an egg and a sperm, each of which is a single cell. Mating combines the two single cells to produce an entirely new organism. The new organism develops and, as it does, it, too, contains either eggs or sperm. Therefore, the new organism can continue the life cycle and contribute its genetic material to yet another generation.

Getting to know gametes

Gametes are the sex cells. A sperm is a gamete, and an egg is a gamete. Each egg and each sperm contain half the number of chromosomes that are normally present in the whole organism. Then, when the gametes join, the organism has all the chromosomes it needs.

Gametogenesis is the process by which gametes form. The process results in haploid cells — either eggs or sperm. *Haploid* refers to the cell having half the number of chromosomes, and it is represented as "N." *Diploid* is the term for having a full chromosome count; it is written as 2N. For instance, the normal number of chromosomes in each human cell is 46. So, 46 is the diploid, or 2N, number. That means that each human egg and each human sperm contain half of 46, or 23 chromosomes (N = 23).

The process of gametogenesis is controlled by hormones, those protein-containing substances that start, stop, and alter many metabolic processes. To explain gametogenesis to you, I'm going to use humans as the organism of focus. But, the process of producing sperm and eggs is pretty much the same in all animals that undergo sexual reproduction. The differences lie in how the animals mate and transfer their gametes.

Spermatogenesis: Making the little guys

Sperm are the male gametes. Spermatogenesis is the process that results in the production of haploid sperm cells. Beginning at puberty, human males start producing millions of sperm every day. The sperm survive inside the male for only a short time, which is why they need to continually be produced. The process by which that happens is outlined here:

Spermatogonia are cells that contain 46 chromosomes. They are the starting point for spermatogenesis, which occurs in the *seminiferous tubules*. The spermatogonia line the walls of the seminiferous tubules and go through mitosis. *Mitosis* is the stage of cell division that replicates a diploid cell. So, each spermatogonium (the single version of the plural spermatogonia) produces a cell called a *primary spermatocyte*, which also contains 46 chromosomes.

The primary spermatocyte goes through meiosis. A primary spermatocyte produces two *secondary spermatocyte*, which contain 23 chromosomes.

Secondary spermatocytes go through meiosis a second time — called a second meiotic division — and produce four *spermatids*, each of which also contains 23 chromosomes.

Each spermatid develops into a *spermatozoa*, which is just the technical name for what you know as a sperm. To become a full-fledged sperm, the spermatid must mature to the point where it has a tail, middle piece, and a head. The *tail* is a flagellum that moves the sperm through body fluids — the tail allows the sperm to swim. The *middle piece* contains many mitochondria, which supply the energy for the sperm's tail to move, and the *head* contains the 23 chromosomes inside its nucleus.

Oh, Oh: Oogenesis

No, *oogenesis* is not pronounced the way it looks — it's not "ooh" genesis. It's o-o-genesis, and it's how eggs are made inside females. A human female is born with all the eggs she will ever have. From birth until puberty, the eggs lie dormant in the ovaries. The hormones generated during puberty kick-start the eggs into the menstrual cycle, which continues monthly from puberty until the woman begins menopause.

When a female is a developing fetus, oogenesis begins in the ovaries. An *oogonium* is the initial cell, and it contains all 46 human chromosomes. It grows in size, eventually maturing into a *primary oocyte*. The primary oocyte begins going through meiosis, but the process is paused until the female enters puberty. Once the hormones of sexual development start flooding the ovaries, the menstrual cycle begins.

At the midpoint of the menstrual cycle, *ovulation* — the release of an egg from an ovary — takes place. To prepare for release, and possible fertilization, the primary oocyte continues on through meiosis. The first meiotic division produces a daughter cell (secondary oocyte), which receives most of the cytoplasm (so it is large), and the first polar body, which has minimal cytoplasm (so it's tiny). Both the daughter cell and the polar body contain 23 chromosomes. The second meiotic division results in the daughter cell producing the egg and a second polar body, while the first polar body produces two more polar bodies.

So, out of one original oogonium, only one functional egg is produced; the three polar bodies that also are produced just wither away. And, think about this: In human females, the meiotic division that pauses in the oocyte remains paused for 40 years or more! From before a woman is born until her ovulation ends during menopause, oocytes are in hang time just waiting to develop into an egg and get fertilized. Of the thousands of oocytes that a female is born with, only about 500 develop into eggs during the woman's lifetime.

The daughter cell and egg are large, and the polar bodies are small, for a very specific reason. It's not that Mother Nature is being unfair in dividing up the cytoplasm. The splitting of the cell — called cytokinesis — is unequal so that the egg ends up full of cytoplasm. The polar bodies basically serve as receptacles for genetic material that is being discarded. Twenty-three chromosomes are in the egg, but the other 23 chromosomes have to end up somewhere. The egg gets most of the cytoplasm so that it can hold plenty of nutrients, as well as the organelles to turn the nutrients into fuel, for a developing embryo.

Mating rituals and other preparations for the big event

This section is not about what cologne to wear on a date, how to seduce a partner, or what birth control to choose. Those are all human conventions. Mating in humans can take place whenever a man and a woman are in the mood to do so. However, most animals follow more rigid reproductive cycles.

Suppose that you were a female oyster in the ocean. Every season, you would release 60 million eggs into the water — yes, you read right — 60,000,000. Your mating process is really left to chance. If one of your eggs happens to bump into a male oyster's sperm, then fertilization would occur. Leaving the continuation of the oyster species to chance is one reason that oysters release such huge numbers of eggs in the first place. Obviously, most of the eggs never bump into Mr. Right, or the ocean would be overflowing with oysters (and pearls wouldn't be so expensive!).

But what if you release all the eggs because you feel like it, but the male oysters in the area aren't in the mood and do not release any sperm? You'd be alone in your shell with nothing to show for it. That's why reproductive cycles and specific mating seasons for animals exist. Animals of the same species need to be in sync to have successful reproduction.

Most species mate when the time is right — that is, when conditions are optimal. Optimal conditions vary for different species, however. Often, the mating season occurs so that the birth of the offspring will occur at the time of year when there is the best chance for survival. For example, in deer, the *gestation period* — the time of development of the fetus inside a female mammal — is approximately five or six months. The best time for a fawn to be born is in the spring. Food is plentiful then, temperatures are a bit warmer, leaves are on the trees and shrubs to provide cover for the animals, and it gives the fawn the longest period of time to develop before conditions get harsh the next winter. So, backing up six months from spring puts mating season for deer around October or November. And that's exactly when you can see bucks competing for does. The strongest deer — supposedly with the strongest genes — gets to mate and pass on its genetic material to continue the species. Nature is so cool.

If you are an aquatic (water-living) organism that lives in the desert, you have to work quickly. These animals reproduce when the scarce desert rainfall produces a temporary pond. During the dry season, the animals are in *diapause*, which is a dormant state. In diapause, the metabolism of the animal is very low, and extreme heat and dry weather do not affect the animal, which allow it to survive through drought or heat waves. When rain does come, the animals become active immediately, quickly breed, and have offspring that develop all before the pond dries up. Then, the new generation gets to sit in the desert in diapause waiting for the next thunderclouds to appear in the sky.

Human reproductive cycles

Humans can reproduce all year-round, but cycles are still involved. Human males can discharge sperm capable of fertilizing an egg on any given day. However, human females are able to have their egg fertilized for only a few days out of every month. Reproduction is controlled by the monthly ovarian cycle and menstrual cycle, both of which occur hand-in-hand and are controlled by hormones.

The *ovarian cycle* controls the development of the egg in the ovary. An oocyte needs to complete meiosis and mature into an egg before it can be released by the ovary. Believe it or not, the brain runs this process. Deep inside the brain are two very small glands that really control most of the important processes of the body: the hypothalamus and the pituitary gland.

The *hypothalamus* keeps a check on how much of the hormones estrogen and progesterone are floating through the bloodstream. When the levels decline, the hypothalamus secretes a hormone called *gonadotropin-releasing hormone (GnRH)*, which, as you probably can guess, prompts the gonads (in females, the ovaries) to get in gear.

When the hypothalamus secretes GnRH, the GnRH heads straight for the *pituitary gland*, and it stimulates part of the pituitary gland to secrete *follicle-stimulating hormone (FSH),* and *luteinizing hormone (LH).* The oocyte that is suspended in meiosis is also called the *follicle*. FSH is the hormone that kick-starts meiosis again and continues development of the follicle so that it can release an egg. FSH also causes the follicle to release estrogen.

As the level of estrogen rises in the bloodstream, the hypothalamus can detect it. The hypothalamus then releases more GnRH, which causes LH to be released about the middle of the ovarian cycle. LH stimulates the release of the egg from the follicle in the ovary — ovulation.

When ovulation occurs, the egg secretes the hormones estrogen and progesterone. These hormones prepare the body for pregnancy. Once ovulation occurs, the egg has a great chance of fertilization. So the body prepares for a possible pregnancy by thickening the lining of the uterus. Estrogen and progesterone are responsible for making sure that the uterus is ready for possible implantation of a fertilized egg. The tissues lining the uterus develop thicker blood vessels, which brings more nutrients into the uterus.

Once the uterus is ready for implantation, the levels of estrogen and progesterone have reached a certain level in the bloodstream. If the egg has been fertilized and implants in the lining of the uterus, an *embryo* begins development. Upon implantation, the embryo immediately starts to secrete the hormone *human chorionic gonadotropin (hCG)*, which is the hormone detected by pregnancy tests. The presence of hCG ensures that estrogen and progesterone production continues so that the lining of the uterus remains nourished by larger blood vessels. Once the *placenta* — a blood-filled, nutrient-rich temporary organ — has formed, the embryo gets its nutrients and blood supply through the *umbilical cord* connecting the embryo to the placenta, which is connected to the mother's blood supply. Therefore, the production of hCG by the embryo declines once the placenta is up and running.

If the embryo does not produce a sufficient amount of hCG, the pregnancy would not continue, and the embryo would abort (a *spontaneous abortion* is another term for a miscarriage). So, many more eggs are fertilized than you may realize. Not every fertilized egg results in a bouncing baby. If the hormone levels are not right from the start, a fertilized egg may never implant or may implant but not secrete enough hormones to maintain the pregnancy. Often, an unusually heavy menstrual period that started a few days late is really the spontaneous abortion of a fertilized egg that didn't work out.

Overstaying its welcome

Once a follicle develops and releases an egg, the empty follicle is called a *corpus luteum*, which is Latin for "yellow body." If the egg becomes fertilized and implants in the uterus, the corpus luteum hangs around to help out with the beginning stages of pregnancy. It secretes progesterone for a few weeks until the placenta is fully developed and can secrete progesterone on its own. The progesterone helps to keep the lining of the uterus rich with blood and nutrients for the developing embryo. Sometimes, the corpus luteum sticks around for a few months. Normally, it eventually shrinks and withers away sometime during pregnancy. However, about 10 percent of the time, the corpus luteum hangs out in the ovary for far longer than it should. Sometimes it remains even if the woman is not pregnant. Then, the corpus luteum can turn into a cyst, aptly called a *corpus luteum cyst*. Usually, this cyst on the ovary is not a problem, unless it continues to grow, twist, or rupture. Only then would it be removed surgically. After the first trimester of pregnancy, the surgery rarely threatens the developing fetus.

If fertilization does not occur, the hypothalamus can detect when the levels of estrogen and progesterone have reached the point where the lining of the uterus is ready for implantation. But, if there is no fertilized egg, there is no impending implantation. So, the hypothalamus causes the pituitary gland to stop producing FSH and LH. The lack of FSH and LH stops the production of estrogen and progesterone, which causes the lining of the uterus — the *endometrium* — to stop receiving all that extra nourishment. The endometrium then starts to disintegrate and eventually sloughs off and is carried out of the body by the menstrual flow.

The first day of menstrual flow is the first day of the *menstrual cycle*, which is also sometimes called the *uterine cycle*. The menstrual cycle consists of the shedding of the endometrium if there is no implanted embryo (the menstrual period), and thickening of the endometrium to prepare for a possible implantation, which occurs simultaneously with the ovarian cycle. The tasks of the ovarian cycle include development of the follicle; the secretion of hormones, culminating with the release of estrogen; ovulation; and the secretion of estrogen and progesterone are secreted from the egg. The biggest fluctuations in hormone levels occur toward the end of the ovarian cycle and *before* the menstrual cycle begins, hence the name *premenstrual* syndrome (PMS). And, I'm sure at one time or another you've experienced someone with the symptoms of that!

Finding a partner

The birds and bees may reproduce by sexual reproduction, but they don't fall in love. They don't suffer angst over whether the other bee loves them and will remain committed. They don't worry about whether their partner will be faithful. Bees "do it" solely for the purpose of creating more bees. They don't plan lifetimes together. All those emotions and feelings, although created in the brain, are not part of the reproductive system. Love is the subject for an entirely different book — check out *Dating For Dummies* by Dr. Joy Brown followed by *Making Marriage Work For Dummies* by Steven Simring, M.D., M.P.H., Sue Klavans Simring, D.S.W. with Gene Busnar (both published by Hungry Minds, Inc.).

This section in this book is dealing solely with the physical aspect of reproduction. Although love, may not be a requirement for fertilizing an egg, attraction is involved. And I'm not talking about just humans here. Many animals exhibit certain behaviors or have certain characteristics that help them attract a mate.

The mating rituals of several species of birds have been well studied. Doves are used as wedding symbols for a reason: They form pairs for life. They are committed, loving, faithful creatures, and they have a ritual. The reproductive cycle of doves lasts about 45 days. First, the male dove struts around bowing and cooing to the female, trying to win her over. But before they mate, they build a home (sound familiar?). The doves work together to choose a place for their nest (no Realtor involved), and they work together to build the nest. During the period of time that they are building the nest, they take a break and have, well, sex. The technical way of saying it is that they *copulate*. A few days later, the female lays two eggs in the new nest, and when the chicks hatch, both parents feed them. When the chicks are old enough to start feeding themselves — in about two to three weeks — the adults begin the reproductive cycle again and start courting (how romantic!).

What is interesting, though, is that if a female dove is put into a cage by herself and is given twigs and straw to make a nest, she will not do it. If both male and female doves are put into a cage but are not given material to build a nest, they will not copulate. Hormones are responsible for bringing everything together. Hormones cause the male dove to start his courting behavior, and the courting behavior actually causes the level of estrogen to increase in the female. While her estrogen is rising, they build the nest. Then, when the nest is done — and only when the nest is done — she ovulates.

In many other animals, the *secondary sex characteristics* are what attract mates. *Primary sex characteristics* are the obvious ones: male reproductive organs and female reproductive organs. Secondary sex characteristics develop as the animal matures. For example, in humans, secondary sex characteristics include hair growth and distribution (beards in males), deepening of the voice (in males), increase in muscle mass (males), increase in amount and distribution of fat (in females), and development of breasts (in females). Male deer grow antlers, male lions grow manes, and male peacocks develop a fan of beautiful tail feathers.

Notice how the males develop the characteristics to attract the females? Females tend to have their choice of who they mate with; males compete to be the ones to pass on their genetic information. For example, the brighter red that a male cardinal is, the more likely a female cardinal is to mate with him. High-quality secondary sex characteristics tend to indicate good genes, which attract the females. But keep in mind that this rule does not always apply to humans! Women do not select a mate based on the length of a beard or the deepness of a voice. Just think how much easier it would be in a singles bar if the best-suited genetic mate could be detected by a visual cue like a brightly colored peacock tail!

The act of mating: The big event

Okay. You've been waiting to read this part. Don't read too quickly, now!

The way that organisms reproduce depends on the organism, of course. Flowers go through sexual reproduction, but it is very specific based on the structures of a flower such as stamen and pistil (Chapter 12, "Making More Plants," gives you the details). But, keeping in mind that the purpose of reproduction is to create a new generation that contains the genetic information from the previous generations, members of different species cannot reproduce. Members of the same species have common characteristics within the same taxonomic level (see Appendix A). So, although humans and chimpanzees are closely related (see Chapter 17) and are of the same kingdom, phylum, class, and subclass, the genus and species names are different because of differing characteristics. Humans are in the same subclass as antelopes, too, but are very different.

Different species contain different numbers of chromosomes, and those chromosomes contain differing genes. For instance, humans carry 46 chromosomes in each cell, whereas chimpanzees have 48. Cell divisions would not be equal, and a theoretical offspring (I won't even go there) probably would not be viable (meaning able to live and survive). Actually, eggs are surrounded by a layer of proteins on top of the plasma membrane that contains receptor molecules only for sperm of the same species. In human eggs, the zone that prevents fertilization by a different species is called the *zona pellucida*. Only human sperm can crack the code to get into the egg.

So, I'll stick to how babies are made in humans. If you've never had this lecture before, the section on intercourse will get your parents off the hook. Or maybe you know what happens externally but aren't so sure of what's actually happening inside your body. I'll set you straight.

Intercourse

The male reproductive organs include the penis, the testes, and the seminiferous tubules where sperm are made. The female reproductive organs include the vagina, the uterus, the ovaries, and the fallopian tubes.

When a man is sexually aroused, the penis becomes erect when the erectile tissue within the penis fills with blood. *Erection* allows the penis to stiffen so that it can remain inside a woman's vagina during intercourse. When a woman becomes aroused, the erectile tissue within the vagina fills with blood, and the increased pressure causes drops of fluid to be squeezed out of the tissue. This *lubrication* prepares the vagina for intercourse so that the erect penis can easily be inserted.

During intercourse, in males, sperm travel from the *epididymides* (tubules in the scrotum that store produced sperm) to the vas deferens. The *vas deferens* are the tubes that carry sperm from the scrotum to the urethra so that they can be ejaculated. The vas deferens also are the tubes that are cut and tied off during a *vasectomy.*

In females, the *clitoris* — which is equivalent to the penis — is the sexually sensitive organ. The clitoris has erectile tissue and a glans tip, just as a penis does. The *cervix*, which is the bottom end of the uterus, extends down into the vagina. Sperm must travel through the cervix to get into the uterus.

When the penis is fully inside the vagina, the tip of the penis is as close as possible to the cervix. The actions that occur during intercourse serve to bring the man and woman to the climax of stimulation, which is followed shortly by orgasm.

Orgasm

Don't let this take the fun out of everything for you, but orgasm serves a physiological purpose. As the sexual stimulation of a male intensifies, the sperm move from the vas deferens into the urethra, and secretions from three glands — the seminal vesicles, the prostate gland, and the bulbourethral gland — all add their fluids to create semen (seminal fluid). Even the semen has a purpose. The fluid contains the following "ingredients" that help promote fertilization: the sugar *fructose*, which gives the sperm energy to swim upstream; *prostaglandins*, which cause contractions of the uterus that help propel the sperm upward; and a pH of 7.5, which provides the basic solution in which sperm can live and helps to neutralize the acidic conditions of the vagina, which would otherwise kill the sperm.

When orgasm occurs in a male, a sphincter muscle closes off the bladder so that no urine gets in the urethra. Shutting out urine allows the urethra to be used solely for ejaculation at that time. (In males, the urethra is shared by both the urinary tract and the reproductive tract; in females, the urethra is solely part of the urinary tract.) At the height of sexual stimulation, orgasm occurs. *Orgasm* is signaled by muscular contractions and a pleasurable feeling of release. The muscular contractions cause semen to be expelled from the penis, which is called ejaculation. The average amount of semen expelled during one ejaculation is less than 1 teaspoon, but it contains more than 400 million sperm!

In females, the height of sexual stimulation also causes intense muscular contractions and a pleasurable feeling of release. The fluid released inside the vagina helps to create a watery environment that the sperm can swim in. The muscular contractions of the uterus slightly open the cervix, which allows sperm to get inside the uterus and also help to "pull" sperm upward toward the fallopian tubes. Fertilization does not occur in the uterus. The sperm have quite a bit of swimming to do before they find the egg.

Sperm have to travel from wherever they are deposited in the vagina at ejaculation through the muscular cervix, up through the entire uterus, and up through the fallopian tubes until they reach the egg. Fertilization — the joining of the sperm and egg — actually occurs in the fallopian tube. Because a human egg lives no longer than 24 hours after ovulation and human sperm live no longer than 72 hours, intercourse that occurs in the three-day period prior to ovulation or within the day after ovulation is the only chance of fertilization during a given month.

If the sperm does find its way to the egg, it must penetrate the egg in order to supply it with its 23 chromosomes. However, human eggs have several layers of cells and a thick membrane surrounding it (women just don't make it easy!). To get through all that, the sperm produces enzymes in a structure near its nucleus called the acrosome. These acrosomal enzymes digest the protective layers of the egg; the sperm basically "eats" its way into the egg. The egg, once activated by the sperm, also helps the sperm to get inside by going through physical and biochemical changes. Once inside, the sperm joins with the egg, creating a cell containing the full chromosome count of 46.

With all the timing involved and all the steps that have to occur for fertilization to happen, it's amazing that the human species has continued and that babies are born every day.

How Other Animals Do It

Humans obviously are not the only animals that mate and undergo sexual reproduction. If that were the case, humans would be all alone on the planet with the plants and the animals that reproduce by asexual reproduction. Here's a look at the birds and the bees. Table 13-1 outlines how some other animals mate.

Table 13-1 Mating Styles of Selected Animals

Animal	*Mating Actions*
Sea urchins	Males and females look exactly the same on the outside; both have a ring of genital pores at the center of their bodies. Males discharge sperm, which look just like human sperm, through their pores into the water; females discharge eggs through their pores into the water. Fertilization is left to chance, but is helped by the fact that sea urchins live in close contact. The eggs have a sticky coat to which sperm adhere. The ejaculation by any one of the sea urchins signals the others to ejaculate too. They are all ripe and ready to go. This helps increase the odds of fertilization.
Planarian worms	These freshwater flatworms can reproduce asexually by constricting their bodies and literally splitting in two. The missing halves then grow. They can reproduce sexually as well. All worms are male and female (called hermaphrodites), so each has male and female reproductive organs. Sperm is exchanged between two worms, so each worm uses its male organs to secrete sperm; then they both use their female organs to create zygotes. The zygotes develop into small worms that then hatch and mature into adults.
Earthworms	Earthworms are hermaphrodites, too. They contain ovaries and testes, along with seminal vesicles, vas deferens, and seminal receptacles. When earthworms copulate, they face in opposite directions and put their citella together. A citellum is the external smooth, nonsegmented part of an earthworm found in the middle of the body, as well as at segments 9 and 10 down from the head. The citellum secretes mucous and help the sperm get from the vas deferens of one worm to the seminal receptacle of the other worm. Cocoons form and are protected by a mucous sheath created by the citellum. The sperm and eggs are fertilized inside the cocoon, and the zygotes stay enclosed in it until they hatch in the soil.

How the birds do it: It's a yolk

Birds copulate. The male bird deposits his sperm inside the female bird. The egg becomes fertilized, but then is deposited outside the female bird's body to continue developing until it is time to hatch.

Humans have a yolk sac, too

In birds and reptiles, the yolk sac consumes the yolk and provides nutrients through blood vessels in the embryo. However, in animals that are attached to a placenta during development, the yolk sac is empty. Perhaps, like the coccyx that used to be a tail, it is a vestige leftover from evolution?

When you crack open an egg, the yellow part — the yolk — is where the developing embryo would reside. The white part — the albumen — serves to nourish the embryo throughout its development.

Just after fertilization, one spot on the yolk of an egg goes through a series of divisions called *cleavage*. At the end of the cleavage divisions, an *embryonic disk* is created on one side of the yolk. (Have you ever noticed a reddish spot when you crack open an egg?) The embryonic disk is called the *blastoderm*. The blastoderm is the initial cell tissue that begins to develop into a chick. The blastoderm separates into an *epiblast*, which is the top layer, and a *hypoblast*, which is the bottom layer. The epiblast cells migrate down into the hypoblast, along a line in the yolk called a *primitive streak*, to create the mesoderm, which goes on to develop the rest of the chick.

How the bees do it: Parthenogenesis

A queen bee receives all the sperm she will ever be impregnated with during her nuptial flight. She never mates with the drones again, but rather stores the sperm cells inside her body. Then, she totally controls when fertilization will occur. When she lays eggs and releases the sperm, the eggs become fertilized. Those fertilized eggs develop into females; some are worker bees, which never reproduce, and a few are new queens. When the queen lays eggs but withholds sperm and prevents fertilization, the unfertilized eggs develop into male drones.

Drones are male bees that never have a full chromosome count; they are born haploid instead of diploid. All the cells of their body are haploid, including the cells that develop into sperm. Therefore, the cells in their testes do not go through meiotic divisions to reduce the chromosome number. The cells in the testes just develop into gametes that are received by the new queen (which technically is one of their sisters!). The worker bees are born diploid but never produce gametes.

The process by which bees reproduce is called *parthenogenesis*, meaning "virgin production." (*Partheno-* is Greek for virgin [as in the Parthenon], and *genesis* means production.) Parthenogenesis is a form of sexual reproduction used by some plants, invertebrate animals, and certain lizards, as well as honeybees and wasps.

Developing New Beings

Sexual reproduction involves the production of gametes and the act of mating to join the gametes so that fertilization occurs. Once fertilization occurs, the term *development* describes how the fertilized egg becomes another new organism.

From single cells to blastocyst

The sperm is a cell, and the egg is a cell. Both contain half the number of chromosomes that are found in an organism. When fertilization occurs in the fallopian tube, the diploid number is restored, and the full set of chromosomes is needed later in development. Once the nucleus of the sperm and the nucleus of the egg fuse, fertilization is complete, and the cell is referred to as a zygote. Figure 13-1 illustrates these steps.

The zygote begins to travel down the fallopian tube, heading for the uterus where it can implant in the lining. The zygote divides into two cells, which divide into four cells, which divide into eight cells, which divide into 16 cells. At this point, the zygote is called a *morula*.

Cell division continues, but the morula becomes filled with liquid, which pushes the increasing number of cells out toward the periphery of the embryo's membrane. The flattened cells form the *trophoblast*, the fluid-filled cavity is called the *blastocoel*, and the sphere of larger, rounded cells is called the *blastula*. The entire cell is called a *blastocyst*. Separation of cell types in the blastocyst is the first step in forming specialized tissues.

Ectopic pregnancy

If the zygote remains in the fallopian tube and implants in the tissue there, the pregnancy that occurs is called an ectopic pregnancy. Eventually, a zygote growing in the small fallopian tube, instead of the larger uterus, causes intense pain early during the pregnancy. To preserve fertility in that tube, the zygote must be removed before the fallopian tube bursts.

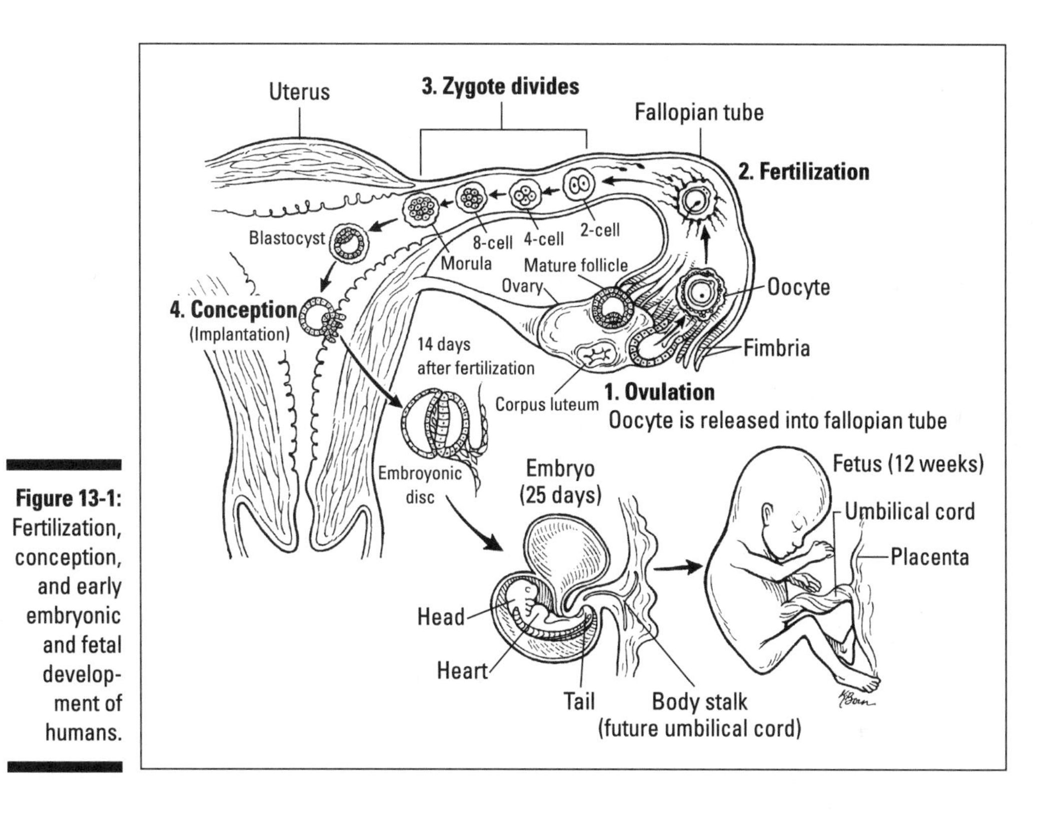

Figure 13-1: Fertilization, conception, and early embryonic and fetal development of humans.

The blastocyst must implant in the wall of the uterus, or pregnancy will not take place. *Conception* is said to occur when the blastocyst successfully implants itself. Conception is not the same as fertilization. The egg can be fertilized, but a pregnancy is not conceived until the blastocyst is rooted where it can develop further.

The trophoblast cells secrete an enzyme that helps degrade the endometrium of the uterus. Once the blastocyst "eats" its way into the wall of the uterus, it sinks into the wall and is implanted. If conception occurs, the trophoblast cells of the blastocyst go on to form the chorion, which becomes part of the placenta. Once implanted, the developing mass of cells undergoes *gastrulation*. Gastrulation occurs when the cells of the blastocyst move inward and form a two-layered *embryo*.

Go, go, embryo

You get more done in the 12 weeks or so that you are an embryo than at any other time in your life. During the embryonic period — essentially the first trimester of pregnancy — every organ in the body forms.

Fraternal "twins" really are different

When you hear that somebody has a twin, you immediately might think that the "twin" is the spitting image of the first person, maybe even with the same type of personality. However, only *identical twins* have the same exact genetic information. During the blastocyst stage, every cell in the mass is genetically capable of becoming a human being. Identical twins are created when the cells of the blastocyst separate or when the inner cell mass splits into two. Then, two identical embryos start developing. After birth, the twins look exactly alike and may even think alike.

In contrast, *fraternal twins* develop when two eggs are released during ovulation. Then, of the hundreds of sperm that make it to the fallopian tubes, one sperm fertilizes one egg, and one sperm fertilizes another. The "twins" are indeed brother and sister, or brother and brother, or sister and sister, just like any other siblings are. The only difference is that they happened to be born at the same time instead of years apart. They are totally individual because their genetic makeup is different. Each sperm in a man is not identical, and every egg in a woman is not identical. If they were, then two parents would create the same offspring every time they created a pregnancy. And, as you know, your brother or sister can have completely different looks, feelings, likes and dislikes than you have. It is even possible that fraternal twins can be half brothers or half sisters! It has been known to happen that a sperm from one man can fertilize one of the eggs, and the sperm of another man can fertilize the other! Talk about genetic differences!

First, cells grow in size and begin to specialize, forming three layers: the ectoderm, mesoderm, and endoderm. The specialized cells of each layer begin migrating toward other cells with the same specialty, which gives an embryo a shape and is called *morphogenesis* (*morph-* is Greek for form or structure).

Outside the embryo, specialized membranes develop. The chorion combines with tissues created by the mother to become the placenta. The *placenta* is filled with blood vessels and provides a large surface area for exchange of gases, nutrients, and wastes.

The *allantois*, in humans, forms off of the center cavity of the blastocyst. Eventually, it becomes the body stalk and then the *umbilical cord*, which connects the fetus to the placenta. In birds and reptiles, the allantois is used to store wastes.

The *amnion* surrounds the amniotic cavity, which fills with *amniotic fluid*. This fluid protects the developing embryo and fetus. The fluid cushions movements created by the mother and protects the developing organism from bumps. Amniotic fluid also is thought to contain *surfactant*, which, as the fetus drinks in the fluid, coats the internal surfaces of the lungs so that after birth the lung tissues do not stick together and prevent the infant from breathing. Cells in the amniotic fluid can be tested during *amniocentesis*. The genetic material in those cells matches those of the developing embryo, so it is possible to see whether the embryo will have a genetic defect. Interestingly, amniotic fluid may be an evolutionary link. Every animal develops in a watery environment, even those who, after birth, live only on land.

After the cells and outer membranes of the embryo are in place, differentiation starts to occur. *Differentiation* is the term for the biochemical processes that make a cell an eye cell instead of a toenail cell. As cells differentiate, organ systems begin to form. In humans, the first system to begin developing is the nervous system, which makes sense because the brain — the center of the central nervous system — controls everything.

First, the *notochord* begins to form from the mesoderm layer. The notochord becomes the vertebrae of the backbone. Then, the *neural tube* forms from the ectoderm. The neural tube spawns the central nervous system, creating the brain and spinal cord. *Neural tube defects* cause problems such as *spina bifida*, which is a hole in the spinal cord. Evidence exists that a woman who consumes adequate levels of folic acid (a B vitamin) prior to conception can prevent neural tube defects. Many breakfast cereals have increased the amount of folic acid they contain in an effort to help prevent neural tube defects.

Throughout embryonic development all other systems and structures of the body form as well. Cells that split off the neural tube and form the neural crest become the teeth, bones, skin pigments, and muscles of the skull, for example. At the end of the embryonic period, a human embryo is about an inch and a half long, and it starts to look less like a lizard and more like a human.

Fetuses

In humans, the fetal period encompasses the last six months of pregnancy, or the second and third trimesters. Fetuses are completely differentiated, meaning that the cells have migrated and formed organ systems. All fetuses do in the uterus is continue to grow and develop features such as hair and nails. As the fetus gets stronger, longer, and heavier, it looks more and more like a newborn baby. When it is ready to be born, it is thought that the baby itself initiates the event. It is known that prostaglandins and the hormone oxytocin cause the uterus to contract. But what causes the production of these hormones to start is thought to be initiated by a chemical produced by the fetus. The fetus is thought to release an as yet unknown chemical that may prompt the mother to start producing the hormones that initiate labor.

When labor does not begin naturally, prostaglandin suppositories and/or synthetic oxytocin — called pitocin — is given to the mother. Upon the birth of the fetus, the organism is called a neonate, meaning newborn. A life begins, and development continues. Go to Chapter 15 to find out what develops in a human. For more information about what happens during pregnancy, check out *Pregnancy For Dummies* (Hungry Minds, Inc.).

Chapter 14

Making Mendel Proud: Understanding Genetics

In This Chapter

- Finding out how you inherit certain traits
- Coming to understand how DNA carries genes
- Figuring out how the information contained in genes turns into a protein used by an organism
- Taking a look at some great achievements in the field of genetics

Gregor Mendel, the father of genetics, would be amazed at how far genetic knowledge has come in the fewer than 150 years since his simple crosses of pea plants. He was probably pleased as punch when he came up with the Laws of Heredity. I doubt he ever imagined that every gene in the human gene pool would be mapped and sequenced and that animals could be cloned or that genes from bacteria can be placed in a plant to *prevent* disease. He would be fascinated. Make him proud. Find out a little about genetics in his honor. After all, genetics — the branch of biology dealing with heredity — certainly knows a lot about you!

Jumping Into the Gene Pool: Some Definitions to Springboard You

If you can really grasp these definitions, you'll be in good shape to understand the more detailed processes that take place in the nuclei of your cells.

- **Chromosomes** are paired structures that are made up of strands of *chromatin*, which contains DNA and protein. In humans, there are 46 chromosomes (23 pairs) in the cell nucleus of regular cells of the body — called *somatic* cells — as opposed to the *gametes*, which are sperm and egg cells that contain only 23 unpaired chromosomes. A chromosome has a short arm and a long arm. The arms are held together by a centromere.
- **DNA** — deoxyribonucleic acid — is the chemical molecule that serves as genetic material. A strand of DNA is a long chain (a polymer) of nucleotides. Each nucleotide of DNA contains a nitrogenous base, a sugar with five carbon molecules called deoxyribose, and a phosphate group. There are four different kinds of nitrogenous bases in DNA: adenine, thymine, cytosine, and guanine. The nitrogenous bases (and therefore the nucleotides) can be and are different throughout the long chain of DNA. DNA exists inside the chromosomes.
- **Genes** lie along the chain of DNA. They are made up of sections of nucleotides. Some genes can have many nucleotides; others just a few. Humans have thousands of different genes, which reside on different chromosomes, but on the same chromosome in all people. For example, the gene for cystic fibrosis is always found in the same location on gene number 7 in all humans. The gene for Huntington's disease is on chromosome number 4 in all humans. The gene for sickle cell anemia is on chromosome 11 in all humans. However, not all humans have all diseases. Some genes are *expressed* (they show the effect), whereas other genes are *repressed* (they do not show an effect).
- **Alleles** are the different forms of a trait. For example, the gene for hair color resides in a certain location on a certain gene in all humans. However, humans can have many different shades of hair color; the different shades are represented by different alleles. Say that the gene for hair color is on gene number 2 (I don't know that it is, I'm just giving you an example). You have a pair of number 2 genes — one from your mother and one from your father. Suppose that your father has dark brown hair, and your mother has blond hair. Brown hair color is more *dominant* than blond hair, so I'll call it *H* — H for hair, capital H for dominance. On one of your number 2 chromosomes, you have a pair of H alleles at the location of the gene for hair color. But you also have your mother's genes. I'll call her hair color *h* — h for hair, but lowercase to represent a less dominant (or *recessive*) color. On your other number 2 chromosome, you have a pair of h alleles at the location where the gene for hair color lies. Both *H* and *h* are alleles of a gene that controls a certain trait. Table 14-1 gives you a few examples of some dominant and recessive human diseases.

Human traits aren't just dominant or recessive

Many of the measurable traits, such as height and weight, as well as hair color and eye color, are *polygenic traits*. Polygenic means "many genes," and many genes are involved in determining how tall you will be or what number you should see on the scale. Other traits, such as baldness, are called *sex-linked traits* because even if both sexes carry the gene for the trait, only one sex usually shows the effect. Females often carry the gene for male-pattern baldness, but they are rarely bald. But, if you are a male with a bald grandfather on your mother's side, you might want to save some money for future hair-weaving, a toupee, or some retinol products.

Table 14-1 Some Dominant and Recessive Human Diseases

Dominant	*Recessive*
Huntington's disease, which causes degeneration of the nervous system	**Sickle cell anemia**, which results in abnormally shaped red blood cells
Marfan syndrome, which is a disorder of connective tissue that affects the skeletal and cardiovascular systems, as well as the lens of the eye	**Thalassemia (Cooley's anemia),** which results in unusually small red blood cells
Cowden disease, which results in multiple lesions in several types of tissues and organs that often become malignant	**Cystic fibrosis,** which results in chronic respiratory infections and thick mucus secretions

"Monk"ing Around with Peas: Mendel's Law

Gregor Mendel was a monk in Austria in the mid-19th century. To pass his quiet time, he watched pea plants grow (nobody said being a monk was exciting). As he observed subsequent generations of pea plants growing, he noticed subtle changes and wondered how they occurred. He observed the color of the seeds, flowers, and unripe pods; the shape of the seeds and pods; the length of the stem; and the position of the flowers. Most importantly, he kept accurate records and an accurate count of what plants showed what traits.

Of course, Mendel had to wait for each pea plant to mature to see whether his predictions (hypotheses) came true. Lucky for him, pea plants grow pretty quickly. Lucky for you, monks are patient people with lots of time on their hands.

During Mendel's time, the general thought was that the traits of a father blended with the traits of a mother. So, a tall father and a short mother were expected to breed average-size kids. Traits in offspring were expected to be averages of the traits in the parents.

Well, one day Mendel crossed a tall pea plant and a short pea plant, expecting to get average-size pea plants. However, three tall pea plants and one short pea plant grew. He crossed tall with short again and again and again. Same thing. What he discovered was that offspring carry all the traits of the parents, but each offspring is capable of expressing different ones. He figured that these traits were passed from generation to generation by something that he called *factors*. What Mendel called factors are what you call *genes*.

After studying 28,000 or so pea plants and reviewing his data, Mendel was confident enough in his research to state that heredity followed specific patterns. First, Mendel stated that traits are inherited independently of each other — this is called Mendel's *Law of Independent Assortment*. This law states that each trait or characteristic is found on separate factors (genes), that each factor (or gene) comes in pairs, and that each pair separates on its own.

Mendel also came up with the *Law of Segregation,* which states that during cell division, each allele of a gene pair will randomly move to different gametes. The section called "Jumping Into the Gene Pool" gives you an example of alleles using hair color. (If you didn't read it, go back and read it now. I'll wait.) On one of the number 2 chromosomes you have two *H* alleles; on the other number 2 chromosomes you have two *h* alleles. You have four alleles for hair color (H, H, h, and h) at the locations of the genes for hair color. When you create an egg or sperm, whichever is appropriate for you, only half of your genetic material goes into each gamete (remember, gametes have only half the chromosomes because they eventually meet up with another gamete that contains the other half). So, you produce two gametes containing *H* alleles and two gametes containing *h* alleles. But, each of those alleles separated randomly into the four gametes.

There are multiple genes and alleles for hair color. Two of the genes may be next to each other, at a particular position on a chromosome, and they will not separate randomly. Their alleles on their paired chromosome will separate randomly, though.

Bearing Genetic Crosses

When a geneticist writes a genetic "equation," the alleles are represented by letters. The *dominant traits* are usually capitalized; the *recessive traits* are usually lowercase. Writing the letters that represent the alleles of a gene is called a *genotype*. A genotype can be written to represent a phenotype. A phenotype is the physical result of the expression of a gene. So, if *H* stands for brown hair and *h* stands for blond hair, *HH* is the genotype for a person with the phenotype of dark brown hair; *hh* is the genotype for a person with the phenotype of light blond hair; and *Hh* is the genotype for a person with the phenotype of light brown hair.

Phenotypes are created by matings — or crosses — between different organisms. If you cross a rose bush that creates red flowers with a rose bush that produces white flowers, you would get a rose bush that has red or maybe even pink flowers. If you were crossing the rose bushes explicitly to see the result of flower color, you would be performing a *monohybrid cross*. A monohybrid cross examines just one trait. (Fittingly, a *dihybrid cross* examines two traits.) The phenotype for the first rose bush is "red flowers." The genotype would be *RR*. The phenotype for the second rose bush is "white flowers." The genotype would be *rr*. Figure 14-1 uses *Punnett squares* to determine the results of two monohybrid crosses.

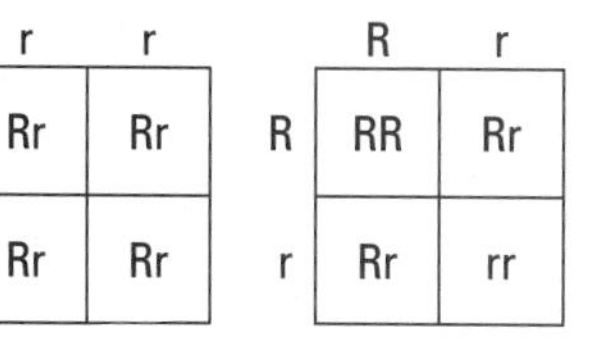

	r	r
R	Rr	Rr
R	Rr	Rr

	R	r
R	RR	Rr
r	Rr	rr

Figure 14-1: Punnett Square of a monohybrid cross. RR = red flowers; rr = white flowers; and Rr = dominant red or possibly pink caused by incomplete dominance.

Now, the results of the monohybrid cross in the top part of Figure 14-1 are all Rr. Because R is dominant, most of the flowers on the new rose bush would be red. However, pink flowers are distinctly possible because of something called *incomplete dominance*. Incomplete dominance results in an intermediate phenotype, one that is a blend of the two original phenotypes. If two Rr plants are crossed, one pure red and one pure white flower will be produced, as well as two more Rr flowers.

Copying Your DNA: Can't Wait to Replicate

Your DNA doesn't copy itself — the technical term is *replicate* — only when you create gametes and mate. Every cell in your body needs to be replaced periodically. Cells don't last nearly as long as humans do. Cells never stop working and eventually wear out. Cell turnover, as it is called, happens constantly. Blood cells need to be replaced every 120 days. But not every blood cell is replaced every four months. On any given day, your body can be replacing some blood cells, some skin cells, some hair cells, and some mucous cells. Whatever is on the body's to-do list, the process of *DNA replication* is the same.

DNA looks like a twisted ladder, with the nucleotide bases forming the "rungs" of the ladder (see Chapter 4, Figure 4-3). During replication (see Figure 14-2), the DNA strand must "unzip" so that the rungs are split apart with one nucleotide on one side and one nucleotide on the other. Each side of the original DNA strand becomes a *template strand* upon which the new *complementary strand* forms. The unzipping of the DNA helix is initiated by the enzyme *helicase*.

The entire DNA strand does not unzip all at one time, however. Only part of the original DNA strand opens up at one time. When the top part of the helix is opened, the original DNA strand looks like a Y. This partly open/partly closed area where replication is going on is called the *replication fork*.

Note on Figure 14-2 the numbers 5' and 3' (read "5 prime" and "3 prime"). These numbers indicate the direction in which replication is occurring. Each template strand of DNA is "read" in the 3' 5' direction. And, because the bases that are complementary (opposite) to the template strand are added, the complementary strand "grows" in the 5' 3' direction.

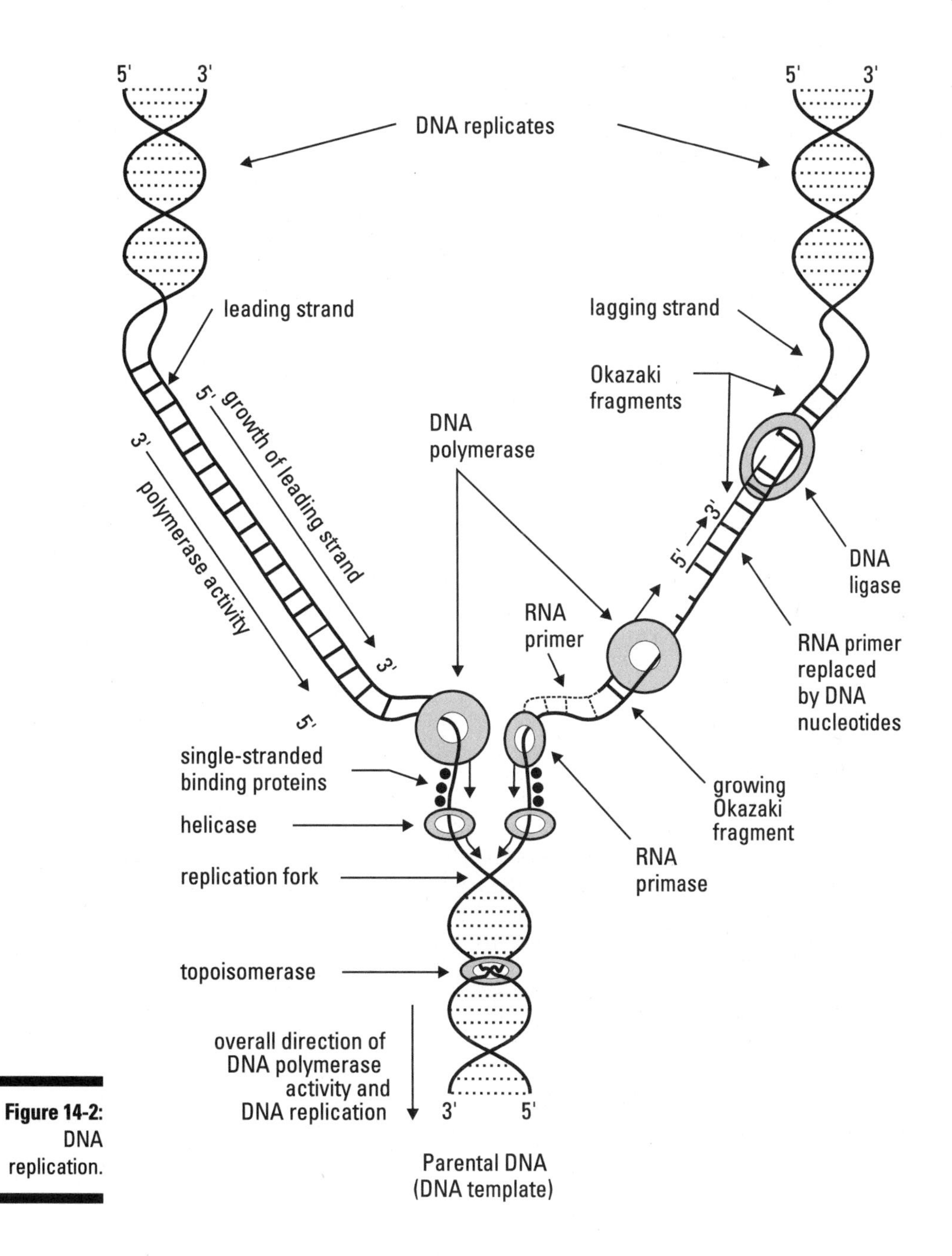

Figure 14-2: DNA replication.

The template strand tells some great stories, so you'll want to know how to read it. The nitrogen bases that make up each nucleotide along the strand of DNA include adenine, guanine, cytosine, and thymine, which are abbreviated as A, G, C, and T, respectively. In a molecule of DNA, adenine (A) always pairs with thymine (T), and cytosine (C) always pairs with guanine (G): A–T and C–G.

As the enzyme DNA polymerase moves along the template strand, if a base says A, then a T is added to the growing complementary strand. If a base on the template strand says G, then DNA polymerase adds a C to the growing complementary strand.

The order of the bases is important because the order of bases delineate the genes, and the genes dictate what amino acids are produced, and the amino acids determine which proteins are produced, and proteins are needed in every cell of your body. Proteins make up cell structures themselves, as well as enzymes that initiate cellular processes that keep you alive. It all starts with the DNA, though.

Refer to Figure 14-2. See how the left side is opening and growing smoothly? The DNA polymerase works continuously on that side, and that side is called the *leading strand*. The other side looks a little messier because the process does not occur smoothly. On that strand, called the *lagging strand*, the DNA polymerase reads the template strand and assembles the new bases in fragments. These fragments are called *Okazaki fragments*, and they are then joined together by the enzyme *DNA ligase* to form the new complementary strand.

Now that you know what the nitrogenous bases of DNA do, I'll tell you what the phosphate and sugar molecules of DNA do. The replicating DNA strand needs energy to go through the steps of reading the template, producing the complementary base, and joining the base to the growing strand. The molecules of the sugar deoxyribose provide that energy. The phosphate bonds that are broken apart when the original strand of DNA "unzips" provides the chemical energy needed to get the whole process started. Nature certainly is well organized, isn't it?

Mistakes Can Happen

Believe it or not, the newly created DNA strands in cells are proofread before cell division is finalized. If a mistake is detected, it's back to the template strand. The nucleotide that was inserted in error is removed, and the correct one is put in. If the *proofreading* function goes awry, *mismatch repair enzymes* are available to shore things up. Error recognition and repair mechanisms exist in organisms with eukaryotic cells — those with a true nucleus, like you, me, and mountain lions — but the details of how they function are not as well understood.

If a mistake in a new strand of DNA goes undetected or unrepaired, the mistake becomes a *mutation.* A mutation is a deviation from the original DNA strand. The nucleotides are not in the same sequence. Although mutations can and do cause serious defects, all mutations are not bad. (In Chapter 15, you can see how mutations help with the development and evolution of species.) In the following list, I explain how mutations, which usually are caused by certain chemicals (such as formaldehyde) or radiation (such as ultraviolet light, x-rays), affect humans.

- **Substitutions:** These types of mutations occur when the wrong nucleotide is put in for another nucleotide. For example, if the code for a particular gene read 5'-A-T-C-G-T-C-A-G-3', the correct complementary sequence for the code on the new strand of DNA would be 3' -T-A-G-C-A-G-T-C-5'.

  Genetic code is written in a specific direction. Because DNA is a double helix in which two strands intertwine, confusion can easily be created when trying to keep track of the ends of the strands. To avoid confusion, one strand of DNA is labeled 3' (3 prime), and the other is labeled 5' (5 prime). The convention is to read the strand in the 5' to 3' direction.

  A substitution mutation occurs if the newly created sequence reads something like this: 3' -T-A-C-C-T-C-A-G-5'. The third base over should be guanine (G) instead of cytosine (C). That base could have been passed over during the "reading" of the strand of DNA, or a new C could have been put in instead of the G. In either case, it's wrong, so it's a mutation. Because it is just one base, it's called a *point mutation*. Chances are that the protein that gene creates would not be affected. If so, the mistake is called a *silent mutation*.

- **Deletions:** If, during the creation of a new strand of complementary DNA, a nucleotide is read but the complementary base is not inserted, the complementary strand is missing a nucleotide. This type of mutation is called a deletion. Deletions can cause serious diseases. Cystic fibrosis is a disease that causes the lungs to continually fill with thick mucus, which can harbor bacteria and cause serious cases of pneumonia, as well as other problems. People with cystic fibrosis often do not live past their 20s or 30s. Cystic fibrosis is caused by a teeny little deletion on chromosome number 7. Duchenne muscular dystrophy, the most common type of the disease that the Jerry Lewis Telethon raises money for every Labor Day weekend, also is caused by a deletion on a chromosome. Muscular dystrophies are genetic defects that lead to muscle deterioration, which can be quite serious. People lose their ability not only to walk, but Duchenne muscular dystrophy eventually leads to severe weakening of the muscles that allow breathing and of the heart muscle.

- **Insertions:** If an extra nucleotide (or many extra nucleotides not in multiples of three to maintain a codon) is slipped into a newly developing complementary strand, the rest of the strand is read wrong. This type of mutation is called a *frameshift mutation* because the reading of the "frames" of genetic code is shifted (think of each nucleotide as a frame on a piece of film). One well-known disease that is caused by the addition of nucleotides is *Huntington's disease*, in which the sequence C-A-G is inserted up to 100 times into a normal gene. Although the sequence is a multiple of three (so technically not a frameshift mutation), the abundance of these insertions screw up the reading of the normal genetic code, causing abnormal protein production or a lack of protein production. In people with Huntington's disease, the nervous system degenerates starting when a person is in his or her 30s or 40s. Another disease caused by an insertion mutation became well known when a movie was made about a person affected with the disease. The movie, "The Elephant Man," gave an account of a man affected with neurofibromatosis, which causes deformities. The cause of this disease is the insertion of DNA sequences that do not code for anything right into the middle of DNA sequences making up a gene that does code for certain proteins. When these noncoding sequences are stuck into the gene, the code for the normal gene cannot be read, and errors in protein production occur.

Producing Proteins Promptly

I need to introduce to you another type of nucleic acid. *Ribonucleic acid (RNA)* is very similar to DNA except for these differences: RNA is single stranded (instead of a double helix), it contains the sugar *ribose* instead of the sugar deoxyribose, and it uses uracil (U) as a nitrogenous base instead of thymine (T). Therefore, the nucleotides in RNA pair up as A–U and . RNA bases can pair up, even though the RNA molecule is single stranded, because RNA has a secondary structure and can fold up and base pair with itself where complementary. RNA molecules are important for the production of proteins. And, just after DNA is replicated, the complementary strands head out to produce proteins. This is the story of what happens.

You know that the DNA harbors the genes that code for what proteins will be produced in your body. But the code buried in segments of DNA is not what initiates protein production. First, the DNA must be "rewritten" into a strand of RNA, and the RNA carries — therefore it is called a messenger — the information out of the cell's nucleus to the ribosomes. At a ribosome, the original message is translated, and then the appropriate protein can be produced. Protein synthesis is initiated on ribosomes that exist free in the cytoplasm. The ribosomes that are attached to endoplasmic reticulum make proteins to be secreted or transported to other organelles. That's it in a nutshell. Figures 14-3 and 14-4 detail the individual steps for you.

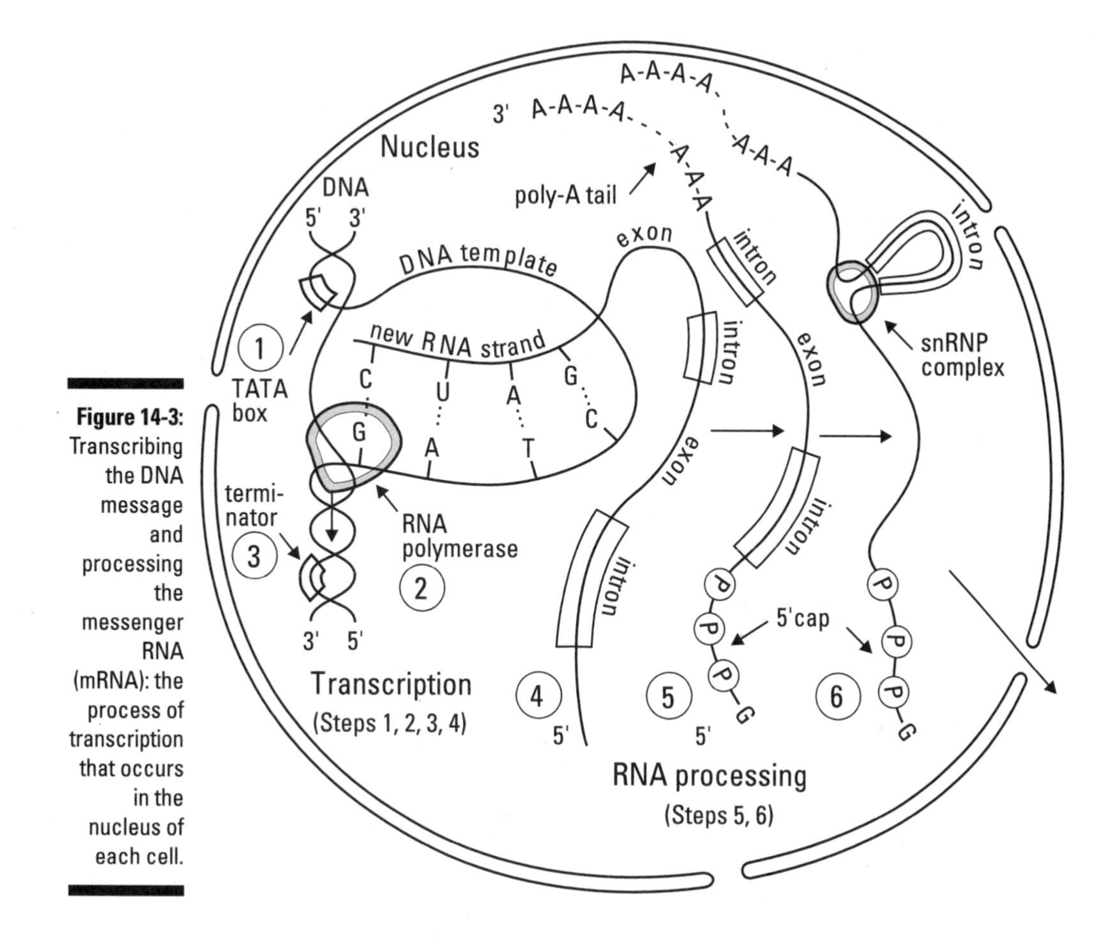

Figure 14-3: Transcribing the DNA message and processing the messenger RNA (mRNA): the process of transcription that occurs in the nucleus of each cell.

Rewriting the DNA's message: Transcription

When an original strand of DNA is unzipped during DNA replication, its nucleotide bases are used as templates for the production of new complementary strands of DNA. Those new complementary strands are used during transcription as the templates from which a strand of RNA is produced. The type of RNA molecule produced from the transcribed message is called messenger RNA (mRNA) because it then carries the DNA's message outside of the nucleus for processing.

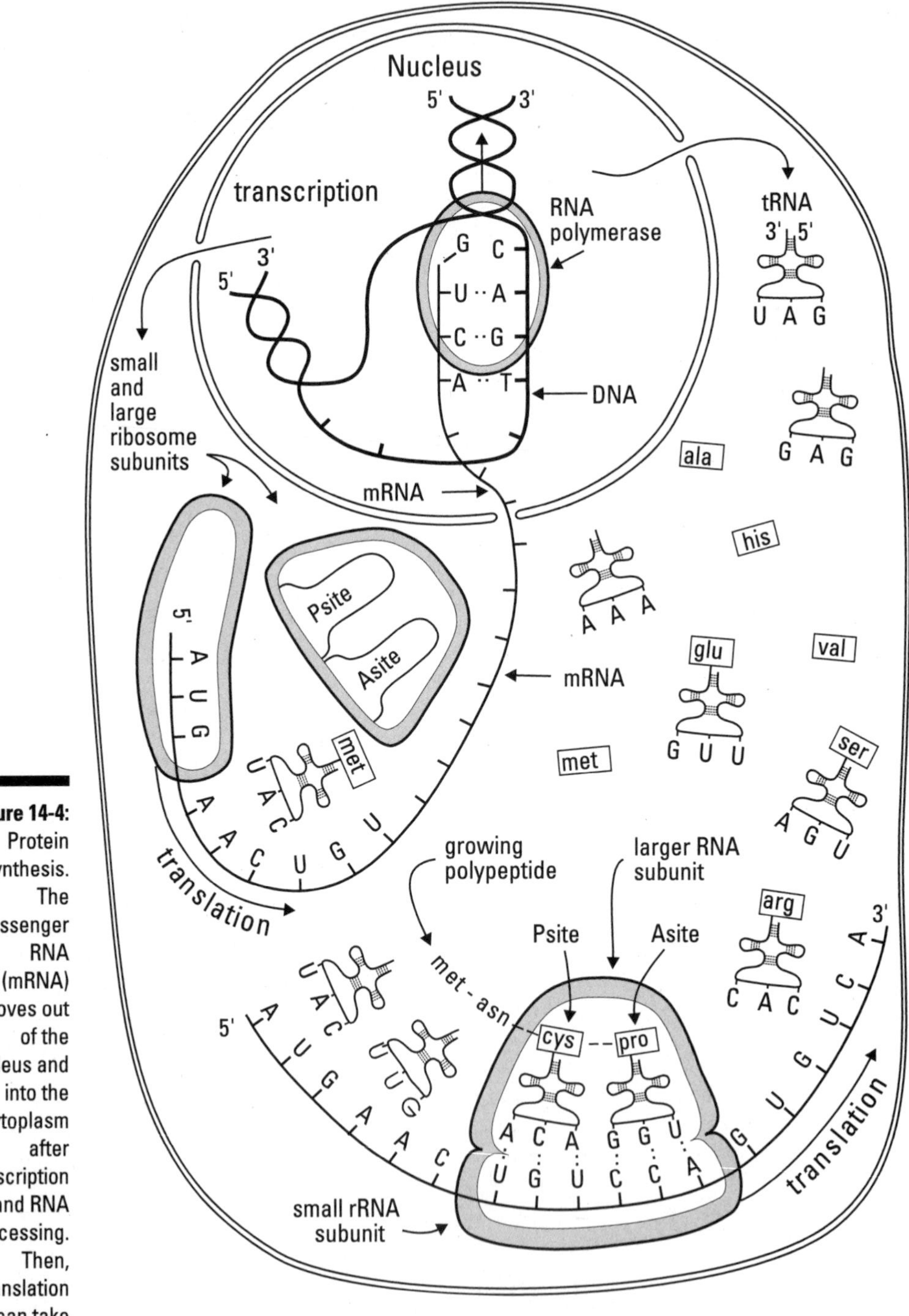

Figure 14-4: Protein synthesis. The messenger RNA (mRNA) moves out of the nucleus and into the cytoplasm after transcription and RNA processing. Then, translation can take place.

The complementary strand of DNA contains certain sequences (such as T-A-T-A and C-C-A-A-T, called TATA or CAT boxes) that exist on the strand just before the spot where transcription should start. Transcription begins at a region on the mRNA that is untranslated. This region, aptly called the 5'-untranslated region (UTR), serves as a signal to the ribosomes to begin translation at the appropriate codon that reads A-U-G. The nucleotides that follow those indicators are "read" and transcribed into the corresponding bases that are appropriate for an RNA molecule. So, A is the template for the addition of a U. T is the template for the addition of an A, C is the template for the addition of a G, and G is the template for the addition of a C. An original DNA template message of 5' -A-T-G-C-T-G-C-A-3' becomes a complementary DNA message of 3' -T-A-C-G-A-C-G-T-5'. Then, that message serves as a template to create the following mRNA sequence: A-U-G-C-U-G-C-A.

As transcription occurs down the template, certain areas are read and transcribed. A large precursor is made called heterogeneous nuclear RNA (hnRNA), which has both intervening sequences (called *introns*) and expressed sequences (called *exons*). The introns are removed, and the exons are joined by a process called splicing. (Splicing takes place only in eukaryotes, not prokaryotes). When all the nucleotides have been read, exons have been spliced together, and the messages have been transcribed, the end of the material to be transcribed is indicated by a termination codon. The termination codon — either T-A-A, T-A-G, or T-G-A — is like a stop sign along the strand. Transcription of a message completely ends at the 3' UTR (untranslated region), which can be quite a long sequence.

Processing the RNA

Once the DNA message is transcribed, and the mRNA is produced, the mRNA gets a cap and a tail. The cap and tail help to stabilize the mRNA molecule as it transports the important messages. The nucleotides that did not code for anything — the introns — are deleted out, and the important nucleotides — the exons — are joined together (refer to Figure 14-3). Think of when you use a word processing program, and you highlight a sentence in the middle of a paragraph and select the Cut command. The sentences before and after the one you deleted then join up to make the paragraph flow smoothly. Once the cutting and splicing is done, the mRNA is mature enough to leave the nucleus and head out into the cytoplasm on its own with the important genetic information (refer to Figure 14-4). You wouldn't trust an unlicensed 15-year-old to drive a moving van filled with your most treasured belongings, now, would you? Maturity is important even at the cellular level.

Putting the code into the right language: Translation

Once a mature mRNA leaves the nucleus of a cell, it heads for a ribosome in the cell's cytoplasm (some head for a ribosome attached to the endoplasmic reticulum). As the strand of mRNA, which is carrying the ever-so-precious genetic information, slides through the two parts of a ribosome, the code it is carrying is read three bases at a time. The group of three bases is called a *codon*.

Genes reside on a strand of DNA, and genes direct the production of amino acids, which are then put together to form a protein. Well, *translation* is the part of the process where the information from the gene is used to create the amino acids and then the protein (see Figure 14-4). To perform this amazing feat, however, the ribosome must be able to take the nucleotide bases and equate it with codons that specify amino acids. Each amino acid is represented by certain codons. The "language" that bridges the gap between gene and amino acid is the *genetic code*.

The genetic code was deciphered within the last 25 years or so and earned the men who cracked it a pair of Nobel Prizes. The genetic code uses the four nucleotide bases — thymine, cytosine, adenine, and guanine — to "spell" out what amino acid should be produced. But, codons usually are written as the RNA sequence, so uracil is used in place of thymine. And because each codon is "spelled" by three bases, the genetic code contains 64 (4^3) different codons (see Figure 14-5). Amazingly, every organism uses the same genetic code. Each organism produces a different number of amino acids, however. Humans use only 20 amino acids, so several codons specify one amino acid.

The mRNA strand moves through the ribosome, bringing codon after codon into position so that it can be read. As the codon is read, a transfer RNA (tRNA) molecule brings the appropriate amino acid to the ribosome. Amino acids are then joined together by peptide bonds to form proteins. Following translation, protein molecules may need to be modified a bit before they become functional. But, once the protein is created, it is soon capable of being expressed within the organism.

Suppose that you are getting too much sun (or unnatural ultraviolet rays). To protect yourself, the cells on the tip of your nose want to darken a bit. If the original template strand of DNA contained the gene for depositing melanin (skin pigment) on the tip of your nose, the complementary strand of DNA created during DNA replication also contained that information. Then, when the bases were transcribed for mRNA during the process of transcription, the mRNA was handed that genetic information. The mRNA carried the info to the ribosomes, and the codons on the mRNA were read during translation.

During protein synthesis, the tRNA brought in the amino acids for melanin (in the appropriate order, of course), and peptide bonds joined the amino acids together to form the protein melanin. The genes directed that the melanin be deposited on the tip of your nose, so lo and behold, you develop a freckle in a very obvious spot! The protein was created and is now expressed, in an attempt to protect you from damaging ultraviolet rays, whether you like it or not.

Figure 14-5: The genetic code. One letter from the vertical column on the left, one letter from the horizontal column, and one letter from the vertical column on the right are the codon. The amino acids that the codon codes for are given. Note that the same amino acids are coded for by more than one codon. For example, UUU codes for phenylalanine, as does UUC.

First Letter ↓	Second Letter: U	C	A	G	Third Letter ↓
U	phenylalanine	serine	tyrosine	cysteine	U
	phenylalanine	serine	tyrosine	cysteine	C
	leucine	serine	STOP	STOP	A
	leucine	serine	STOP	tryptophan	G
C	leucine	proline	histidine	arginine	U
	leucine	proline	histidine	arginine	C
	leucine	proline	glutamine	arginine	A
	leucine	proline	glutamine	arginine	G
A	isoleucine	threonine	asparagine	serine	U
	isoleucine	threonine	asparagine	serine	C
	isoleucine	threonine	lysine	arginine	A
	methionine & START	threonine	lysine	arginine	G
G	valine	alanine	aspartate	glycine	U
	valine	alanine	aspartate	glycine	C
	valine	alanine	glutamate	glycine	A
	valine	alanine	glutamate	glycine	G

Exploring the Unknown: Genetic Pioneering and Genetic Engineering

People have been wondering why they look like their parents for centuries. Observations of nature over the past few millenia have led people to ask "Why?" and "How?" repeatedly. The search for answers have led to fascinating findings.

Genetic pioneering

Mendel's pea plant experiments jumpstarted the field of modern genetics. Once it was known that heredity was based in the cells and that genes carried the hereditary information, scientists built upon each other's work, adding more and more knowledge to the base.

Tiny animals that didn't take up too much laboratory space, didn't eat much, and that could create successive generations quickly were used to try out theories and learn more. Fruit flies (the genus and species name is *Drosophila melanogaster*) and mice or rats are most often used. To produce a new generation, which is the only way to see whether a mutation or trait is passed on, you must wait for the parental generation to mature to the point where they can reproduce. Then, you must wait through the gestation period once the parents have mated. Mice and fruit flies mature quickly and have short gestation periods. Humans, though, are not capable of reproducing until after puberty occurs during the teenage years, and then the gestation period is nine months. That's quite a wait for results!

Since Mendel's days in the monastery garden, DNA was found, and its structure was figured out. When James Watson and Francis Crick figured out that DNA was a double helix, scientists were able to determine that it split apart to be replicated. Once scientists knew how DNA was copied within the cell, they could figure out the genetic code. Knowing the genetic code allowed them to determine what amino acids and proteins were produced. And that led them to the Human Genome Project.

Mapping ourselves: The Human Genome Project

I'm sure you've heard about this project on the news, even if you didn't know what a genome was at the time. (By the way, a *genome* is the total collection of genes in a species.) In 1988, laboratories all across the world, but heavily based in the United States, began determining the DNA sequences of human DNA. If you are wondering why the Human Genome Project is a big deal, think of it this way. If you were a researcher and you wanted to study a specific human gene, first you would have to know what chromosome it "lived" on. To provide the "address" of each human gene, researchers set out to build a map of the nucleotide sequences in the DNA of each human chromosome.

Sounds like a daunting task, doesn't it? Well, the process of DNA sequencing became automated, and with several laboratories around the country all working toward the same goal and sequencing different pieces of DNA using really sophisticated computer programs, the project was largely completed several years ahead of schedule — how often does that happen?

Now armed with a roadmap of where every gene is located, researchers can turn their attention toward making good use of that information. Knowing where each gene resides in the chromosomes, the "bad" genes — the ones that cause disease or cancer or other undesirable traits — can be sought out. Gene therapy research is trying to prevent the bad genes from having their undesirable effect or to convert them to good genes. It is predicted that the future of medicine will heavily use gene therapy to prevent the occurrence of diseases rather than medicines to treat diseases that have already taken hold.

However, now that research is dealing with human genes, plenty of controversy is peppering the positive results. An uproar in the 1980s occurred when a genetically engineered strawberry was created. As geneticists, biochemists, and molecular cell biologists discover more about what can be done with genetic information, others are worried about the implications of such technology. Even after gene therapy has been successfully used, people just are not sure how to approach the future.

Should gene therapy and cloning be regulated by the government? What would happen if genes being inserted into a patient went to the wrong chromosome? If plants and animals are altered, will the balance of nature be disrupted? Will "designer" babies be created? What do you call your mother if she's your clone, and therefore also your twin sister? These questions have been asked not only by researchers, but also by government officials, journalists, and people sitting around their dining room tables. It's a brave new world out there, best to approach slowly, I guess. But, with so much promise in developing genetic techniques, it is hard to contain enthusiasm. Researchers know that they can help people now. Why wait?

Table 14-2 gives you just a few examples of what is being done now with genetically engineered products. Table 14-3 shows you what is on the laboratory benches now. By the time this book is updated, I bet these tables will grow much longer. Any takers?

Table 14-2 Genetically Engineered Products

Product	*Benefit*
Alpha-interferon	Normally produced in small amounts in the body, has very important immune function. Genetically engineered bacteria can cause the body to create lots of alpha-interferon. Now used to shrink tumors, as well as treat hepatitis B and hepatitis C.
Beta-interferon	Also a naturally occurring protective protein that is produced in small quantities. The genetically engineered variety is used to treat multiple sclerosis, which is a serious autoimmune disorder in which the body attacks its own nerve fibers, eventually causing an inability to move.
Humulin (human insulin)	In the past, pigs were used to create insulin that was used in humans with diabetes. However, because it was pig insulin, and not human insulin, some serious side effects could occur. Now, *Escherichia coli*, a very common intestinal bacteria, can be inserted with the gene for human insulin and turned into little human insulin factories. The insulin they make causes much fewer side effects and is much safer.
Monoclonal antibodies	Antibodies are cells in the immune system that fight off invading organisms. Monoclonal antibodies are made by combining B lymphocytes (cells from the immune system) from mice with cancer-causing cells. These hybrid (mixed) cells start to produce antibodies against the cancerous cells. Monoclonal antibodies are used instead of chemotherapy in patients with a form of bone cancer.
Tissue plasminogen activator (tPA)	This protein is the body's clot buster. It occurs naturally in the body to keep blood flow moving. What scientists did was genetically engineer the substance so that it could be produced outside of the body and in larger quantities. The genetically engineered product is given to patients who just had a heart attack or stroke to dissolve blockage that was the culprit.

Table 14-3	Genetics in the Works
Product/Research Area	***Expected Effect***
Functional genomics	Study of certain DNA sequences in an organism and how they function, taking into consideration all the DNA of the organism.
Microarray analysis	Instead of studying one gene in one organism, the techniques associated with microarray analysis may allow thousands of genes to be studied at one time or in many different organisms at once.
Antisense therapy	Hopes to stop bad genes from functioning, which would prevent the protein they produce from having a negative effect.
Creating new chromosomes	Could possibly create entire human chromosomes that would contain genes to cure certain diseases. These could be inserted into people with a disease so that their body would replicate the good genes instead of the bad ones.

Part V

Ch-, Ch-, Ch-, Changes: Species Development and Evolution

"For the last time— pregnant vegetarians do NOT give birth to Cabbage Patch Dolls"!

In this part . . .

Like a preschooler, life doesn't stand still. And compared to the age of the planet, living organisms are like preschoolers. They really haven't been around all that long. It took a long time for plants and animals to evolve from primitive cells that formed in the primordial soup of gases and elements. And just because humans are here now, who says evolution is finished? It's not. Cells continually divide and replace themselves. While they're doing that, changes can take place. Environments change. Habitats are lost. Living organisms must adapt, or they will die off, just like the dinosaurs.

In these chapters, you find out how cells become specialized, how you continue to develop long after you are born, and how species continue to evolve long after they are established.

Chapter 15

Differentiating Differentiation and Development

In This Chapter

- Figuring out how a cell "knows" where to go and what to do
- Exploring the development of different animals
- Hearing about totipotency
- Finding out what hormones control differentiation and development

If human cells contain all 46 chromosomes, how do the cells know what genes to express? That is, how does a hair cell express the gene for hair color in your hair and not on your elbow? How do your toenail cells "know" to produce keratin — the protein nails are made of — and not the blue pigment found in your eyes? When you were a developing embryo, how did your leg cells end up at the bottom of your body and your arm cells at the top?

You discover the answers to these fascinating questions and more in this chapter. You find out how some genes are turned on, and some are turned off, as well as how hormones regulate when the expression of a gene is on or off. Here's another question: Because hormones are proteins, and genes direct the synthesis of proteins, do certain hormones regulate their own production? It's getting deep, now. I'll just cut to the heart of the matter here.

Defining Differentiation and Development

Differentiation is the specialization of cells that occurs during development. Differentiation determines what the structural and functional aspects of the cell will be. *Development* is the overall process of an organism going through stages of differentiation. Over time, the changes occurring during development at the cellular level become visible. *Determination* occurs when a cell commits to developing in a certain manner — that is, when it is destined (or determined) to become a brain cell instead of a bone cell.

How these processes are controlled and what prompts them to start is still a little fuzzy. But I'll share what *is* known with you. At least what is known now is a vast improvement over the *homunculus theory*. Back in the 1600s, biologists actually thought that inside every sperm was a preformed human — a tiny little man called a homunculus. The theory was, as you might guess, that when a sperm fertilized an egg, the tiny little human was freed and able to develop. Some biologists back then even argued that the homunculus actually contained sperm that held members of future generations that were even smaller than the homunculus itself! Remember, this was before they knew how traits were inherited. They were right that inherited factors came from the sperm; they were just a little off in how it occurred.

Then, in the early 1700s, the preformation theory continued but took a different slant. Biologists in that century thought that the egg contained a miniature body and that the seminal fluid stimulated the unfolding and expansion of body parts. Finally, in 1759, biologists got on the right track when a scientist studied chick embryos and determined that layers became organized from simple granular material (cells, maybe?) and folded into a body. This hypothesis was called the *theory of epigenesis*. Once the microscope was invented, studies in this area moved along a little more quickly.

Think about this: You started out as one tiny cell. When the nucleus of your father's sperm fused with the nucleus of your mother's egg, a single cell was created containing all the genetic information you would ever have for the rest of your life. That single cell held the details about how tall you would be; what color hair and eyes you would have; the tone of your skin color; the diseases you would be susceptible to or be capable of passing on; what you would look like at age 5, 35, and 65; and possibly even your intelligence, talents, and personality. Amazing, isn't it?

That single cell began to divide, and the genetic material began to replicate. Those processes continued for a while until the zygote was ready to implant in the lining of your mother's uterus. At that point, the mass of cells needed to know what end was up. Plants also orient themselves before development continues. As a seed begins to develop, the root hair cells gravitate to the bottom, and the cells that will become the stem and leaves migrate toward the top. Once the organism is oriented correctly, the cells begin to become different from one another and specialized. This stage is when the cells of various organs begin to develop, and it is called *organogenesis*. Table 15-1 outlines differences among a human, frog, and bird when it comes to this stage in embryonic development.

Table 15-1 Differentiation During Embryonic Development in Humans, Frogs, and Birds

Humans	*Frogs*	*Birds*
A blastocyst forms from cell divisions in the zygote, and it contains the trophoblast and embryonic disk.	A gray crescent forms after a frog's sperm penetrates a frog egg, causing the cytoplasm of the egg to reorganize. If only a portion of the gray crescent is in a cell, it can develop into a frog. Without the gray crescent, development does not ensue.	The blastodisc is a flattened area of cells that sits on top of the massive yolk. Cell cleavages (divisions) do not occur throughout the entire yolk; they occur only in the blastodisc.
Trophoblast embeds into uterus and produces human chorionic gonadotropin (hCG) to maintain pregnancy. Trophoblast forms chorion, which becomes part of placenta.	Gastrulation occurs when cells cover the top edge of the blastopore, which once was the gray crescent.	At the beginning of gastrulation, the blastodisc invaginates (folds inward) along a line called the primitive streak.
Inner cell mass forms in cavity created by trophoblast. This mass presses against one end of the developing cell and becomes the embryonic disk. A primitive streak forms, gastrulation — the formation of the three primary layers from which all tissues develop — occurs, and development of organs and extraembryonic membranes begins.	Developing frog eggs contain yolk. A yolk plug forms near the top edge of the blastopore.	As the cells move downward into the primitive streak, the crevice that develops becomes an elongated blastopore from which the bird embryo develops and cells differentiate.

Genetic equivalence of the nucleus and totipotency

A more recent theory as to how differentiation might be initiated in cells is the idea that during development the nuclei of some cells lose their ability to express all the genes they contain. Of course, every cell contains all the genetic material because every cell contains a full set of chromosomes (except for the gametes). But not every gene is expressed all the time.

So, what takes away the ability for all genes to be expressed? Say that a cell in the iris of the eye is being created. The idea is that only the genes relating to eye color would be expressed, and all other genes that are not important in an iris are ignored. But what controls that?

Two researchers — Robert Briggs and T.J. King — tested this theory on tadpole cells. They found that up until the blastula stage, at which point the organism contains 8,000 to 16,000 cells, a single cell retained the capacity to develop into an entire organism. One of those cells could develop into a whole new tadpole. However, when they used cells from a later period in development, they were not successful. Then, another researcher — J.B. Gurdon — had success in "growing" normal frog embryos from adult frog skin cells that were transplanted into eggs that had the nucleus removed. As long as the egg into which the cells were transplanted was not past the critical point in development, the tadpoles developed. These studies led to the cloning of adult animals, like old Dolly the sheep.

Although animals seem to have a point at which cells become determined to differentiate into certain types, plants are much more flexible. You know how you can grow a whole plant from a cutting of another plant? The cutting does not contain roots, but it is able to grow them. It is also possible to take a few plant cells and grow an entire new plant from them. The ability for single plant cells to develop into whole plants is called *totipotency*. This refers to the fact that every cell in a plant has the potential to grow into a whole new plant. Each plant cell, like each animal cell, contains all the genetic information relating to the entire organism. But something in animals changes during development, whereas in plants, the totipotency remains intact.

Factors That Affect Differentiation and Development

When embryonic development begins, all the cells in the embryo have developed from that one single cell created by the fusion of the gametes. The gametes are the sex cells — egg, sperm, pollen grain, and so on.

Something has to occur to start making the cells turn into cells of the nervous system or cells of muscle tissue, or cells of the heart or lung, and so on. That "something" is differentiation, and several influences occur early in embryonic development to initiate differentiation.

Embryonic induction

Embryonic induction is the influence of one group of cells on another group of cells, causing the recipients of the influence to change their course in development. Cells that wield this power are called *organizers*, and they exert their influence by secreting certain chemicals.

Embryonic induction occurs when the lens of the eye is developing (and not just in humans, but in all vertebrate animals). When the eyes begin to develop, they start out as bulging outgrowths (*optic vesicles*) on the sides of the early brain. When the optic vesicles touch the ectoderm — one of the germ layers created during gastrulation — the ectoderm thickens into what is called a *lens placode*. The lens placode then develops into the curved lens of the eye. Induction causes these structural changes, and it also causes the biochemical changes that allow the production of many special proteins that refract light. These special proteins, called crystallins, allow vision to occur.

Interestingly, if the optic vesicles are removed from the ectoderm prior to induction occurring, regular skin develops in that spot, and an eye is not created. Induction causes a type of chain reaction. An inducer molecule binds with a receptor molecule in a manner similar to that of hormones binding to cells. However, inducers have their effect long before the cells are differentiated enough to produce hormones.

Cytoplasmic factors

Another method of causing cells to develop in a certain way is thought to involve the cytoplasm found in the developing cells. During cell division, the cytoplasm often is divided unequally among daughter cells. Unequal division of cytoplasm happens in birds, in which one cell gets most of the cytoplasm to form yolk. It also happens in frogs, in which the gray crescent is specialized cytoplasmic material. And it happens in humans. During oogenesis, the process that produces eggs in a females, one daughter cell gets most of the cytoplasm, and the other three cells are tiny polar bodies with very little cytoplasm. The cell with most of the cytoplasm goes on to become an egg.

What exactly causes the variations in the division of cytoplasm is still unclear. However, it is known that the factors involved do not travel through the developing blood vessels. In an experiment by E.P. Volpe and S. Curtis, two frog tadpoles were surgically connected so that they shared blood vessels. Then, the *primordial germ cells* — special cells stemming from special cytoplasm in the yolk of a frog egg that become the gonads — of one of the tadpoles was irradiated with ultraviolet light. The treated tadpole did not develop any gonads, but the untreated one did. Now, because the two tadpoles shared a circulatory system, the irradiated cells should have traveled to the untreated tadpole, if the primordial germ cells traveled through blood vessels. But that was not the result. So, how do the primordial germ cells become the gonads? How do they get there? If you are looking for a career in biology, here's one area that needs more research.

Homeotic genes

Homeotic genes are special genes that turn other genes on or off, kind of like a key turning a switch. When certain genes are turned on, certain proteins are produced that contribute to development. When certain genes are turned off, the protein normally created would be withheld so that it cannot affect development. These actions would control what substances would be present or absent in a developing embryo, thereby controlling the development of that embryo.

This effect is seen in fruit flies. Yes, fruit flies. If you read Chapter 14, you might remember that fruit flies are a staple organism in a genetics laboratory. They are easy to reproduce, they don't eat much, and they don't take up a lot of room. And almost all their genes are mapped. Therefore, they are excellent animals to use for genetic research. Anyway, if the homeotic genes in fruits flies are mutated, legs can appear where antennae should be. Body parts end up in the wrong places.

Several years ago, a stretch of DNA that is about 180 nucleotides long (read that as not very big) was found in most of the homeotic genes in many species. This short segment of genes is called a *homeobox*, and, yes, even humans have homeoboxes. A homeobox is the genes in the homeobox that remain unchanged generation after generation. The genes in the homeobox help to control development, so can you imagine the implications of mutations in the human homeobox? I picture a Picasso drawing.

Hormonal control

Of course, hormones, those ever-important regulatory proteins, eventually do play a role in development. But apparently, they do not step in until most of the vital organs (heart, lung, kidneys, liver) are formed, and limbs and other appendages are where they belong. Then, the hormones seem to control the actual appearance of the body.

Hormones control growth, and they also continue to control differentation and development that occur throughout an organism's life. In embryos, however, hormones are able to control changes that occur in body form.

Metamorphosis

Frog tadpoles go through some stunning changes during metamorphosis — the process that causes a tadpole to grow long legs and webbed feet and lose its long tail. A frog tadpole looks very much like a fish, but eventually, a frog appears. The hormone *thyroxine*, which is produced in the thyroid gland, is responsible for the changes. If the thyroid gland is prevented from secreting any hormones during metamorphosis, the tadpole remains, and the frog never develops. However, the tadpole will keep growing in size, just not differentiating and developing. Giant tadpoles can be created.

Metamorphosis in insects changes the egg into a larva, the larva into a pupa, and the pupa into the adult insect. Hormones also control this cycle in insects. Insects produce a brain hormone in their brains that travels along nerve cells until it is released into the blood and carried to the *prothoracic gland* (gland that appears before — *pro* — the body — *thorax* — appears). The prothoracic gland then releases the hormone *ecdysone*, which is called the *molting hormone*.

Ecdysone promotes the molting of a larva into a pupa, and it helps develop a pupa into an adult. When the insect is in the larval stage, another hormone, called *juvenile hormone*, is secreted in response to the ecdysone. The two hormones work together — synergistically — to keep the insect in the larval stage. The insect will molt into another larva as long as juvenile hormone is being secreted. Once the insect is in the last larval stage, juvenile hormone secretion drops off, and the ecdysone causes the larva to molt into a pupa, eventually becoming an adult.

In experiments, when juvenile hormone was removed early in the larval stage, the ecdysone had its effect and produced a tiny pupa that turned into a tiny adult. When juvenile hormone is added at the point when it naturally would be turned off, an extra large pupa and giant adult are created. (Maybe excess juvenile hormone in the water is creating all those monstrous moths you see fluttering around light posts in summer?)

Gender differentiation in humans

Humans are no strangers to the effects of hormones during development. In fact, human males and females are identical organisms until the time that sexual differentiation occurs.

At very early stages of development, the reproductive system of humans (and other vertebrate animals — you know, the ones with backbones) have two sets of ducts: one for the female reproductive system, and one for the male reproductive system. When both sets of ducts are present, the stage of development is called the *indifferent* stage because there is no difference yet between male and female. Humans remain in this stage until about seven weeks after fertilization (about the end of the second month of pregnancy), which is why an ultrasound any earlier than this time cannot tell the sex of the developing embryo. It's not just that the embryo is too small to see (that's part of it, of course), but more importantly, the sex is not yet evident.

The two sets of ducts are the *Wolffian ducts*, which eventually become the male vas deferens, epididymis (on the testes), and seminal vesicles; and the *Müllerian ducts*, which eventually become the oviducts, uterus, and vagina.

Inside the cells, the chromosomes can tell whether the embryo will develop into a male or a female. Of the 46 human chromosomes, the last pair — the two chromosomes 23 — either are two X chromosomes or an X and Y chromosome. Two X chromosomes indicate female; an X and Y chromosome indicate male.

If two X chromosomes are inside the cells of the developing reproductive system, the Müllerian ducts will develop, and the Wolffian ducts will disintegrate. If an X chromosome and a Y chromosome are in the cells, then the Wolffian ducts will develop, and the Müllerian ducts will disintegrate. Sex-determining genes on the Y chromosome stimulate the development of testes. If the sex-determining gene on the Y chromosome is absent, the primary gonad develops into ovaries, but two X chromosomes are required for the ovaries to be maintained.

However, one other factor, besides what chromosomes make up pair number 23, determines male or female gender. In mammals — you know, the vertebrate animals that nurse their young with mammary glands — if the male sex hormones are being secreted by the developing gonads, the Wolffian duct will develop into the vas deferens, and the embryo will become male. If the male sex hormones are withheld — it's not that female sex hormones are secreted — then the Müllerian duct structures develop.

One reason that this type of hormonal control occurs in mammals is because mammals develop inside a female's body. If the embryo began producing female hormones, it may affect the mother and the pregnancy, too. (Estrogen and progesterone must be at certain levels to maintain a pregnancy.) By using just small amounts of male hormones as an on/off switch for gender development, the desired effects can be achieved without disturbing the mother's hormone balance.

Hormones also direct the development of internal and external genitalia. The tubules necessary for ejaculation of semen are complete at about 14 weeks gestation (beginning of second trimester of pregnancy). At about that time, the penis, testes, and scrotum develop from the urogenital tubercle, urogenital swellings, and urogenital folds. The urogenital tubercle becomes the glans penis in the male, the urogenital folds become the shaft of the penis, and the urogenital swellings become the scrotum. In males, these structures develop under the influence of the hormone *dihydrotestosterone (DHT),* which is produced by cells in the newly formed testes.

In females, with the absence of DHT, the urogenital tubercle becomes the clitoris (equivalent to the glans penis), the urogenital swellings become the labia majora, and the urogenital folds become the labia minora. However, ovaries do not produce a hormone that stimulates the development of the female external genitalia. It is simply the absence of DHT that spurs on this development. In fact, the female external genitalia will develop even if the internal genitalia failed to develop. External female structures are completed around the same time that male structures are: between 14 and 16 weeks gestation. The first ultrasound of a normal pregnancy is usually performed at 16 weeks gestation, allowing visualization of the sex of a developing fetus. Of course, visualizing the sex of the baby is more accurate the later that the ultrasound is performed. I'm sure you know of people, as I do, that were told they were going to have a girl and then had to re-paint the pink nursery and return all the little dresses when the bouncing baby boy was delivered.

Errors can occur in the hormonal stimulation of genitalia. The complex process of sexual differentiation involving genes and hormones is not without error. Turner syndrome is a genetic disorder that can occur in two ways. First, a genetically female (XX) individual may be missing part or all of one of the X chromomsomes, resulting in an XO individual that is neither completely female nor male. Second, an embryo may have an X and a Y chromosome, which normally indicates male, but a deletion occurs in the region of the sex-determining gene on the Y chromosome. This deletion means that the Wolffian duct structures do not develop into a testes, so no DHT is produced. Then, the Müllerian duct structures develop into internal female genitalia, and in the absence of DHT, female external genitalia develop as well. Remember, ovaries are not required for the development of other female genitalia. So, people with Turner syndrome are genetically male (XY) or partially female (XO) with female genitalia, but they are unable to reproduce because there are no ovaries.

Embryos that have an abnormal androgen receptor cannot bind the DHT necessary to produce male genitalia. Therefore, they may be male genetically (XY) but have female external genitalia. In embryos that oversecrete adrenal androgens (hormones that are involved in the normal synthesis of DHT and testosterone), a genetic female may have masculinized external genitalia complete with a penis but have normal ovaries and other female internal reproductive structures. Or, a genetic male may be undermasculinized. Either condition creates a hermaphroditic organism; that is, both male and female gonadal tissue are present.

Developing Beyond Birth and Throughout Life

Development doesn't stop after you pack up and move out of your mother's womb. You certainly develop the most from that one-cell stage to when you were a wee embryo and then a fetus. And you develop fast when you are a newborn, toddler, and child. Puberty certainly adds many new features and develops children into young adults. But whether you're 16 or 60, you are still developing. Aging is a normal developmental process. Aging doesn't happen only to "old" people; it happens from the minute you are born and throughout every minute of your life. Perhaps "aging" should just be called "changing" because change, like growing older, certainly is one of the constants of life.

The life expectancy of humans has increased over the years, but the average life span has not. That means that people should be capable of living well past 100, but as yet they are not because of diseases like heart disease and cancer. From a biological perspective, the human species receive no advantage from having people live so long after their reproductive years. Once they pass on their genetic information and raise their offspring, older members of the species just compete for resources (such as food and water) with the younger (reproducing) members of the species.

Aging is the process that decreases the ability to sustain life. The process is long and slow, but over time, functional and structural changes decrease the ability of an organism to maintain homeostasis and fight off infections. When a man is 70 years old, the weight of his brain is only 56 percent of what it was when he was 30 years old. His maximum breathing capacity is only 43 percent of what it was at 30 years of age. Even the number of taste buds on a 70-year-old man's tongue is only 36 percent of what were there when he was 30, which I guess is a blessing since the basal metabolic rate at 70 years old is only 84 percent compared with 100 percent at age 30. But most telling of the aging process is that if a 70-year-old man had a metabolic disturbance that altered his blood pH (which is extremely important to maintain in the normal range), his capacity to adjust it to regain or maintain homeostasis would be only 17 percent, compared with 100 percent when he was 30. Eventually, the physiological processes of the body decline beyond the point where they can be recovered. Death is the point of reaching the body's genetic and physiological limits.

Some aging studies have shown that cells grown in a laboratory under optimal conditions go through about 50 cell divisions before dying. Cells are replaced in the body every day throughout your life, but the theory that stems from those laboratory studies is that maybe aging cells run out of the ability to replace themselves. Eventually, too few "fresh" cells exist, and they are unable to keep important processes running at the proper levels. However, this idea is just a theory, but it does tie in to the *immunological hypothesis of aging.*

Aging and ailing: Changes in the immune system

In the immune system of humans (and birds as well), two important groups of lymphocytes (protective white blood cells) exist: *T cells*, which become differentiated in a gland called the thymus, and *B cells*, which, in birds anyway, become differentiated in a sac called the bursa. (Mammals don't have a bursa, but the name "B cell" stuck so scientists don't have to continually refer to them as "not the T cells, the other ones.")

During the really busy, trying time of puberty, the thymus starts to get smaller and smaller and smaller until it is virtually gone later in adulthood. Without the thymus, T cells are not differentiated as well or as often. Production of B cells starts to wane as well. As the cells in the bone marrow that are precursors to the T cell and B cells go through a slowdown in production, the bone marrow is less able to put cells through cell division. This weakening of the immune system may possibly explain why people toward the high end of the average life span become more prone to infection.

Older people also have an increased risk of *autoimmune diseases*, which are diseases in which the body's healthy cells are attacked by cells of its own immune system. One example of an autoimmune disease that is much more common in older people is arthritis. In people with arthritis, the cells of the immune system can attack the cells lining the joint spaces, causing inflammation and deterioration (not to mention pain and swelling).

Growing wild: Cancer and aging

Cancer is much more prevalent in people at the high end of the life span for several reasons. First, some tumors are caused in part by viruses. If the immune system of an older person is deteriorating, a virus is more likely to take hold. What some viruses do (and you can read about this in Chapter 18) is insert themselves into the genes of a host, such as a human. Then, the virus makes the host go along for the ride of virus reproduction because viruses are incapable of reproducing on their own. This mechanism is how a tiny rhinovirus makes you miserable with a cold. Your immune system tries to fight off the invading viruses by increasing mucus production and causing inflammation in tissues such as your throat and sinuses. But if a cancer-causing virus takes over genetic material, it can cause the host's cells to reproduce like there's no tomorrow, and the immune system can't fight back. Although the rate of cell division doesn't increase, the cells become unable to stop dividing. This process happens in animals (Rous sarcoma virus, Kaposi's sarcoma associated with human immunodeficiency virus) as well as plants (crown gall tumor).

Radical, man: Antioxidants eliminate free radicals

When your body is going through metabolic processes and digestion, oxygen molecules are given off. These oxygen molecules are free, but they do not like to be solo. So as these oxygen molecules travel throughout the body, they pick up electrons from other molecules. This bad behavior on the part of these free oxygen molecules earned the molecules the name *free radicals.* As free radicals steal electrons from other molecules, those molecules take electrons from still more molecules. The chain reaction of electron transfer damages cells and tissues found in organs and arteries. This damage can contribute to atherosclerosis (see Chapter 7) and cancer, among other problems.

Research performed back in the 1980s led to more studies in the 1990s, and now it is known for certain that substances called *antioxidants* take care of free radicals within the body. Antioxidants give the lonely oxygen the electron for which it is searching. So, instead of the oxygen scavenging for an electron, the antioxidants scavenge for free radicals. And, with the free radicals under control, the body can heal itself instead of undergoing damage.

Antioxidants are found in fruits, vegetables, fish, and certain plant-based oils. However, cooking and processing foods can deplete the natural antioxidants. Recently, antioxidants such as vitamins A (beta-carotene), C, and E have been added to juices and other foods. However you take in antioxidants, whether it is through fresh fruits and vegetables (always preferable), drinking fruit juice with added antioxidants, or taking antioxidant supplements, just get it into your system to minimize the damage and maximize the body's natural ability to heal itself.

Another factor involved with increased cancer rates in older people is that cell division does not work as well or as cleanly as it does in younger people. Normally, cell division moves through the cell cycle of growth and duplication of DNA, growth and preparation for mitosis, mitosis, and more growth. As a person ages, more and more cells drop out of going through the cell cycle repeatedly. They may just hang out until they are worn out, but they don't replicate genetic material and divide.

Throughout life, tiny genetic changes may occur that you don't even realize — a tiny mutation caused by an x-ray here or another tiny mutation caused by breathing in polluted air there. Some mutations are repaired; other just exist in the body's cells without much notice. After genetic changes accumulate over time, however, the cells may not be able to repair a mutation, and the cells may become transformed. Transformation may throw cells into an uncontrolled cycle of dividing and growing, which causes a mass of cells called a *tumor* to form. A disease state occurs as these abnormal cells become more numerous than normal cells. The immune system of older people may be too weak to control this neoplastic ("new tissue") growth.

Chapter 16

Changing World, Evolving Species

In This Chapter

- Discovering what people used to believe about how organisms got here
- Figuring out what people have learned and how it changed their beliefs
- Looking at how Darwin came up with his theory of evolution
- Examining the proof of evolution
- Finding out what species preceded the current human species

You know that you have changed over the course of your life and will continue to do so. Do you see yourself as part of the great circle of life? Do you recognize the fact that all living things participate in a continual process of birth, growth, death, and renewal?

I'm sure you've seen fossilized bones or tools from ancient ancestors. Look at how humans have changed and expanded their knowledge over the millenia. It's not so hard to believe that humans have evolved. But what was the starting point of evolution? From what did they evolve?

This chapter tells you about the beliefs of ancient people regarding evolution, how Darwin studied geology to come up with his theory of evolution, and what the current thoughts are on the origin of species. You read about studies that provide proof for evolution and find out some pretty fascinating facts about how things used to be and how they've changed.

What People Used to Believe, Believe It or Not

From the time when ancient Greece was the cultural hotspot until the early 1800s, philosophers, scientists, and the general public believed that plants and animals were specially created at one time and that new species had not been introduced since then. You could call this way of thinking "fundamentalism."

People also thought the earth itself never changed, but this belief was held not long after Christopher Columbus proved that the world wasn't flat by going out to sea and not falling off the edge of the earth.

Prior to the discovery of the printing press, people received most of their information from and had their beliefs shaped by their religious leaders. When Johann Gutenberg developed the printing press, he printed the Bible so that even the common people had access to it. People could read the Bible for themselves, and then they started to question what they had been told to believe. People sought out education, they wanted to travel the world to see what was really out there, they wanted to learn why, how, what, where, and when — about everything. Science exploded. People studied plants and animals, performed experiments, and published their findings. Colleges and universities sprang up, and educated people would get together to debate such heady issues as heredity, evolution, theology, mathematics, and natural science. They wondered what was in outer space. Planets were being discovered, and people learned that the sun rather than earth was the center of the universe. This time in history was the periods of the Renaissance and the Reformation, and it was certainly interesting.

The printing press had such an incredible impact on science and technology that it spurred on investigations that, over the years, led to research still going on today. Mendel's pea plant experiments probably would not have come about if people hadn't developed a strong interest in science and nature. And without Mendel proving that traits were passed on from one generation to the next through genes, there would be no field of genetics.

Humans certainly have learned a lot of information since the golden age of Greece.

How Charles Darwin Challenged the Thinking of the Mainstream

Charles Darwin, the English country gentleman who set out on a seafaring journey on the HMS *Beagle* in 1831 (when he was only 22 years old), actually set out to keep a journal of "God's work." He was hired by the captain of the ship to be a naturalist. He was no less religious than others of his day, nor was he lacking the fundamentalist thinking that was typical of that time. But he was very seasick during his trip, so he spent most of his time on the ship lying down. While supine, he read a book called *Principles of Geology*, which was written by Charles Lyell.

In the geology book, Lyell proposed that the earth was not a stagnant mass that never changed. The geology author gave his evidence as to how the continents of the earth had been worn away by erosion, and how rocks that were transported by creeks and rivers eventually were deposited in the ocean, where they became sediment. The geologist proposed that because all continents of the earth were not at sea level due to erosion, a force within the earth must push sediment upward, creating new land masses.

What struck Darwin was that if it held true in geology that the earth was dynamic and going through a process of gradual change, why wouldn't it hold true for biology that living organisms also go through processes of gradual change? So, while on his five-year voyage around the world making notes about nature, he gathered plenty of evidence that supported his hypothesis.

Darwin traveled to Australia, Africa, and South America. Along a riverbed in Argentina, he discovered the fossil bones of three extinct species. Darwin started to form his theory of organic evolution right then and there. The extinct species resembled three living species, so Darwin began to wonder whether the prior species were ancestors to current species. If so, then evolution was a certainty. The extinct species had changed, and the new species evolved from the old.

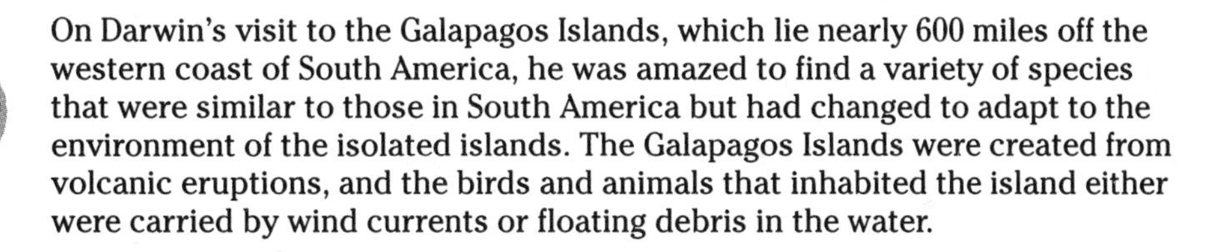

On Darwin's visit to the Galapagos Islands, which lie nearly 600 miles off the western coast of South America, he was amazed to find a variety of species that were similar to those in South America but had changed to adapt to the environment of the isolated islands. The Galapagos Islands were created from volcanic eruptions, and the birds and animals that inhabited the island either were carried by wind currents or floating debris in the water.

One species that Darwin found fascinating was the giant tortoise. After observing the shells of the tortoises, Darwin found that the shape of the tortoise shells varied from island to island. Also, the tortoises that lived in dry areas of the islands had longer necks than those that lived in wet areas. Darwin came up with a theory that the tortoises had ideal conditions on the islands, including no natural predators. So they were able to breed often and suffered few losses, quickly populating the islands. However, because the turtles were isolated on different islands, the tortoises also quickly adapted to the variations in environment. These changes happened through the generations of tortoises, but provided Darwin with a pure model of how evolution could work.

Then, Darwin noticed a group of birds living on the Galapagos Islands. The birds were finches from South America, and like the giant tortoises, they quickly established a large population on the islands. In South America, this type of finch ate only seeds. However, because no predators of the finch were on the Galapagos Islands, the little birds thrived. Soon, they overpopulated themselves, and there were not enough seeds for every bird. To remedy the competition for food, some finches began to eat insects instead of only seeds. Now, some finches on the Galapagos Islands eat only insects and no seeds. Some even evolved into cactus eaters. Although the finches on the Galapagos Islands all have a common ancestor that either flew or floated to the newly formed land masses, they became geographically isolated, which meant they were reproductively isolated from each other. Because the finches on separate islands did not mate, the island birds now are genetically different and varied from the original ancestor.

What Darwin observed in the tortoise and finch populations in the Galapagos Islands is referred to as *adaptive radiation.* Adaptive radiation happens when members of one species get into environmental niches and have very little competition for resources at the outset. This ability allows them to become rooted in a new environment and increase their population. As the population increases, competition for resources begins, and the original species breaks off into several new species that adapt to different environmental conditions. The changes in the tortoise and finch populations, as well as the discovery of extinct fossils, led Darwin to develop his *theory of organic evolution.*

Darwin's theory of organic evolution

In short, Darwin's theory of organic evolution goes like this: If a species does not adapt to changing environmental conditions, it becomes extinct. Species that adapt to change, branch off into new species from an original ancestor. Darwin believed that every species of plant and animal, whether it was extinct or still living, was descended from a previously existing species and then modified to fit current environmental conditions.

It took Darwin 23 years after his return from the sea to finally publish his book *The Origin of Species.* And it's a good thing he didn't wait any longer. A young naturalist named Alfred Wallace was quickly coming to the same conclusions that Darwin had come to on the Galapagos Islands, and he almost beat Darwin to the punch. When *The Origin of Species* rolled off the printing press, it was an instant sellout (literally, all copies sold the first day!), but the ideas in the book elicited sharp condemnation from the church. However, before Darwin died, the church ended its attack and conceded that it was entirely possible to believe in God and in the theory of evolution at the same time.

Darwin's evidence

When Darwin was traveling the world observing nature and taking notes, he collected three main categories of information. These categories turned out to provide the evidence for the evolution of species.

- **Life forms are diverse.** Darwin found that environments around the world varied greatly. What amazed him was that for every little ecological niche he found, several species of animals and plants were well suited to live there. Darwin wondered what made living things adapt so well to their environments. Now, people involved in the study of *biogeography* search for answers and more clues as to how organisms adapt to their surroundings.

- **Life forms are similar.** Although this sounds like a contradiction of the preceding category (how can life forms be diverse yet similar?), it really is not. Although Darwin found a great variety of life forms seemingly specialized for their environments, he also found that overall all plants are very similar to each other. All plants are rooted in the ground and have stems and leaves, for example. And all flowers and seeds have the same structures; only the size, color, and shapes differ. Likewise, all animals have basically the same anatomic features. The arms of a human, the front leg of a dog, and the flipper of a seal all contain the same bones — an upper "arm," an elbow, a lower "arm," and five "fingers." The only differences are in size and shape. Humans, worms, insects, and bunnies all have similar digestive systems. Humans, horses, goats, and frogs all have hearts and circulatory systems. A neuron in a human is no different than a neuron in any other animal. Looking at the similarities among different animals gave rise to the field of *comparative anatomy*. And, if you read Chapter 14, you find out that the genetic code — the code that states which codon (order of three bases from DNA) signifies which amino acid should be produced — is exactly the same for all species on earth. Many findings from the field of genetics provide proof for evolution.
- **The fossil record tells an important story.** When Darwin saw for himself that different layers of rock contained different fossils (he had read this fact when he studied geology), he was struck by how different the older fossils at the bottom were from the "newer" fossils at the top. The simplest, and now extinct, species were at the bottom, more complex forms in the middle, and forms resembling current species at the top. Darwin noticed that the oldest, simplest forms disappeared as you moved your way up the rock formations. Of course, one rock formation does not contain all species that ever existed. Fossils from layered rocks all over the world must be examined. Studying fossils made him realize that something caused species to develop from simple organisms and then change. As the changes became more complex, the simple forms became extinct and other diversifications became evident. These changes took place over millions of years, of course, but the impact of the information is clear: Species came into being as simple organisms, changed over time, and became more complex, leading to present-day species that have in common that simple ancestor from billions of years ago. Interest in this area of science spawned the field of *paleontology*.

Darwin's Other Theory: Survival of the Fittest

Darwin did find himself having to defend his observations. The skeptical fundamentalist society he lived in did not easily come around to his way of thinking. They did eventually, but it wasn't easy at the beginning. Darwin had to explain how species could have changed over time. He came up with the theory of *natural selection,* which still holds water today. His theory of natural selection is often described as the "survival of the fittest" theory.

Darwin explained that an organism that had traits superior to another organism of the same species was simply showing better adaptation to its environment. This superiority increased the organism's *fitness*, which Darwin described as an organism's ability to survive and produce offspring. In Darwin's time, fitness had nothing to do with washboard abs and a low percentage of body fat.

When a trait improves the survivability of an organism, the environment is said to favor the trait or naturally select *for* the trait. Selection acts *against* unfavorable traits, which are considered to show a lack of adaptation to the environment. The most fit individuals are "selected" to survive (survival of the fittest). Darwin had several strong arguments for his theory of natural selection.

First, a species has the potential to increase the population by leaps and bounds, but population sizes remain about the same because not every baby that is born survives. Not every baby survives because of several factors related to natural selection. The amount of available resources does not increase with the size of the population, which means that as the sources of food and water (and light, in the case of plants) remain the same but a population size increases, increased competition for resources occurs. The competition is survived only by the most fit individuals because they have adapted traits best suited to the environment and are able to outcompete other individuals for resources. As the improved traits increase, the most well-adapted individuals produce offspring who inherit the improved traits. And those offspring produce offspring that have inherited the adaptive traits. Over time, the adaptive traits become more dominant, and evolution occurs. Table 16-1 summarizes the four main types of natural selection.

Table 16-1 Types of Natural Selection

Type of Natural Selection	*Effect*
Stabilizing selection	Eliminates extreme or unusual traits. Individuals with the most common traits are considered best adapted, which maintains the frequency of common traits in the population, and over time, nature selects against extreme variations of the trait.
Directional selection	Traits at one end of a spectrum of traits are selected for, whereas the other end of the spectrum are selected against. Over generations, the selected traits become common, and the other traits become more and more extreme, eventually being phased out.

Type of Natural Selection	*Effect*
Disruptive selection	The environment favors extreme or unusual traits and selects against the common traits. One example is the height of weeds in lawn grass compared with in the wild. In the wild, natural state, tall weeds compete for the resource of light better than short weeds. But in lawns, weeds have a better chance of surviving if they remain short because grass is kept short.
Sexual selection	Females increase the fitness of their offspring by choosing males with superior fitness; females are concerned with quality. Males contribute most to the fitness of a species by maximizing the quantity of offspring they produce. Because males are concerned with quantity, competition between males for opportunities to mate exists in contests of strength. Therefore, structures and other traits that give a male an advantage in a contest of strength have evolved, including antlers, horns, and larger muscles. Because females choose the mate, males also have developed traits to attract females, such as certain mating behaviors or bright bird feathers.

Artificial selection is performed by humans (therefore, it doesn't fall under "natural" selection). Humans may have a desire to breed animals with certain traits or cross plants for certain effects; this practice often is seen in the farming industry. During the Roman Empire, wild dogs were domesticated by humans. Since then, humans have created many of the current dog breeds by mating two types of dogs with desirable traits.

Mitochondrial DNA Provides a Link to the Past

Evolution is even apparent in molecules. Certainly, comparative anatomy has allowed scientists to find out how vertebrate limbs evolved from amphibious (water-living) creatures. But with the technology available now, Darwin would have been amazed to learn that almost all living things contain DNA as their genetic material, and that there is one genetic code contained in the DNA of all organisms.

In addition, all living organisms convert food sources to energy in the form of ATP. Whether the organism is a flower taking in nutrients and water from the soil and light from the sun, a lion chomping down a wildebeest, or a human consuming a gourmet meal cooked by Emeril Lagasse himself, the foods are broken down by a digestive system, and the nutrients are transported by a circulatory system to every cell in the organism, where they are converted to the molecule ATP in organelles called the mitochondria. ATP is then used to provide energy for cellular processes. In the cellular processes of any organism, ATP fuels the production of proteins, and the production of proteins is directed by the genes on strands of DNA.

Biochemists have analyzed proteins in an attempt to learn more about how genetic material is expressed in organisms. If organisms produce similar proteins, they must contain similar genes. These types of studies have been valuable in studying the evolutionary relationships between species and determining common ancestors. The more recently that species have evolved from a common ancestor, the more proteins (and genes) they will have in common. This sameness is referred to as *homology*. Likewise, the fewer homologous proteins and genes that organisms share indicate a longer period of time since they branched off the evolutionary tree.

Proteins associated with DNA molecules are called *histones*. In one of the histone proteins, the amino acid sequences produced are identical in many, many organisms. This fact points to a single organism serving as the ancestor to all the organisms currently populating the earth. Also intriguing is the fact that these proteins and gene sequences have not changed in millions of years; they are said to be *highly conserved* sequences.

One of these highly conserved sequences produces a protein called *cytochrome c*, which is part of the electron transport chain that occurs in mitochondria. Humans and chimpanzees have exactly the same amino acid sequences in their cytochrome c proteins, which indicates that humans and chimpanzees branched off the trunk of the evolutionary tree very recently ("recently" in evolutionary times is still quite a long time, about 6 million years in this case). The cytochrome c protein in rhesus monkeys differs from humans and chimpanzees by just one amino acid (out of a total of 104).

One thing that can alter what protein is produced is a mutation. (The details of DNA replication and protein synthesis are given in Chapter 14.) Although mutations sound like horrible things, they actually occur quite regularly. They occur so regularly, in fact, that mutation rates for genes exist. Mutations are what allow organisms to change slowly. Yes, sometimes mutations can cause disastrous results, but most of the time they go unnoticed. However, they do fuel selection for certain traits.

Scientists can use mutation rates of genes to calculate how long it has been since two species diverged. And to further prove Darwin's theories, the amount of time calculated through mutation rates of genes matches the dates that paleontologists have come up with by dating fossils.

So, Who Were Your Ancestors?

To guide you through this discussion, I must mention Charles Darwin. Besides *The Origin of Species*, he wrote other bestsellers, including one called *The Descent of Man*. In his book about the evolution of humans, Darwin suggested that humans descended from ancestral species found in Africa. This idea was held for a while, but fell out of fashion in the late 19th and early 20th century.

In 1891, researcher Eugene Dubois discovered a few bones found in Java, Indonesia, a large island off the southeast coast of Asia. Calling the discovery Java Man, Dubois thought he found the link between ape and man. What he found certainly was an ancestor to modern *Homo sapiens* — the genus and species name for humans — but it was not apelike. What he found was actually a member of the species *Homo erectus*, one of the earliest walking hominids. Other Homo erectus bones have been found in China, but they have also been found in Africa. The controversy, or rather the question posed, is did the *Homo erectus* species evolve in Africa and then travel to Indonesia and Asia, or was it the other way around? What makes this question so difficult to answer is that the age of the bones of both Java Man and the bones found in Kenya, Africa, are 1.8 million years old. They are very close in age, but were found 10,000 miles apart! So, who did the walking?

During the 1930s, a researcher named Raymond Dart examined a small skull that was found in Taung, a town in South Africa. After studying the bone structure and realizing that inside the skull was a petrified brain, Dart came to the conclusion that the skull belonged to a child who was about 6 years old and was a member of a human ancestral species. The remnants were called Taung Child. Dart thought *he* had found a missing link between apes and men.

Others disagreed. Dart was ridiculed for suggesting that a human ancestor was "out of Africa," when the thinking at that time was that the first human species was from Asia because of the hullabaloo surrounding Java Man.

But Dart persevered, even though his belief was unpopular at the time. He classified his skull as *Homo habilis*, meaning handyman, because crude stone tools were found near the bones. Eventually, Dart's views were vindicated.

In the 1930s, Louis and Mary Leakey began excavating in Olduvai Gorge in Tanzania, Africa. In the early 1960s, the Leakey's son, Richard, noticed the jaw of a saber-toothed tiger sticking out of the archaeological site. Digging continued in the area, and eventually pieces of three skeletons were unearthed. The skeletons were also classified as those of *Homo habilis* and were dated at about 2 million years old. The Leakeys continued their work at Olduvai Gorge well into the 1980s, and in 1984, they unearthed a spectacular find: the first (and still the only) full skeleton of a *Homo erectus*. The *Homo erectus* bones that the Leakeys dug up were dated at 1.6 million years old.

A 1.5 million-year-old *Homo erectus* specimen that Richard Leakey discovered in 1974 revealed information about the socialization of human ancestors. On the leg bones of this female skeleton were signs of disease that were later determined to be caused by vitamin A poisoning.

Vitamin A poisoning occurs in people who eat too much liver, because the liver filters vitamin A. Symptoms include loss of hair and teeth, aching joints, and blood clots that calcify into lumps of bone. When the lumps of bone form, an affected person is unable to walk.

Because the lumps of bone were so prevalent on the Homo erectus bones, researchers determined that someone took care of this woman — bringing her food and water and protecting her from predators — until she died of natural causes, most likely the vitamin A poisoning. So, about 1.5 million years ago, humans were forming social bonds. As evolution continued, physiological changes required the formation of stronger social bonds, leading to family units. Hang on, the changes are astounding.

- As human ancestors began to walk upright, they soon began to hunt. So, instead of being herbivores as they were in the forests, they became carnivores consuming animals.
- One factor that led to this development was climate change. As the earth began to warm up, some of the forests disappeared and became open savannas. In an open savanna, it's much easier to see prey (especially if you are standing). So, the predecessors became successful hunters and ate plenty of meat.
- Eating plenty of meat — with the fats and proteins that meat contains — made their brains bigger, and bigger brains were "selected for" over time. As their brains got bigger, they developed tools and began to wonder what was across that wide-open savanna. The ancestors, armed with their tools, began to travel. One hominid alone did not cover the large distances between sites where bones were found. It took thousands of years for populations to develop across the miles. One estimate is that it took 25,000 years for human ancestors to populate their way from Africa to Indonesia and Asia.
- As the shape and size of brains changed and enlarged, ancestral females had to give birth earlier so that the head could fit through the pelvic bones. Because babies were born earlier, they were much more dependent on their mothers for a longer period of time. This change meant that the mother could not contribute to hunting, but she still needed adequate nutrition to make milk to breastfeed her baby. So, the father and other members of the clan had to help the mother by bringing her food. The mother having to rely on others for her survival and that of her baby led to close ties with the others.

In the years since Dart and the Leakeys, several more fossilized bones have been discovered all along the southeastern coast of Africa, giving Africa the nickname the "cradle of civilization." The oldest remains of a hominid — an upright-standing, walking human ancestor — belong to Lucy, a 3.2 million-year-old skeleton of the *Australopithecus afarensis* species. She was found in Ethiopia in northern Africa in 1974 by Don Johanson.

Richard Leakey's wife, Maeve, headed upstream from Olduvai Gorge to Kanapoi in northern Kenya, and found a 4.2 million-year-old ancestor. The lower jaw was complete, and the teeth are surprisingly like those of a human today. However, the shape of the jaw is like that of a chimpanzee. Pieces of lower leg bone were also found, and indicated that the creature walked on two legs. Maeve named it *Australopithecus anamensis*. This 4.2 million-year-old creature was an ancestor of the species to which Lucy belonged.

There are bones dating back 16–20 million years, but those bones indicate more apelike features, such as all four limbs being of equal size. Humans are very similar to chimpanzees. In fact, 99 percent of the DNA in humans matches that in chimpanzees. But, only 97 percent of the genetic material of humans is shared with gorillas. Today, the current line of thinking is that there was an apelike species 10–20 million years ago that branched into a line of gorillas around 7 million years ago, then two lines around 5–6 million years ago: one that led to chimpanzees and the other that evolved into humans. Table 16-2 gives you an overview of changes that occurred from apes to humans. Table 16-3 reviews the changes in brains that have occurred in human ancestors.

Table 16-2 Changes from Apes to Humans Over the Past Few Million Years

Anatomic Structure	*Changes*
Arms	Because apes walk on all four limbs, their front limbs do not straighten completely, or they would suffer dislocations. So apes do not have elbows, which allow the arm to be straightened, but humans do.
Brains	Modern humans have prominent foreheads. The size and shape of the skull has changed as the size and shape of the brain changed. Human brains are now larger and more rounded than those of ancestral species. And the bony ridge above the eyebrows of humans has significantly shrunk in comparison with the predecessors.
Feet	Now that humans walk upright, the shape of the heel has changed to absorb the impact of the foot hitting the ground differently.

(continued)

Table 16-2 *(continued)*

Anatomic Structure	***Changes***
Hands	The hand of a human and a chimpanzee are amazingly similar. The anatomic structures are the same; the differences lie only in the fingerprints. Humans and other primates have prehensile (grasping) thumbs, which allow the gripping of objects. (Think about it; it wouldn't be easy to grasp without a thumb.) Prehensile thumbs appeared in human predecessors about 18 million years ago.
Jaws	The human jaw and teeth have shrunk. Now that humans cook food (instead of eating it raw), teeth do not have to tear and grind as much. Instead, humans have developed chins to help support the thinner jawbone. The changes in the jaw and flattening of the face have allowed humans to produce language.
Knees	The knee allows humans to walk upright. The ability to straighten the leg supports the weight of the body, and because the knee is positioned under the pelvic bones (instead of in front of them), humans don't waddle during movement. Waddling slows a human down, and humans need to run at times.
Tails	Apes do not have tails, and humans no longer have them, either. This anatomic feature disappeared about 25 million years ago. However, the remnants of a tail are evident in what is the coccyx bone at the end of your spine.

Table 16-3 Evolution of Brains from 4.5 Million Years Ago to Today

Genus and Species Name	***Developments***
Australopithecines anamensis, Australopithecines afarensis	Brains were about 400 centimeters in size, comparable to that of chimpanzees or gorillas. Could walk on two legs (bipedal), but were intellectually apelike.
Homo habilis	Brains were about 650 cc. Used stone tools.
Homo erectus	Brains were 850–900 cc. Began socialization.

Genus and Species Name	Developments
Homo neanderthalensis	Brains of 1300 cc, but also larger bodies, which required larger brains for control.
Homo sapiens	Brains between 1200–1600 cc. Larger frontal lobes and broader foreheads because of increased brain capacity.

Although there is just one human species now, several hominid species supposedly populated different areas of the earth at the same time millions of years ago. Currently being studied are bones from a 4.4 million-year-old skeleton found in Ethiopia. It is called *Ardipithecus ramidus* and, as the oldest known ancestral fossils, they are being studied to determine if this organism was in fact a direct ancestor to humans. Studies are also being done on bones found in North America. As scientists discover more about where humans came from and where they have been, you and I just may find out where the species — creationists and evolutionists alike — is headed. Stay tuned, and continue that brain development, please.

Keeping Up with the Joneses

If one species evolves because it adapts to environmental changes, it may develop to the point where it can escape a predator. But, then what happens to the predator if it can't catch its prey? It becomes extinct if it can't adapt to the change. In the pattern of *coevolution*, predators and prey evolve along with each other, each trying to outdo or adapt to change in the other. As plants evolve, plant-eating insects evolve. As flowering plants evolve, so do bees. As the immune system of animals evolves, so do the bacteria and viruses that attack them.

In *divergent evolution*, two or more species develop from a common ancestor, like chimpanzees and humans developing from an ancient line of apes. Divergent evolution can occur when there is a *geographic barrier*, such as a large body of water that cannot be crossed or a mountain range. When the populations cannot get together to mate, they evolve into separate species. Divergent evolution also can occur when there is no geographic barrier but populations become *reproductively isolated* for other reasons. One example would be if hybrids form and the genes of the hybrid allow it to adapt even better than the parents. Or, divergent evolution can be caused by *adaptive radiation,* which is rapid evolution of many species from a single ancestor. When the original species spreads out geographically, environmental conditions may change, causing the spread-out populations to evolve differently. Did you read about Darwin's finches?

Okay, So How Did the World Form in the First Place?

Evolution certainly occurs among species, and change is happening all the time. But, how did life form to begin with? Where did it all start? This branch of evolution is called *chemical evolution*. You may have heard of the Big Bang Theory of physics. With biology, you get the *heterotroph theory*, which states that the first forms of life were heterotrophic cells. Heterotrophs are organisms that cannot make their own food. So, with a will to live and a desire to eat, heterotrophs sought out food to survive. Here's a quick overview of the timeline.

Lamarck's advice: Use it or lose it

Jean Baptiste de Lamarck, a contemporary of Thomas Jefferson, was a little off the mark when he suggested that traits acquired by parents during their lifetime were passed onto offspring. For example, Lamarck explained that giraffes had long necks because as parents stretched to reach high branches, their "reach" was passed to their offspring. As the offspring stretched to reach higher, the ability to reach higher than their parents was then passed onto their own offspring. It just doesn't work like that. Lamarck called this his *theory of natural transformation*. It has since been proved incorrect.

But, at the time — a good 50 years prior to Charles Darwin's *The Origin of Species* and Gregor Mendel's pea plant experiments — scientists were trying to explain how traits were inherited and how evolution could occur.

Lamarck was on the mark, however, when he suggested that adapting to one's environment played a role in the origin of new species and in evolution. And, he was correct when he stated that changes were very gradual.

But, Lamarck is most often remembered for his theory on the use and disuse of body parts. Lamarck's theory is that body parts, such as muscles, develop the more they are used and atrophy (or weaken) if they are unused. If you perform any type of exercise or play any sport regularly, and then stop for a long period of time, you know this is true. The more you repetitively use muscles, the stronger and more energy efficient they become. This includes not only skeletal muscles but the heart and lungs as well. Even the digestive tract functions better when it is filled with fiber that makes it work harder.

If you do not use your body well, its capability of functioning at optimal capacity declines. Apparently, this theory holds for brain cells as well. Recent research has shown that people who keep their mind active, such as by playing games, reading, playing musical instruments, or doing crossword or jigsaw puzzles have a lower risk of developing Alzheimer's disease or senility. So, use your body well, or lose its amazing capabilities.

1. **The earth and the atmosphere formed** (this is where the Big Bang Theory comes in to play) from volcanoes releasing lots of gases but no oxygen.
2. **The primordial seas formed** as the earth cooled down from the explosions. The gases previously produced formed water and minerals (inorganic molecules).
3. **Complex organic molecules formed** when energy from the sun, lightning, heat, and radioactivity struck the inorganic molecules and water. Some of these organic molecules were amino acids. Oxygen was not present, or these reactions would not have occurred.
4. **Polymers of amino acids formed and molecules started to replicate themselves.** A polymer of amino acids is a protein, and proteins were created in the organic primordial soup. And, when proteins formed, DNA formed.
5. **Organic molecules became concentrated in cell precursors.** Chemical reactions occurred in the probionts (the cell precursors), but the probionts could not reproduce. Borders, like cell membranes, formed to separate where the reactions occurred and the place to where the products of the reactions went.
6. **Hetrotrophic prokaryotes formed.** These true cells consumed (ate, ingested) organic substances. These cells reproduced by dividing, eventually increasing the competition for the organic material that was available. The "survival of the fittest" theory held even way back in the beginning of world. The cells that were the best at obtaining food went on to reproduce; natural selection had begun.
7. **Autotrophic prokaryotes formed.** A mutation occurred that gave a heterotroph the ability to feed itself; that is, it became capable of producing its own food. This mutation led to photosynthesis, in which plants create their own food by absorbing water and minerals from the soil and absorbing light energy from the sun.
8. **Oxygen was produced by photosynthesizing bacteria.** As the autotrophic bacteria began producing their own food, they took in carbon dioxide and released oxygen as a waste product. The oxygen accumulated in the atmosphere, creating the ozone layer when oxygen and ultraviolet light mixed together. Because the oxygen and subsequent ozone layer absorbed the ultraviolet light, there was not enough ultraviolet energy to create molecules from the organic "soup," and the primitive cells died off.

9. **Eukaryotic cells formed through mutually beneficial relationships among prokaryotic cells.** According to the *endosymbiotic theory*, mitochondria, chloroplasts and other organelles may have been early prokaryotic cells. They contain structures that allow them to produce energy. When they entered other prokaryotic cells (maybe they were ingested?), a *symbiotic* (mutually beneficial) *relationship* was formed. The mitochondria and chloroplasts then produced energy for the prokaryote that was housing them. The beneficial relationship is that the larger prokaryotic cell gained energy from the smaller "prokaryotic cells," and the smaller prokaryotic cells gained protection and a host that would provide raw materials from which energy could be produced. The arrangement worked well, so it stuck. What formed from that relationship is eukaryotic cells — the type that that is in every plant and animal on earth today. Natural selection, mutations, and evolution all took place over billions to develop the diverse *flora* (plants) and *fauna* (animals) on the planet now.

Part VI
Ecology and Ecosystems

In this part . . .

Even if you are a loner who prefers solitude, you must realize that you never are actually alone. You share a world with, to paraphrase one of my favorite scientists — the late Carl Sagan — billions and billions of other organisms. All the other people, plants, and animals that you see on a daily basis are just a tiny fraction of the amount of planet-mates you have. Your own body is not even completely your own! You have bacteria that use you as a host and live on your skin, in your mouth, under your fingernails, and in your intestines. I hope for your sake that they are proper houseguests! Even tiny mites live in your bed and eat the skin cells you slough off at night — yes, bedbugs.

In this part of the book, you read about how the multitude of organisms manage to live in relative harmony on earth. You find out about all the tiniest organisms: bacteria, viruses, and insects. And, you take a look at what some biologists are working on now, including looking for organisms on other planets and some current genetic research.

Hopefully, you will come to understand that bacteria, viruses, and insects are not all bad when you realize that you actually need them to survive as much as they need you. Even though it may not seem like it at times, the natural world really is a cooperative place. Please, keep it healthy and try not to damage your environment. Oh, yeah, one more thing. Don't let the bedbugs bite!

Chapter 17

Sharing the Globe: How Organisms Get Along

In This Chapter

- Seeing how organisms are distributed around the world
- Finding out how organisms interact with their environment
- Figuring out how habitats support many organisms
- Examining how predators are just a natural part of the big picture
- Seeing how organisms help each other to survive
- Helping to keep the world in balance

You've lived on this planet a while now. I'm sure that you have noticed that different parts of the world have different climates. Arizona is a far cry from Antarctica. A cornfield is vastly different than a beach. The mountain forests of Pennsylvania are much cooler than an African savanna. You get the picture.

Now I'm sure that you have also noticed that different animals inhabit those different areas: lions, elephants, and antelope on the savanna; black bears and blackberries in the temperate forests; prairie dogs, scorpions, and cacti in the desert; and penguins and polar bears in Antarctica.

How do such diverse organisms live together in the world? *Ecology* is the branch of biology that deals with this question, and it's the focus of this chapter.

Populations Are Popular in Ecology

Technically, *ecology* is the study of how many organisms live in certain areas, how the organisms interact with other organisms, and how they relate to their environment.

A *population* is a group of members of a certain species living in the same area. So, it is possible to measure how many members of the *Homo sapiens* — that is, human — species are living in Buffalo, New York, and it is possible to measure how many buffalos are living on a prairie in Wyoming.

When several populations occupy the same area, a *community* develops. For example, when my family and I lived in the tiny village of Germania, Pennsylvania, there was a small population of only about 120 humans; larger populations of deer, bear, turkey, coyotes, bobcats, raccoons, and skunks; and small populations of elk and eagles. But all of these species occupied several square miles of a community among beautiful mountain forests and farmland. And quite a community it was!

The word *ecosystem* describes how the varied populations in a community get along in the physical environment. Ecosystems take into consideration the *habitats* of organisms, which are the type of place that the organisms prefer to live (such as desert, woodlands, rocky mountainous regions), as well as the actual quality of the environment (such as freshwater or saltwater, type of climate or soil).

Population ecology

Ready for some math in a biology book? *Population ecology* is the branch of ecology that deals with determining how many organisms exist in a specific group. And it uses lots of statistics to do it.

The size of a population is looked at in terms of its *density* — that is, how many organisms occupy a specific area. For example, Atlanta, Georgia, contains approximately 4.5 million people, which gives you the population of the general area. But the population density can be determined for smaller areas around metro Atlanta. For example, according to data supplied by the Atlanta Regional Commission, in the suburbs of north Fulton County, 1,164.8 people live in one square mile. Further south in Fulton County, 4,992 people live in one square mile. The same kind of measurements can be determined for how many bees or ticks exist in a two-block radius or how many deer live within five square miles.

Looking at how a population is distributed throughout a certain area is called *dispersion*. Organisms may be dispersed in a *clumped* fashion, like bees in a hive or ants in a hill. Dispersion can be *uniform,* like grapevines in a vineyard or rows of corn plants in a field, or it can be *random*.

Several statistics are looked at to determine *population growth*. *Age structure* provides the number of organisms for each age group, and *survivorship curves* explain at what stage of life most of the organisms of a species die. Humans have a type I survivorship because most individuals survive to middle age (about 40 years) and beyond. Type II survivorship curves show randomness in which death can occur at any age. Mice have type II survivorship; they never know when the cat or trap will strike. Species with type III survivorship have few members that survive to reproductive age. Type III organisms die young. Species such as frogs that produce offspring that must swim on their own as larvae fit into this category. Other animals eat many of the larvae before they reach the adult stage and can reproduce. These different levels of survivorship keep the populations of organisms on a fairly even keel. If all the larvae produced survived, the oceans would be filled overpopulated with organisms. If all mice lived to a ripe old age, you'd see them a lot more often. You wouldn't want that, would you?

The maximum growth rate of a population under ideal conditions is called the *biotic potential*. This factor must be looked at to determine the maximum population size that could be produced by a species. The biotic potential assumes that no competition for resources such as food or water exists, and that no predators or diseases affect the growth of the organisms. Other factors involved in determining the biotic potential include the age of the organisms when it is able to reproduce, how many offspring are usually produced from one successful mating, how often the organisms reproduce, how long a period of time they are capable of reproducing, and how many of the offspring survive to adulthood.

Taking all those factors into consideration, it has been determined that if two elephants mated under ideal conditions, and all the preceding factors mentioned were fulfilled, that in just 2,000 years, the weight of all the descendants of those two original elephants would be more than the weight of the entire earth. Good thing elephants don't live 2,000 years! Death ensures that populations do not exceed the earth's capacity to support them; in the case of elephants, I mean that literally!

Zero population growth, in which populations of organisms remain stable, occurs when the birth rate and death rate are equal. If the death rate were higher than the birth rate, the population would be declining. If the birth rate is consistently higher than the death rate, overgrowth can occur, which increases the competition for resources.

The *carrying capacity* is the maximum amount of organisms of a single population that can survive in one habitat. If the carrying capacity is exceeded (think of an elevator having too many people and too much weight in it), the population goes down. Nature wins in the end. The habitat can be damaged if too many organisms are living in it, which further reduces resources and leads to death of some organisms. This situation decreases the population so that the carrying capacity is met; however, with a destroyed habitat, the carrying capacity is lowered even further, necessitating more deaths to restore balance.

Human population growth is beginning to exceed the available resources, as you can read about in the next section.

Humans are increasing exponentially

Up until only 1,000 years ago, human population growth was very stable. Food was not as readily available as it is now, no antibiotics fended off invading bacteria, no vaccines fought against deadly diseases, and no sewage treatment plants ensured that water in rivers and streams was safe to drink. People didn't shower or wash their hands as often, so they were capable of spreading incurable diseases. These factors increased the death rate and decreased the birth rate. In addition, people didn't have electricity or natural gas for heating and cooking, and they didn't have insulated homes and synthetic fibers to keep them extra warm and toasty. Therefore, they were not able to survive in some environments, so people had fewer habitats available to populate.

But now, especially in the last 100 to 200 years since the scientific and industrial revolutions, the food supply has increased, people are able to survive in very hot and very cold environments due to technology, and hygiene and medicines have reduced deaths due to common illnesses and diseases. So more people are born, and more of them survive well past middle age. The population of humans has experienced exponential growth, which means the reproductive rate of humans has increased, and the size of the population has risen quickly.

What does this mean for other organisms in the world? What does it mean for humans? Will there be competition for resources soon so that Darwin's "survival of the fittest" theory will be tested? Will humans be forced to adapt to a changing environment, leading to further evolution of the species? Ecologists try to answer these questions.

When a population is *stressed,* meaning when overcrowding has occurred, and the competition for resources has increased, certain physiological reactions can occur in some of the members. Certain hormones that are released in response to stress can harm immune systems. The weakened immune system then cannot fight off disease, and death due to illness rises. Females may not be able to reproduce, and, if they do reproduce, they may not be able to nurse or care for their offspring, causing the offspring to develop abnormally and reducing their ability to reproduce. All in all, a stressed population starts to reduce its own numbers. And ponder this question a while: Aren't some of these characteristics apparent in the human population now?

Another way that animals respond to overcrowding is to *disperse*. If a population is increasing beyond the carrying capacity, some of the members of the species will leave and move to a new habitat. The members who survive in the new habitat start a new population. Dispersal is thought to have played an important role in the evolution of humans. When the clans of human ancestors began to get too big for their own savanna, some of them moved on to new locations. Populations sprang up along the trail of their nomadic wandering, as is evident in the fossilized skeletal remains of human ancestors found along the coast of east Africa, up and across into Asia, back to Africa, and over to what is now Europe (see Chapter 16). Eventually, humans made it to North America, South America, and Australia, and populations continued to grow. But now, humans cover the entire planet. Not much room is left for dispersal.

Ecosystems, Energy, and Efficiency

When scientists study an ecosystem, what they are trying to learn is how well energy is produced and used in the system. Ecosystems involve all organisms in a community — plants, animals, and bacteria — and the way that each member of the ecosystem uses and produces energy must be taken into account.

To look at an ecosystem accurately, the organisms must be divided into *trophic levels*. Trophic levels (or food chains) describe what the organisms consume as fuel. The word *trophic* stems from the Greek word for *nutrition*.

Autotrophs, which create their own food, harness energy from the sun and turn it into chemical energy. Plants and photosynthetic bacteria fall into this category. They make energy available to get *food chains* started, so they are called *primary producers*. They are not consumers because they do not consume any other organisms; they take in only the ingredients from which food is made (such as water, carbon dioxide, sunlight, and minerals).

Primary consumers are the organisms that eat the primary producers. Because the primary producers are mainly plants, primary consumers are *herbivores* (plant-eating animals).

Secondary consumers eat the primary consumers. Because the primary consumers are animals, secondary consumers are also called *primary carnivores* (meat-eating animals).

Tertiary consumers eat the secondary consumers. Food chains usually do not go much beyond this point because the amount of energy available in the food is reduced with each level of consumer. Past the tertiary consumer level, too much energy is depleted from the food for it to be of good use to any organism. The amount of energy at each trophic level in proportion to the next trophic level is called *ecological efficiency.*

Detritivores and *decomposers* aren't very picky eaters. They get their energy from consuming dead plants and animals, which comprise the *detritus.* Detritivores eat dead organic matter, such as vultures and earthworms. Decomposers are bacteria and fungi that help to break down dead organic matter.

Decomposers and the Biogeochemical Cycles

Boy, put biology, geology, and chemistry together, and you get *biogeochemical!* When you talk about the "circle of life," the circle to which you are referring is a *biogeochemical cycle.* The plants and animals that live and then die are the *bio* part; the earth that they decompose into comprises the *geo* part; and the process by which organic matter returns to the chemical elements in the earth is explained by the *chemical* part. There are four biogeochemical cycles, and each of them returns to the earth important elements that are required in living organisms.

The hydrologic (water) cycle

Plants absorb water from soil, and animals drink water or eat animals, which are made mostly of water. When plants go through the process of transpiration (check out Chapter 7), they give off water. When animals create perspiration, they release water, which is evaporated into the atmosphere. Water also is released from plants and animals as they decompose. Decomposing tissue becomes dehydrated, which is what causes the dried-out tissues to break down and fall off into the soil. As water evaporates into the air, wind moves air over bodies of water, and precipitation (rain, snow, sleet, hail) releases water into larger bodies of water such as lakes, rivers, oceans, and even glaciers. Water from precipitation and decomposing tissue also gets into groundwater, which ultimately supplies larger bodies of water.

The carbon cycle

Plants take in carbon dioxide for photosynthesis. Animals consume plants or other animals, and all living things contain carbon. Carbon is what makes organic molecules organic (living). Carbon is necessary for the creation of molecules such as carbohydrates, proteins, and fats. Plants release carbon dioxide when they decompose. Animals release carbon dioxide when they decompose or respire. (Animals take in oxygen and release carbon dioxide when they breathe.) Carbon dioxide also is released when organic matter such as wood, leaves, coal, or oil are burned. The carbon dioxide returns to the atmosphere, where it can be taken in by more plants that are then consumed by animals. Decomposing animals and plants leach carbon into the ground, forming fossil fuels such as coal or oil. Peat also forms from the decomposition of organic matter. Some carbon is stored in the form of cellulose in the wood of trees and bushes.

The phosphorus cycle

ATP, that ubiquitous energy molecule created by every living thing, needs phosphorus. You can tell that by its name; *triphosphate* indicates that it contains three molecules of phosphate, which requires phosphorus. DNA and RNA, those ubiquitous genetic molecules present in every living thing, have phosphate bonds holding them together, so they require phosphorus, too, as does bone tissue. Plants absorb inorganic phosphate from the soil. When animals consume plants or other animals, they acquire the phosphorus that was present in their meal. Phosphorus is excreted through the waste products created by animals, and it is released by decomposing plants and animals. When phosphorus gets returned to the soil, it can be absorbed again by plants, or it becomes part of the sediment layers that eventually form rocks. As rocks erode by the action of water, phosphorus is returned to water and soil.

The nitrogen cycle

Amino acids have a basic core of $-NH_2$ (an amino group) in their structure. And because amino acids build proteins, nitrogen is pretty important. Nitrogen also is present in the nucleic acids DNA and RNA. Life could not go on without nitrogen. The nitrogen cycle (Figure 17-1) is the most complex biogeochemical cycle because nitrogen can exist in several different forms. Nitrogen fixation, nitrification, denitrification, and ammonification are all parts of the nitrogen cycle.

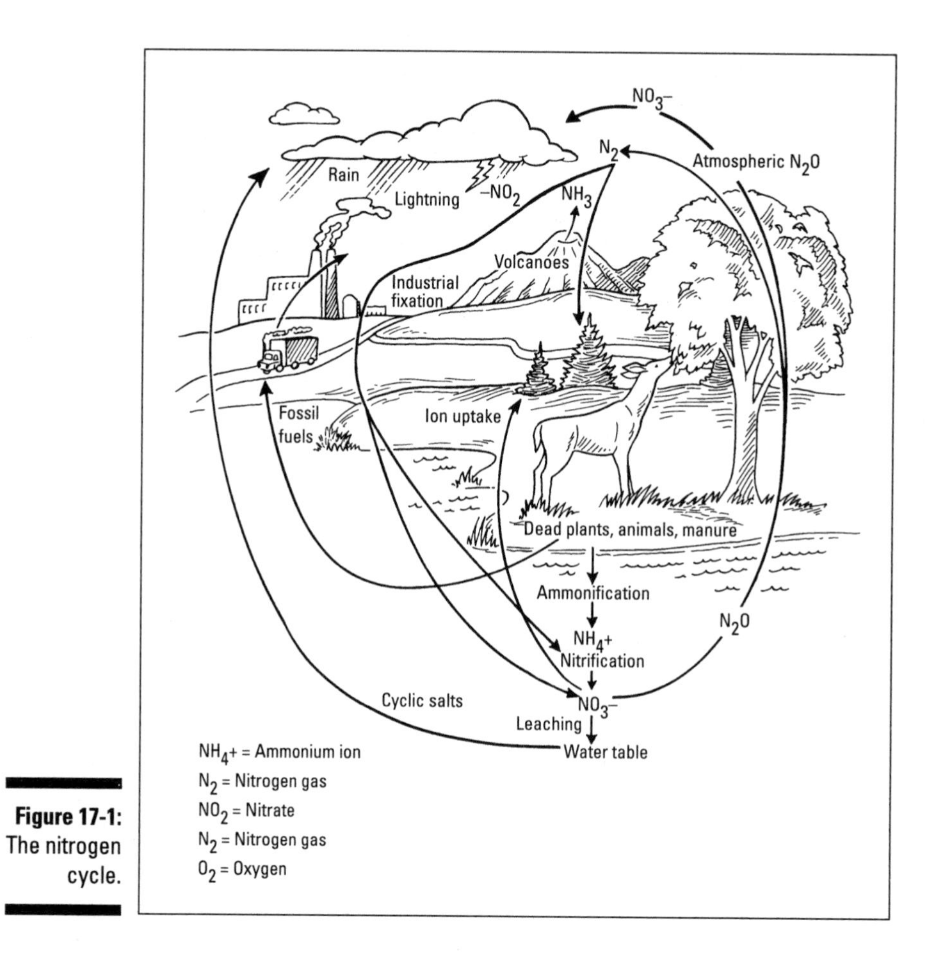

Figure 17-1: The nitrogen cycle.

- **Nitrogen fixation:** In the soil, as well as in the root nodules of certain plants, nitrogen is "fixed" by bacteria, lightning, and ultraviolet radiation. The "fixing of nitrogen" does not mean nitrogen was broken; a better term might be "fixated," because the bacteria put elemental nitrogen (N_2) into a form that can be used by living organisms (NH_4^+ or NO_3^-) and do not allow it to leave that form and revert to elemental nitrogen.
- **Nitrification:** Certain bacteria take the forms into which nitrogen was fixated and further process it. These bacteria oxidize NH_4^+, which changes it to NO_2^-. Oxidation provides energy for the nitrogen cycle to take place — the bacteria that live in soil cannot harness energy from the sun. The energy they use during their work in the nitrogen cycle has to come from somewhere. The NO_2^- is further processed to NO_3^-.
- **Denitrification and ammonification.** Plants absorb nitrates (NO_3^-) or ammonium (NH_4^+) ions from the soil, and turn them into organic compounds. Animals obtain nitrogen by consuming plants or other animals.

Therefore, the waste products of animals contain nitrogen. Ammonium ions, ammonia (NH_3), urea, and uric acid all contain nitrogen. So regardless of what form of excretion an animal has, some nitrogen is released back into the ecosystem through excrement. Dead plants and animals are food for decomposing bacteria. Some decomposers convert NO_3^- to nitrogen gas, which is released into the atmosphere, through a process called *denitrification*. Other decomposers convert organic compounds to NH_4^+ through a process called *ammonification*. The NH_4^+ ions are stored in the soil, as are NH_3, NO_2^-, and NO_3^-.

How Humans Affect the Circles of Life

All the places on earth that contain living things are cumulatively called the *biosphere*. The biosphere does not exclude many places on earth. And many of the things that humans do to survive cause damage to the biosphere.

Humans start fires. Burning wood and fossil fuels produces carbon dioxide. The more carbon dioxide in the atmosphere, the more heat is held in the atmosphere. The increase in atmospheric heat causes the *greenhouse effect*, in which temperatures around the world are increasing. Weather patterns are affected, which can disrupt ecosystems and cause some species to disperse or migrate, further disrupting ecosystems. And changing weather patterns also affect agricultural yields. If less food is produced, more competition occurs for resources. And warmer temperatures cause some of the polar ice to melt faster, which could potentially raise the sea level, eliminating some habitats.

Another effect of the burning of fossil fuels is an increase in air pollution. Air "pollution" consists of sulfur dioxide and nitrogen dioxide — two dioxides that shouldn't be in the air. The atmosphere contains lots of water vapor (it's more noticeable on a humid day, but water is always in the air), and when the water vapor mixes with one of those dioxides, two acids are produced. Sulfur dioxide becomes sulfuric acid, and nitrogen dioxide becomes nitric acid. When precipitation falls, the drops of water (or flakes of snow) — called *acid rain* — take the acids with them down to soil, lakes, and rivers. When there is too much acid, the pH level of water or soil decreases. And some organisms cannot survive under acidic conditions. If some organisms are killed by the acidic conditions, other organisms that consumed those missing organisms are left without food. It's a vicious cycle.

The ozone layer was created when ultraviolet light reacted with oxygen created by autotrophic organisms. This layer protects really strong ultraviolet rays from reaching the earth. Ultraviolet rays (see Chapter 14) are partly responsible for mutations to DNA in plants and animals. When certain chemicals, such as the chlorofluorocarbons (CFCs) used in aerosol sprays and refrigeration units, evaporate into the atmosphere, they erode the ozone layer, which decreases the protection that the ozone layer normally would provide. Holes have even been eaten into the ozone layer, allowing the damaging ultraviolet rays to reach the earth. Who knows what exactly will be mutated, and what the consequences will be? Maybe it would be wise to prevent further erosion of the ozone layer and possibly help to repair it.

In addition to the chemical pollution such as from CFCs, hundreds of other products that humans have created have damaged the environment. Pesticides and fertilizers can remain in the soil for long periods of time, affecting the organisms that live in the soil and in water.

Suppose that you spray your lawn with fertilizer and spray your outdoor shrubs and flowers with pesticides. After all, looking good has become a human necessity, right? Wrong. After you coat your property with chemicals, you eagerly wait for it to rain or you water everything so that the products soak into the ground and really help your lawn and plants. Well, some of the products end up as runoff. They get washed down your driveway, down your street, into a drainage ditch, and head right for the closest stream. If all your neighbors are using chemicals, too, the concentration of chemicals in the water could reach much higher-than-normal levels. Lawn-care or cleaning products that contain phosphates are the most dangerous.

Phosphates that get into water can cause overgrowth of algae. The algae reduce the supply of oxygen in the water. When the algae die, bacteria consume them, which further reduces the oxygen content. Eventually, so much oxygen is removed from the pond, lake, stream, or river that invertebrate animals and fish die. Plants are part of the ecosystem, too. And with the ecosystem so out of balance, the plants also die. The death of the living organisms means that decomposition puts the nutrients back into the soil. This process is called *eutrophication.* It is a normal ecological process when it happens slowly so that population growth rates remain in balance. However, humans have caused eutrophication to increase way too often, and in doing so are seriously damaging ecosystems.

Another way that humans are disrupting nature is by removing habitats altogether. The rain forests that form the belt of the world all around the equator provide an extremely large amount of oxygen to the atmosphere. And they harbor many species of plants. In biology, variety and diversity are good things because they reduce competition among species. But because humans are destroying rain forests (and other forests), many species of plants and animals are becoming extinct. Once they are gone, they'll never come back. They can't be moved out of the way temporarily so that a construction project can take place or a multibillion dollar industry can have its way. Extinction is permanent.

Chapter 18

Living with Little Buggers: Bacteria, Viruses, and Insects

In This Chapter

- Finding out that bacteria provide countless beneficial functions
- Understanding why bacteria can develop resistance to antibiotics
- Finding out that viruses may have been genes once upon a time
- Understanding why the worldwide HIV epidemic has been so hard to control
- Seeing why an insect is always a bug, but a bug may not always be an insect

Humans share the world with all kinds of different organisms. It's the very diversity of life that allows species to thrive. But some of the organisms that share the planet with plants, humans, and other animals can really bug people. As hard as humans try, it's tough to feel good about seemingly beastly bugs like bacteria, viruses, and insects. But long before higher life forms hit the scene, all these organisms had put down evolutionary stakes. They've been here since the beginning of life, and they're not going anywhere.

You may be surprised to find out that many little buggers actually are beneficial to humans and other life forms. Of course, many bugs can really make you sick. Over the last few centuries, scientists have worked diligently to control these bugs.

Who's winning the battle? Judge for yourself.

Bacteria and Viruses: They Really Make Me Sick!

Feeling a little under the weather today? Chances are you're hosting a party for one of two types of microorganisms that can really make you sick. These two bugs that really bug people are bacteria and viruses. Neither are the headaches — and stomachaches and backaches — confined just to humans. Bacteria and viruses attack all kinds of plants and animals. In that way, they are truly democratic little buggers.

And when it comes to squatting rights, both viruses and bacteria have it all over most organisms. Both have been around for billions of years, in some ways making animals the interlopers in their territory.

Here's the other thing to remember about both viruses and bacteria: They are not always harmful. Some types of bacteria, for example, aid the digestive process in both animals and humans, whereas others break down nutrients in soil so plants can digest their own lunches.

Take a look at bacteria and viruses: who they are, what they do, how they affect our lives, and the historical attempts to control them. As recent headlines about "super bugs" show, humans are not doing too well on the battlefield at this moment. In fact, the antibiotics that were once thought would obliterate many disease-causing bacteria have actually helped to create organisms that are increasingly difficult to fight. Because bacteria are living organisms, they have been able to evolve and change to develop new, antibiotic-resistant forms, sending researchers scurrying for new ways to beat them. And, interestingly, researchers are now actually looking at one pre-antibiotic treatment approach as a way of mounting a surprise attack on the new super bugs. More on that in the section "An Old Weapon Revisited."

A Bacteria by Any Other Name Would Still Be a Prokaryote

Bacteria (bacterium, singular) are those single-celled microorganisms that lack a nucleus and have a cell wall composed of a protein-sugar molecule. They belong to the kingdom *Prokaryotae,* a grouping that puts them in a world of their own, really. They're not really animals, but neither are they plants, even though some can actually perform photosynthesis and make their own dinners.

Generally speaking, bacteria range in size from 1 to 10 micrometers (μ) in length. At that size, needless to say, a whole lot of bacteria are around. In fact, bacteria are the most common organisms on earth. (They edge out rude telemarketers by a nose.) At least 1,700 species of these little buggers fall into three categories:

- **Eubacteria.** The name eubacteria means "true bacteria." This category includes those nasty little bugs — pathogens — that really make you sick, along with bacteria that decompose dead organisms or waste material.

- **Cyanobacteria.** The name cyanobacteria means "blue-green bacteria." These bacteria are plant wannabes, performing photosynthesis and releasing oxygen.
- **Archaeobacteria.** The name archaeobacteria means "original or ruling bacteria." These may be the most talented of all bacteria. They can tolerate extreme environments (making some scientists think that there's a good chance they exist in the harsh atmospheres of other planets), and they actually turn inorganic molecules into energy. That would be equivalent to you getting the nutrients you need from an entrée of lead shavings, with a dessert of fluffy silicon mousse.

All bacteria share a number of distinct characteristics. Along with being nucleus-free, they have a genome that is a single circle of DNA. And they produce asexually, which is why you don't see too many happy prokaryotes around. Because the process of reproduction is so simple, daughter cells are produced with an identical, single circle of the parental DNA. They lead simple lives, these bacteria, with none of the muss and fuss of meiosis or mitosis.

Some bacteria move about by secreting a slime that glides over the cell's surface, allowing it to slide through its environment. Others have flagella that they swish around to move through their watery homes. It all sounds pretty unattractive and makes you glad you're not a prokaryote, right? Well, here's the bad news — or the good news, depending on how you look at it. Scientists now think that eukaryotic cells — the more complex cells with a true nucleus that make up higher life forms like humans — probably evolved from early prokaryotes. Here's how.

Larger, primitive cells had small, separate cells for lunch. Then, in a process called *endosymbiosis,* these little cells became part of the larger cells, continuing to function within the larger organism. Such organelles as mitochondria and chloroplasts were once happily swimming around in the primordial glop on their own, researchers believe. (Think of it as a Jonah and the Whale scenario.) Many researchers also believe that this same drama of evolution could be playing out on other planets, making it quite possible that the inhabitants of earth are not alone in the universe.

The good guys

Despite the fact that bacteria have such a bad reputation, they really do provide a number of invaluable functions. Some forms of bacteria decompose dead organic matter or sewage materials into simpler molecules that can be easily recycled. And that yogurt you like so much wouldn't exist without the aid of those little buggers in bacterial cultures. (Beer and wine are made possible by the support of yeasts, which are organisms larger than bacteria and viruses that belong to the fungi family.)

The pharmaceutical industry uses some forms of bacteria to make vitamins and antibiotics — a process that may be just a little too successful, as is evident in the problem of antibiotic resistance. One really interesting use of bacteria has been to help clean up oil spills. As you may remember from news reports of the massive spill a few years ago along the Alaskan coast from the Exxon tanker Valdez, bacteria that can actually ingest and process petroleum were added to the water to help clean up the mess. (One more reason you should be happy you're not a prokaryote!)

Closer to home, bacteria and higher life forms — like humans — have forged a number of mutually beneficial relationships. In fact, it is the norm for most organisms to be covered by populations of bacteria called *normal flora.* If an organism doesn't have it's usual coat of bacteria, it is considered an abnormal situation.

Some bacteria have warm and cozy existences in human intestines, where they produce antibiotics to their harmful cousins — forms of bacteria known as *pathogens*. And a bit of a struggle is going on between the bugs in white and those in black, with the white (that's the good bacteria) sucking up some of the nutrients that might otherwise help the pathogens get bigger and nastier. Then, they aid the digestive process by releasing vitamin K. (They don't need it anyway.) Vitamin K is essential for proper blood clotting, which keeps you from losing too much blood from simple cuts, but you can't get it from foods. Only these beneficial bacteria produce it inside you. Consider it the rent they pay for using your intestines as their home.

Most of the time, humans live in happy harmony with these good guys, ingesting familiar forms along with a hamburger or glass of milk. But, as anyone who has traveled extensively knows, there are times when this harmonious relationship turns ugly. When you take in unfamiliar, local bacteria, it takes a while for the host's guts and gut's guests to get really familiar. The result often is that dreaded traveling companion, Montezuma's revenge (otherwise known as *traveler's diarrhea*).

Then, there are the bacteria that prefer green, leafy companions and in fact provide them invaluable service. These little buggers are capable of metabolizing nitrogen in the atmosphere and making it usable for the plants. (Plants like things to be really simple.) And, of course, who could forget the attractive blue-green bacteria, which scientists believe have been decorating the atmosphere since the early days, when the earth was just toddling out of its dark womb to become the life-supporting planet you now know and love?

The early atmosphere was just crying out for a little oxygen to make it feel a little more like home. So along came the accommodating little, photosynthetic blue-green bacteria that released oxygen into the environment. In fact, these bacteria — which live in colonies, usually in water — are still sending oxygen into the environment. Take a deep breath, say "thank you," and think twice before you refer to these colonies by their unflattering nickname, pond scum. (It's not easy being blue-green.)

In fact, humans could learn plenty about cooperation from these little blue-green pals. Some of the larger cells ingest nitrogen from the environment and break it down into ammonia, which is a form of nitrogen that other, smaller members of the colony can use more readily. To paraphrase former first lady Hillary Clinton, it takes a village of pond scum. . . .

Bacteria in black

Then there's the other side of the fence: those nasty little bacteria that really live up to the bad reputation they've developed over the eons. These are pathogens, those microbes that cause infectious disease.

Several forms of bacteria cause disease in humans. One of the more infamous pathogens is the form of bacteria that caused bubonic plague (The Black Death), which devastated much of the European population during the 13th century. Other bacteria-related illnesses include strep throat, pneumonia, syphilis, and tuberculosis.

It is interesting to note that some pathogenic bacteria don't actually start out being bad. For example, the *Streptococcus pneumoniae* bacteria exist in the throats of normal, healthy people all the time. In fact, the moist tissues inside a human body are their normal habitat. Most of the time, the bacteria are well behaved, just hanging about in warm, dark little crevices. But if the host is weakened by a cold or flu, things get ugly. The bacteria get a little power-hungry. They take advantage of you while you're down. They begin to reproduce rapidly, possibly leading to strep throat or pneumonia in the lungs — or both.

In some cases, bacteria cause diseases by altering the physiology of the host. One particularly nasty example of this is *leprosy.* When *Mycobacterium leprae* invades, the bacterium takes hold of the host's cells very slowly, pretty much going unnoticed until the infection is chronic and very serious. At that point, the bacterium produces sores that eventually spread over the body and kill the tissues. The bacteria are *saprophytic,* which means that they live off of dead tissue. Eventually, if too much tissue or really important tissues are affected, death occurs.

The bacteria that causes leprosy is closely related to the bacteria that causes tuberculosis. And tuberculosis works in a similar manner. It slowly takes hold of cells, usually in the lungs, and then the symptoms appear when the infection has progressed far. The bacteria create *tubercles* inside tissues. Tubercles are lesions formed when the tuberculosis-causing bacteria — *Mycobacterium tuberculosis* — take over cells of the immune system that normally fight off bacteria. Eventually, the tissues die, and the dead tissue then serves as food for the bacteria so that the cycle can continue.

And these little buggers can make you sick by "remote control," without ever taking over the cells in your bodies or causing an outright infection. Here's how they do it: Some bacteria release toxins. For example, if food is improperly processed, these toxins become the secret ingredients in the meals on our tables. One good example of this is *botulism,* which usually is caused by improper canning, allowing the bacteria *Clostridium botulinum* to grow in the nonsterilized food and release their toxins. The toxin, rather than the bacteria, makes you sick. Other toxin-related bacterial diseases include tetanus and diphtheria, which occur when bacteria living in a body release toxins. (Not a very neighborly thing to do.)

And it's not only animals that are bugged by these little buggers. Plants cope with their share of bacterial diseases, including conditions that can reduce the yield of both fruit and vegetable plants, costing farmers millions of dollars every year.

Are humans losing the fight?

In the war against bacteria, the small are proving to be the mighty, marching forward in spite of all manner of weaponry that is tossed their way. First, there were sulfa drugs, soon to be followed by penicillin. Each was touted as a miracle drug in its time, but bacteria made short work of them. And modern antibiotics — which researchers thought would make bacterial infections a thing of the past — have proven no match for the resilient microbes. Here's why:

First of all, bacteria can reproduce at an absolutely fantastic rate. Assuming that they have all the food they need and can eliminate waste in a timely manner, bacteria can grow and divide in as little as 20 minutes, then divide again in another 20 minutes, and then again in another 20 minutes . . . you get the picture. Before too long, you have a massive colony where once there was a single bacterium. (Imagine Custer at Little Big Horn, facing all those Native Americans that just kept appearing on the horizon, and you've got the picture.) And here's the worst part: Natural selection — a basic principle of evolution (go to Chapter 16) — dictates that the antibiotic-resistant strains will develop and reproduce most vigorously. So every time researchers develop a new antibiotic, a new antibiotic-resistant strain of bacteria develops. This problem has been magnified by the overuse of antibiotics for minor infections, making the drugs ineffective for treating major, life-threatening illnesses.

Here's another problem: When bacteria are facing harsh environments, they cope by creating little sleeping cells within their walls, which contain DNA and cytoplasm. These *endospores* are mini-cells with low metabolic rates. When conditions are hostile, they remain asleep inside the protective embrace of the mother cell. Then, when conditions improve, they start to grow and produce toxins. In the case of the toxin that produces botulism, for example, endospores can remain inactive for years and even tough it out through boiling water. In order to actually destroy the toxin and endospores, water must be heated to at least 120 degrees Celsius (that's 280 degrees Fahrenheit; water boils at 212 degrees Fahrenheit), or the environment must be acidic.

An old weapon revisited

So, the darlings of the medical research world, antibiotics, have failed to control bacterial infections — and because they've created a whole class of superbugs, have in fact become weapons against humans. So in the last few years, medical researchers have turned their attention back to an old friend, *bacteriophages.* These bacteria-blasting viruses were first discovered in the early years of the last century at the Pasteur Institute in Paris. Canadian microbiologist Félix d'Hérelle came across the little creatures while looking for means of treating dysentery in Paris. He saw phages take on and completely destroy a whole colony of much large bacteria. Logically enough, he hoped the microbes would help to eliminate some of the world's worst bacterial infections. He and a group of True Phage Believers traveled around the world treating all kinds of bacterial infection.

Until about 1940, these tiny microbes — they're only about 1⁄40th the size of bacteria — were the miracle cure for many bacterial infections. Then, antibiotics came on the scene, and the medical world turned its back on these little creatures. There were several problems with phage treatment, though. For one thing, d'Hérelle and his phage-friends didn't have the technology they really needed to be sure they'd eliminated all possible toxins. Then there's the problem that phages are finicky eaters. You have to find just the right bacteria to appeal to the palate of a specific phage or the deal — and the meal — is off. Antibiotics, on the other hand, are much broader in spectrum. Why go through the trouble of culturing and nurturing these little buggers, when handy-dandy antibiotics did the trick? Or so the medical world thought.

Every year, nearly 100,000 Americans die from hospital-acquired infections (called *nosocomial infections*) related to antibiotic-resistant bacteria. And that's just part of the problem. Infections that humans thought they had under control, including such nasty diseases as tuberculosis and bubonic plague, are rearing their ugly heads in developing countries around the world. That's why researchers — and recently some big pharmaceutical companies — have supported the use of phages once again. This section takes a look at phages and how they work to keep bacteria at bay.

The good news — good for humans and bad for bacteria — is that phages may well be the most abundant organisms around. And — more bad news for bacteria — they're happy in the same neighborhoods that bacteria call home. They blissfully swim around in piles of sewage — just like bacteria — and hide in cozy little corners in your body. They have big heads filled with genetic material, whip-like tails, and long, insect-like legs. (Okay, pretty they're not. But they're good friends anyway.) Their less-than-beautiful bodies allow them to make short work of their prey: bacteria.

A phage wraps its legs around a bacterium and then drills into the cell body with its tail. Then, it shoots genetic material into the interior of the cell. The genetic material fools the bacterium into producing copies of the phage (daughter cells). In fact, the process results in the production of whole armies of daughter cells, as many as 200 per hour. At that rate, it doesn't take long before the rambunctious little phages burst right through the cell wall. Then it's "bye, bye bacterium." And that's hardly the end of it. The infant phages move on to nearby bacteria, making short work of every cell in their paths. Before very long, a whole colony of bacteria is history.

So phages are the way to get rid of the bacteria bugging you; phages could be the cure-all that antibiotics were not. Right? Well, maybe not. A number of problems are still involved in phage therapy. Even though scientists have gotten much better at identifying just which phages like to knock off which bacteria, there's not always time to do the necessary matching. For example, if a bacterial infection is raging through a village killing dozens of people by the hour, there's hardly time to culture the bacteria to see what it is and which phage will take care of it. Scientists also are worried that bacteria could develop resistance to phages, just as they did to antibiotics. And then there's the fact that bacteria are really tricky little devils. They sometimes save their cytoplasmic skins by hiding inside cells that phages can't penetrate.

All those problems aside, phage therapy can be extremely effective. In one experiment at the University of Texas, researchers infected mice with lethal doses of *Escherichia coli* bacteria. Infected mice treated with phages had a 92 percent survival rate, compared with only 33 percent for those receiving the antibiotic streptomycin.

Researchers and pharmaceutical companies are now working to find better ways to match phages with bacteria and ensure the safety of therapy. Many scientists advise conservative use of the technique to help prevent the development of phage-resistant bacteria, but say that this could be a valuable treatment-of-last-resort, to be used when even the most powerful antibiotics fail.

There's nothing like old friends to help you out of a jam.

Viruses: Those Pinheaded Little Buggers

Here's a riddle for you: What has genetic material, exists by the billions, functions like a living parasite, but may not truly be a living thing? Answer: a virus.

That's right, those nasty little bugs that cause diseases ranging from human immunodeficiency (HIV) to food poisoning to the common cold and even some forms of cancer may be the world's most efficient parasites. But, in the strictest sense of the word, they're not really alive. A virus can't exit outside of a host cell, meaning that it is incapable of independent living and therefore not really alive. But for creatures that aren't even really alive, these buggers can sure cause a lot of damage!

Viruses are tiny pieces of DNA or RNA covered with protein as protection. (You can see why they cause so much trouble, being made of DNA and RNA and all.) Because they're so small — a fraction of the size of bacteria — they can't be seen with a light microscope. So they weren't discovered until the end of the 19th century, when more powerful microscopes were beginning to come onto the scene. When scientists did see them, they were pretty creeped out. Although they come in different shapes in sizes — some are round, others rod-shaped, others coiled — they are uniformly ugly little things, often with nucleic-acid filled heads and spindly tail fibers that look like spider legs. Not exactly the beauty queens of the microscopic world!

No one is really sure how viruses developed in the beginning. Some scientists think they were originally intracellular parasites that got so good at what they did — being parasitic, that is — that they were able to make it with nucleic acid alone. Others think they are cellular escapees, genes that ran away from home, but can't replicate until they return to a specific kind of host cell. (You know how some kids just keep coming back because they can't make it on their own.)

And viruses don't knock on the door of just any cell in the neighborhood. They're pretty picky about just whose home they want to ravage. They can attach themselves only to cells that have the right receptors. This pickiness is one of the factors that feed into the theory that they were actually an integral part of cells — genes that split off, but need to return home to add to their numbers.

How a virus attacks a cell

First, a virus attaches to a receptor on a cell, shooting its nucleic acid into the cell. Viral nucleic acid is pretty powerful stuff; it's like a drug that clouds the cell's metabolic mind, convincing it to replicate viral nucleic acid, protein, and the other components new viruses need to be in full dress. Here's the really tricky part: The virus makes use of the cell's own enzymes and ATP (energy) to achieve this feat. Retroviruses (like the one that causes AIDS) are bullies. They carry an enzyme called *reverse transcriptase* into a cell, which forces the host's cell to make a DNA copy of the retroviruses' RNA genome. Pretty rude, isn't it?

Then, the various virus components get themselves together to form mature viruses, eventually causing too much of a crowd for the cell to handle. The cell wall ruptures, and the viruses go on to wreak havoc in other cells in the host's body. The number of viruses that go on attack at this point can range from ten to tens of thousands, depending on the type of virus.

Needless to say, finding ways to control viruses has been difficult. They are difficult to see and isolate, and they are equally difficult to classify. Note that humans are not alone as victims of viral attacks. Virtually every species is subject to attacks by these pinheaded predators.

Actually, your own body is showing scientists the way toward new and increasingly effective treatments for viral infections. In a host, an infected cell produces *interferons* and other *cytokines* (soluble components that help to regulate the immune response), which work like alarms telling other, uninfected cells to batten down the hatches. These neighboring cells then get ready for the intruders, ceasing virus replication. Certain cytokines are now being used in therapies to attack specific viruses. However — like bacteria — the tricky little devils are already building forms of resistance against the drugs.

And here's another difficulty in the development of antiviral therapies: Because the host cell and viruses become so integrated, it's sometimes hard just to target the virus, without destroying or upsetting the mechanics of the cell itself. So in many cases, it's just the symptoms of viral illnesses that are treated. However, antiviral drugs have been developed for a number of viral diseases including the flu, herpes, and HIV.

In many cases, prevention is the most effective way — sometimes the only way — of controlling the spread of viral diseases. And vaccinations (ouch!) have been used to prevent the spread of viral infections since Edward Jenner first held his own form of clinical trials back in late 18th century Britain. Jenner had noticed that milkmaids in England who had contracted the mild cowpox form of the disease from their bovine friends didn't become infected with the smallpox virus. So he injected a number of human subjects with the cowpox virus and proved that they were indeed immune to the disease. Boy, would he be in trouble with the Food and Drug Administration today!

Vaccines now are available for polio, flu, rabies, chicken pox, and a number of other viruses. The principle behind inoculation is that it activates the body's immune system, creating an immune response to the virus, while causing little or no cellular damage. The body forms antibodies against the virus, and those particular antibody-producing cells that make the antiviral antibody remain in the body for the rest of your life. (Sometimes booster shots are needed, though, to keep the troops ready for battle.) Of course, some vaccination programs have been more effective than others. Smallpox has been virtually wiped off the face of the earth, largely because the virus is confined to human hosts — so it can't grow and mutate in animal populations. Other viruses have been more successful at eluding the long arm (or needle) of the World Health Organization (WHO), which works to control the spread of infectious disease around the world.

Poliomyelitis (polio) is one virus that continues to rear its ugly head because it is capable of leaving its human host and mutating, thus rendering vaccinations useless. Then there are flu viruses, which mutate so rapidly that new vaccines are required each year. When a virus mutates, it can continually adapt to new intracellular environments and beat the street to escape from the host immune response. Mutations can change everything from the strength of a virus to its ability to latch on to new kinds of cells types or new animal hosts.

The AIDS virus: Why it's been a bugger to beat

Take the HIV virus (please, as in the style of Henny Youngman, please take the HIV virus!). It's a pretty good bet, but still an unproven theory, that this virus that causes AIDS (acquired immunodeficiency syndrome) originally plagued monkey populations in Africa. The theory is that after going through several mutations, it was passed on to a human host through a scratch or bite. It was first identified as a disease in 1981, and has since infected millions of people, with the virus now reported in every country in the world. In some African countries, the virus has wiped out whole villages.

HIV has an incredible ability to mutate, sometimes creating more than 100 mutant strains from a single parent strain. That makes vaccination very difficult, because a vaccine would have to be developed that would protect against all mutant forms. It's like trying to build a weapon to kill a specific enemy that has specific traits, but before you can even build the weapon, the enemy has added an army of reinforcements — with entirely different traits.

The cells that serve as hosts to the HIV virus have a special marker known as CD4. CD4 markers recognize certain specific antigens (things against which the immune system reacts). The cells that contain the CD4 marker kind of give a heads up to the immune system, letting it know what foe is in the house. Some brain cells and a number of cells from the immune system, including the T-helper cell, which is part of the body's first line of defense against the big bad world, contain cells that have this marker. This faithful marker friend helps defend the body against nasty invaders like toxins, tumor cells, and a whole range of infectious organisms. When the cells with the CD4 markers are destroyed by the HIV virus, the entire immune system is compromised, leaving the host vulnerable to all kinds of infections and cancers.

Currently, researchers are using a number of drugs — sometimes in combination — to combat the effects of the virus. Now, after dealing with the epidemic for two decades, medical science often can help AIDS patients live relatively normal lives. And new genetic research promises to yield treatments that may actually hit viruses like HIV where it hurts — in their genes.

Insects: The Bugs You Can See

If you want to insult someone, there's an easy way to do it; call him an insect. It's the all-purpose, never-fail way to make someone feel absolutely awful. You've just compared him or her to one of those creeping crawling bugs slithering around in the dirt beneath your feet. Yuchh! You've told him that he's worthless, a total failure. Well, if that's the message you want to give the loathsome lout, you should find another insult. In truth, insects are among the most successful animals on earth.

At least one million species (about half of all the animal species that have been discovered) of insects are known. And as anyone who's tried to hide food from ants at a picnic knows, they're everywhere. Insects are incredibly adaptive, able to survive in extreme temperatures and hostile conditions of all sorts. There's one insect — a sort of cricket — that lives high in the Himalayas, where its only company is probably Buddhist monks. The little creature keeps warm by manufacturing a kind of antifreeze. (There are rumors that General Motors has offered the little bugger millions to direct a new research program, but the cricket won't leave the monks.)

Scientists believe that insects and their fellow arthropods wandered up onto beaches about the same time plants were leaving the watery depths and making beachheads of their own. (That was a heck of a long time ago!) At this point, they began to develop their exoskeletons as protection from the harsh elements. This outer protection, which is also waterproof, helps to keep precious moisture inside their bodies. And both insects and spiders developed a system of *malphigian tubes,* which the insect uses in the elimination of waste and re-absorption of water into the body. When insects and other arthropods left their watery abodes behind, they were determined to keep a little bit of home with them.

Most insects are not pretty. That much is for sure. In fact, whenever a Hollywood director is casting the role of a nasty alien from a far galaxy, bet on him using an insect. (Of course, when they are pretty, they're absolute knockouts. Take butterflies, dragonflies, and some beetles, for example.) Insects have segmented bodies with three main parts: *head, thorax,* and *abdomen* (see Figure 18-1). As adults, they generally have three pairs of legs and one pair of *antennae.* And as an added bonus, they often get two pairs of wings, because it's always nice to have a spare pair on you.

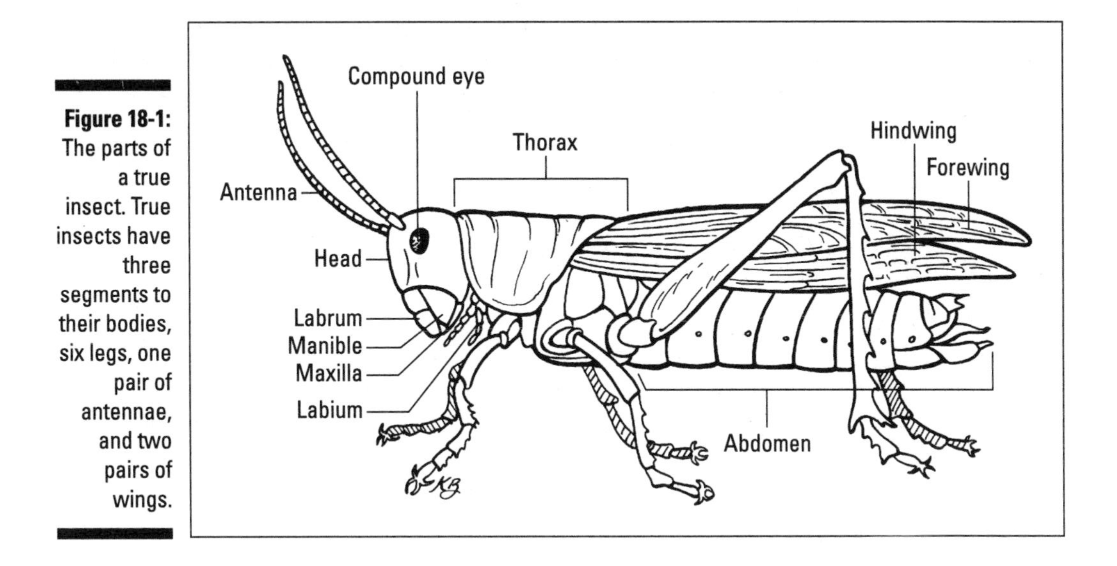

Figure 18-1: The parts of a true insect. True insects have three segments to their bodies, six legs, one pair of antennae, and two pairs of wings.

Here's an interesting point about insects' wings: Insects don't have any muscles in their wings — despite the fact that some species can beat their wings as fast as 1,000 beats per second. Muscles inside the thorax power the wings, allowing the insect to move them up and down and up and down and up and down and up and down and up and down. . . .

As you may have suspected, an insect uses its antennae to scope out the environment. The antennae, which usually reach out from between two bulgy eyes, are really like incredibly long noses. They are lined with olfactory nerves, which insects use to smell food and detect the *pheromones* — odor-carrying, hormone-like molecules — released by potential mates.

Insects breathe air, but they lack backbones, which means they are invertebrates, but not necessarily cowards. They have exoskeletons, which are hard shells that serve as protection and make that sickening little crunch when you step on them. Exoskeletons are neat forms of protection, but they do present one problem — they're really more like suits of armor than living, growing body parts. Exoskeletons don't grow, which means insects have to shed them as they grow in a process called *molting.*

Insects' segmented bodies, exoskeletons, and jointed legs put them into the category of *arthropods,* a designation they share with both crustaceans and arachnids. Crustaceans include some of the tastiest arthropods: crabs, crayfish, and lobsters. On the other hand, the last thing you'd ever call an arachnid is tasty — unless you have skewed concept of gourmet cuisine. Some common arachnids are spiders and scorpions. Yum! People often think that all arthropods are "bugs," but they're not all insects. Don't let a spider try to tell you he's an insect, for example. Tell him he has one too many pair of legs and a missing body segment.

No one will ever accuse an insect of being a dainty eater. Insects consume huge amounts of food — often several times their body weight in a day. They eat everything from the leaves of all kinds of plants to (unfortunately for your new sweater) fabric. Many insects are *herbivores* (plant eaters, such as aphids), but several species are *carnivores* (meat eaters, such as ladybugs), ungraciously munching on fellow insects or you for lunch. These are the guys you want in your garden because they help to control pests that might otherwise make quick work of a field of crops.

Their massive and varied diet is one factor that has helped insects thrive all over the world for so very long. Another success factor is their incredibly rapid rate of reproduction. Even when an insect population is almost entirely wiped out, the few remaining individuals can quickly reproduce and build up the population again. (Of course, they're not happy about taking care of all those babies, but they know that an insect's work is never done.) And in some cases, insects can respond to unfavorable conditions by remaining in the egg or larval stages for extended periods of time, only emerging as adults when conditions improve.

My, how you've changed! Reproduction and metamorphosis

It's an insect and egg dilemma: Do you begin the story of insect reproduction with the adult insect, the egg, or the larva? As with any cyclical life process, you can jump in anywhere and make the circular trip. So here goes.

A few insects give birth to live young, but the great majority of insects deposit eggs. Insects come in separate male and female varieties, with the male fertilizing the female's eggs internally. In the case of external fertilization, the female lays eggs first, which are then fertilized by the male. This activity usually takes place in water.

The female insect lays fertilized eggs that are itty bitty, but there can be lots of them. Ever wonder why so many houseflies are around? Well, a housefly can lay up to 1,000 eggs in just two weeks. Of course, not all of these survive. (Thank goodness! Flypaper budgets couldn't take it.)

Here's where the process of metamorphosis, which basically means change, begins. The animal that slithers out of the egg (and I do mean slithers) often bears little resemblance to Mama Fly or Honeybee. For one thing, it has no wings, probably doesn't have any legs, and lacks its functioning reproductive organs.

There are actually two kinds of metamorphosis: complete and, you guessed it, incomplete.

In some cases, Mama Insect would probably not recognize her own darling baby (well, it's darling to her). In the case of insects that undergo *complete metamorphosis* — including butterflies, moths, and flies — the animals that come out of the egg bear little resemblance to their parents. Just about everything about them is different, in fact. They have different diets and need different environments to thrive. These are *larvae.* Then, after a period of gorging themselves on Gerber Larvae Food (just kidding!), they are finally their full size. At this point, they become a cute little *pupa.* That's when things really change. Under the cover of a cozy *cocoon* (in the case of a butterfly, it's called a *chrysalis* because a butterfly should have a beautiful word for its cocoon), the pupating insect breaks down the larval body and substitutes and adult one. Neat trick. The adult insect then has a coming-out party. It breaks free of the cocoon and tries out its new wings for the first time.

In the case of *incomplete metamorphosis,* the change is not quite so dramatic. Some insects, including dragonflies and grasshoppers, are at least recognizable as such when they wander out of their eggs. These young insects are known as *nymphs* or *naiads,* in the case of dragonflies. Thee young gradually develop adult bodies by changing just a little bit each time they molt.

From that point on, it's Rall about sex and reproduction. Insects are absolutely obsessed with the notion of finding partners and reproducing from the moment they enter adult stage. In fact, the insect world is really like one great big singles' bar. And, just like the inhabitants of a singles' bar, insects are armed with any number of ways to attract the opposite sex. Fireflies, of course, have their beautiful flashing lights, while crickets call to each other with that distinctive sound of summer. (In case you've ever wondered, the male cricket is asking, "What's your sign?")

And — again like the inhabitants of a singles' bar — insects can have some nasty sexual experiences. You've heard the stories of female insects that bite their partners' heads off — well, they're sad but true. While *praying mantises* are copulating, the larger female sometimes literally bites her beloved's head off and then moves on to consume the rest of his body. The male praying mantis could learn from the *empid fly,* which brings along a little insect for his sweetheart to munch on in the marital bed. Serves the praying mantis right for not at least bringing a cheap bottle of wine.

Then there are the feminists of the insect world, who have found a way to reproduce by eliminating the need for male fertilization. This form of reproduction is called *parthenogenesis.* In one form of parthenogenesis, chromosomes simply double inside the egg. The advantage to this form of reproduction is that it allows the insect to choose the right moment, rather than waiting around until she finds Mr. Right. And everyone knows how long that can take!

Learning to Love (or at Least Live with) Those Little Buggers

Face it. None of the bugs that share the planet — bacteria, viruses, or insects — are going anywhere. They've been here long before humans first walked out of caves and will probably still be here long after humans have made the world unlivable for humankind. The key — as with all forms of life — is to learn to live with them in harmony, understanding the good things they give you and modifying the effects of the bad. It's something like marriage, except in the natural world there are no divorces.

Part VII

The Part of Tens

In this part . . .

This part of a *For Dummies* book is saved for some extra-light reading. A *For Dummies* Part of Tens chapter is meant to contain some fun facts or useful information. In this part, I give you both. First, in Chapter 20, you can read about ten really great biology discoveries. Of course, there were more, but I picked a few that I think stand out and contributed the most. Then, in Chapter 21, in case you want to try to work on your own really great biology discovery, you are supplied with ten great Web sites where you can find more information in your biology quest. As a bonus, I've added a list of ten pretty interesting biology facts in Chapter 22. Be sure to check them out. Enjoy!

Chapter 19

Ten Great Biology Discoveries

In This Chapter

- Finding out the whole story behind some commonly known biology discoveries
- Seeing how one discovery leads to another and science information builds

What you read in this chapter is not ranked in Top Ten fashion. And there are certainly many more than ten great biology discoveries. These are ten that came to my mind as being extremely important. Hopefully, you will keep reading about biology and find ten more that you think deserve credit.

Seeing the Unseen

In 1675, an uneducated Dutch shopkeeper, Antoni Van Leeuwenhoek, became the first person to see bacteria. Leeuwenhoek described these bacteria as little animals having "tails" or "legs," which allowed them to move about here, there, and everywhere. (Perhaps he inspired The Beatles as well?) How did Leeuwenhoek see these tiny little critters, you ask? By using a microscope, which, coincidentally enough, he also invented. His achievement just goes to show you that you don't have to have a Ph.D. to discover biological wonders.

Beautiful Loser

Everyone knows (well, maybe not *every*one) that in 1928, Alexander Fleming discovered penicillin, which is made from *Penicillium* mold, earning him the title of "father of antibiotics." But is his title totally accurate, or should someone have been dubbed the "father of antibiotics" 50 years earlier? Actually, in 1875, England's most notable physicist, John Tyndall, was already aware that the *Penicillium* mold he saw under his microscope could destroy bacteria. So why was Fleming given the glamour rather than Tyndall? Well, since Robert Koch didn't discover that bacteria could cause disease until 1882, Tyndall's discovery of *Penicillium* in 1875 was not terribly interesting. In his notes, Tyndall noted only that *Penicillium* was "exquisitely beautiful." It wasn't until

Koch made it known that bacteria could cause disease that Fleming's accidental discovery of penicillin had any meaning. Fleming was studying a strain of staphylococcus bacteria, when his petri dish became inoculated with *Penicillium* spores that had become airborne from the laboratory below Fleming's. To Fleming's surprise, wherever the *Penicillium* spores had landed on the petri dish, the staphylococcus bacteria did not grow.

However, Fleming thought he had discovered only a topical antibacterial substance. The idea of taking penicillin internally did not come about until scientists began to change their way of thinking. At that time, scientists didn't conceive that it was possible to inject an antibacterial substance into people. It wasn't until World War II that two younger, more forward-thinking (and also lucky) scientists — Ernst Chain and Howard Florey — experimented with Fleming's strain of *Penicillium* mold and discovered that it could safely be taken by humans internally. They began producing penicillin and using it in patients in 1941. Fleming, as the "father of antibiotics," may have discovered what penicillin was capable of doing, but Chain and Florey, who really put it to use, could appropriately be called the "fathers of the pharmaceutical industry." Timing is everything.

Eradicating Smallpox: Jenner Had Help

Would you believe that the idea of vaccinating people against diseases like smallpox, measles, and mumps originated in ancient China? Healers there would grind up scabs taken from a smallpox survivor into a powder and blow this dust into the nostrils of their patients. Gross as this may sound, these ancient healers were actually inoculating their patients to help prevent spread of the disease. Edward Jenner is given the credit for popularizing the idea of vaccination to battle smallpox back in 1791. However, British traders reported that the Chinese had been using their own crude method of vaccination to battle this same disease many years earlier.

DNA's Many Discoverers

If you've heard that James Watson and Francis Crick discovered DNA, you only know half the story. In 1953, Watson and Crick's landmark paper on the double-helix structure of DNA was published in the journal *Nature*. Watson and Crick did indeed elucidate the structure of DNA, and that discovery has allowed the realization of how DNA replicates, how genes are made of DNA, and how genes direct the production of proteins. In addition, it ultimately led to the Human Genome Project. Dr. Watson was the first director of the Human Genome Project, which started in the fall of 1988. And just before the calendar turned to 2000, the entire genetic code of humans had been cracked. Nearly every human gene is now in only 12 years!

However, you may be surprised to know that the discovery of DNA itself really should be credited to a 19th century researcher, Friedrich Miescher. In 1869, Miescher completed a paper containing the details of a chemical substance he called "nuclein" and the unknown chemical substance attached to the protein component of nuclein. Miescher's paper on nuclein was finally published in 1871, but it was not until 1943, when Erwin Schrodinger introduced the concept of genetic coding, that the significance of Miescher's discovery was fully realized. So with all due respect to Watson and Crick, the prize for identifying DNA as a chemical entity rightfully goes to Mr. Miescher. But, of course, Watson and Crick could not have figured out the structure of DNA leading to all the good things going on today if Miescher had not discovered its existence.

And Speaking of the Human Genome Project . . .

With the August 24, 1989, announcement that cystic fibrosis was caused by a tiny deletion from a gene on chromosome 7, scientists from Howard Hughes Medical Institute at the University of Michigan and the Hospital for Sick Children in Toronto officially put an end to trying to map the first human gene. This identification of a genetic defect, and the realization that this defect causes a disease, opened the floodgates of genetic research. Having identified the genetic flaw that causes cystic fibrosis, researchers began to focus their attack on the disease using gene therapy.

And, of course, this success bred more success. Now, the genes for other diseases such as Huntington's disease, breast cancer, sickle cell anemia, Down syndrome, Tay-Sachs disease, hemophilia, and muscular dystrophy (there are more, too) have been found. Two genetic tests are available to detect whether an unborn baby has a defective gene or whether two potential parents would likely produce an affected baby. And knowing what causes the diseases enables researchers to focus on ways to possibly cure the diseases.

Mendel and Genes: Like Peas in a Pod

Even if you're not a vegetarian, the story of Gregor Mendel and his garden "laboratory" is intriguing. Mendel, a mid-19th century Austrian monk, performed the fundamental studies of heredity that serve as the basis for genetic concepts to this day. Because pea plants have a number of readily observable traits — such as smooth peas versus wrinkled peas, tall plants versus short plants, and so on — Mendel was able to observe the results of cross-pollinating and growing various varieties of pea plants in the monastery garden. Through his experiments, Mendel was able to establish ratios of the various traits he observed in the pea plant offspring. These ratios helped Mendel to express the occurrence of various genetic factors in terms of statistical probability. Although his work was done before the discovery of DNA and chromosomes, the genetic principles of dominance, segregation, and independent assortment that Mendel originally defined are still used to this day. What Mendel called "factors" you know as "genes." And you thought that peas were only good for making soup!

Darwin's Evolutionary Ideas

Almost everyone has heard of Charles Darwin's trip to the Galapagos Islands. His study of giant tortoises and finches, and his resulting theory of natural selection (also known as "survival of the fittest"), are detailed in *On the Origin of Species,* which was published in 1859. The main point of his theory is that both plant and animal species have some variations that are better suited to the conditions in which they live. These better suited variations tend to thrive in the given area, whereas less suited variations of the same species will not do as well or will even die off.

Darwin's little theory created quite a stir when it was published, and, although widely accepted, it continues to be debated to this day. Darwin's theory was, and still is, debated by Creationists who believe that all species are the same as they have been since the Creator created them. Darwin was not a molecular biologist, nor a genetics expert. These fields did not even exist in Darwin's time, yet his theories are taken as fact to this day. What is even more amazing is that Darwin was only 27 years old when he completed the five-year study on which his theory is based. How about that for productivity?

Schwann's Cell Theory

In 1839, zoologist Theodor Schwann conducted an experiment that would be hailed as one of the most important findings in all cellular research. Schwann demonstrated that although living organisms (bacteria) would readily form in meat broth, they would not form in meat broth that was boiled and then sealed in airtight containers. This proof that air (oxygen) was needed to produce living organisms flew directly in the face of Schwann's contemporaries who maintained that organisms found when plants or animals decayed (as in the case of meat broth left open to air) simply appeared spontaneously. This research was not only critical to Louis Pasteur's work — he was the guy who discovered sterilization and fermentation (for those of you who enjoy beer and wine) — but also lead to further investigations on how cells were produced and reproduced. Schwann's research proved that cells were formed by the division of existing cells. This doctrine — the cell theory — became the basis of all subsequent cellular research. Just think, all of this knowledge came about as a result of some putrid meat broth.

Krebs' Cycle (No, It Wasn't a Harley)

The Krebs' cycle, named for German-born British biochemist Sir Hans Adolf Krebs, is the major metabolic process in all living organisms. This process results in ATP, which is used by both plants and animals as energy to fuel all cellular functions. Defining how organisms use energy at the cellular level opened the door for further research on metabolic disorders and diseases. Although the Krebs' cycle won't result in fuel that will run your Harley, it will produce energy that keeps you running.

Gram's Reputation is Stained Forever

It's a good thing Danish physician Hans Christian Joachim Gram (1853–1938) didn't listen to his mother when she told him to stay clean. Gram devised a method of staining microorganisms using crystal violet, iodine, and a contrasting dye that is used to this day, aptly called Gram stain. Microorganisms that retain the violet color are said to be gram-positive. Those that do not are said to be gram-negative.

You're probably asking why some guy's method of putting colored dyes on things that are not even visible to the naked eye merits being among the top events in biology history. Here's why. Bacteria have many distinguishing characteristics and come in many shapes and sizes. One of the major characteristics of bacteria is whether it takes Dr. Gram's concoction. When physicians need to prescribe an antibiotic, identifying the type of bacteria is key to selecting the correct antibiotic, and knowing whether the bacterium is gram-positive or gram-negative is key to identifying the type of bacteria that has made you sick. For example, penicillin is used to treat most gram-positive bacterial infections. Hats off to Dr. Gram for doing the "dirty" work.

Chapter 20

Ten Great Biology Web Sites

In This Chapter

- Surfing the information superhighway for super biology information
- Navigating through science Web sites
- Looking for cool resources on your new favorite subject — biology, of course!

In this chapter, you can find out where to go on the World Wide Web to find a whole range of biology-related information. Although just a few are listed, some are portals with link to hundreds of other biology and science sites on the Internet. Start with these sites and take a virtual tour of the world of biology. Happy hunting!

Cellbio.com

```
www.cellbio.com/education.html
```

This site offers links to all sorts of fantastic sites. There's even one to colleges and universities that offer biology programs, so you don't have to look up every institution of higher learning! It's really like biology central for the bio-fanatic. The site sends you in any number of cyber-directions, with links leading to sites focusing on topics ranging from genetics to bugs of all kinds. And, if you have a question, you'll love the Ask-A-Scientist feature. You'll be teeming with life!

Tripod.com

```
botany_plant.tripod.com/
```

This terrific site links you up with some of the best e-botany resources available: from the botany Departments of the University of Toronto and the University of Georgia to the Plant Biotech Institute of Canada. Botany.com offers up its encyclopedia of plants. More interested in plant molecular biology and genetics? Check out the link to the agricultural genome information system. Experience a virtual plant. The site also offers the link to PLANTS national database, the Internet Directory for Botany. Looking for the oddities? Try Wayne's World of Unusual and Noteworthy Plants.

Euronet.nl

`www2.euronet.nl/users/mbleeker/folis/bsmain-e.html`

On the site of the Botanic Sorting Machine (BSM), you get an impression of the enormous diversity within the plant kingdom and of how botanists created order here. In the taxonomy chapter, you find descriptions of plant families, remarks about their differences and similarities, and a look at plant evolution. You also find photographs of a number of different plant species and links to Web sites about these species and families. Under plant geography, read descriptions of floral realms, biomes, vegetation types, and climates. The site also links you to related Web sites, and a clickable world map ensures easy browsing.

This site also takes you on a virtual tour through the rainforest in Surinam (SA). Read about plants, animals, Indians, and Maroons. Click the links to look at the photos and listen to the sounds.

Hoflink.com

`www.hoflink.com/~house/index.html`

With more than 3,000 links to 13 areas of biological study, this site checks hyperlinks for accessibility on a weekly basis. The site explores animal anatomy/physiology and behavior; ecology; evolution; human physiology; microbiology; plant physiology; and more. Hoflink.com has received the Web Page Award from the Academic Assistance Center, Link of the Week from Computer Currents Interactive (CCI), and Biology Pick of the week for January 6 to 12, 2000, from About.com, Site of the Day on September 11, 2000, from *New Scientist*, and the Wise Owls Honor Award.

Arizona.edu

`www.biology.arizona.edu/`

Whether it's basic chemistry or toxicology, the University of Arizona's biology Web site offers a wide range of subjects and information. Some of the specialties are offered also in Portuguese and Spanish. The biochemistry area looks at metabolism, enzymes, energy, and catalysis, large molecules, photosynthesis, pH & pK_a, clinical correlates of pH, vitamins B_{12} and folate, and regulation of carbohydrate metabolism. Cell biology studies cells, mitosis, meiosis, the cell cycle, prokaryotes, eukaryotes, and viruses also are included. Chemicals and human health covers basic toxicology, lung toxicology, environmental tobacco smoke and lung development, kidneys, and metals. The site also takes a look at developmental biology, human biology, immunology, Mendelian genetics, and molecular biology.

Biochemlinks.com

`biochemlinks.com/bclinks/bclinks.cfm`

If you need information about biology or chemistry or the all-important Periodic Table, this site has tons of links to university updates, safety guidelines, and discussions of careers in science.

Astrobiology.com

`www.astrobiology.com/`

Stop the world! I want to get off and look at biology elsewhere! Try this Web site. It has many up-to-date articles and archived resources for the biologically curious about exobiology — the study of possible life on other planets.

Madsci.org

`www.madsci.org/index.html\`

Based at Washington University Medical School in St. Louis, Missouri, this site is a great jumping off point for biology and many other related sciences. The premise behind the site and its creative name, The Mad Science Network, is that its bio-focused search engine is like a room full of big-brained scientists ready to answer your questions. You can also click onto a guided tour of the human body and a collection of scientific experiments, for those who want to translate virtual science into real-life experiences. What more would you expect from the "Show Me" state?

Meer.org

`www.meer.org/`

MEER, Marine and Environmental Education and Research, Inc., based in Sebastopol, Calif., sponsors a site dedicated to all kinds of topics related to the watery world. The site offers a search function to help you find information about all kinds of marine life, the field of marine biology, and conservation efforts around the world. You can take virtual field trips, sign up for distance-learning course, order CD-ROMs on the subject, or submit your own questions to the MEER experts.

Discover.com

`www.school.discover.com/`

The *Discover* site offers articles from current and past issues of the popular science magazine, many of which are biology-related. One recent online feature explored the discovery of a 90-million-year-old fossilized tick. The site also offers links to other valuable science sites, an educator's guide, and a look at the Discovery Channel.

Chapter 21

Ten Interesting Biology Facts

In This Chapter

- Winning Trivial Pursuit!
- Impressing your friends!
- Having the "fastest finger!"
- Being the strongest link!

In researching the information for this book, I came across some really fascinating facts that I think you'll enjoy. Consider it a little light reading after all those pathways and cycles. My gift to you is to help you win some trivia games. (Remember me if you get into the hot seat — I'll be your phone-a-friend!) I found more than ten interesting facts, though, so maybe in the next edition. . . .

The Kangaroo Rat Is a Mammal That Doesn't Need to Drink

And it's a good thing, too. These little rodents live in the desert. They avoid water loss due to evaporation by staying deep in their cool, humid burrows during the day and only venturing out at night. They eat high-fat, low-protein seeds, not water-filled succulent plants like many desert creatures do. The water they live on is acquired from the water given off as a waste product when their food is digested. The large intestines of the kangaroo rats reabsorb almost all the water from their fecal matter, and their kidneys produce highly concentrated urine that contains very little water. Because they lose so little water, they don't need to take in any water.

The Lungfish Breathes Air and May Be an Evolutionary Link

Lungfish, which are found in Africa, South America, and Australia, have gills, but they don't get oxygen from the water. They are obligate air breathers, which means they must come to the water's surface to breathe in oxygen from air. Because of this trait, they are thought to be an evolutionary link between amphibians and terrestrial animals. This trait also allows lungfish to survive in really stagnant water. As long as they have water to swim in and they can surface for air, they are very hardy animals. If the water they're living in dries up, the lungfish can form a cocoon around themselves out of mud and a substance they secrete. When the mud dries, the cocoon hardens. Lungfish have been stored on laboratory shelves in their hardened cocoons for years. When the scientist was ready to study it, the cocoon was put into water, and the lungfish emerged to pick up where it left off. Pretty amazing creatures.

Fish Can Drown

The lungfish, because it is an obligate air breather, must surface for air to survive. It doesn't get its oxygen from the water like other fish do. So, if a lungfish is caught in a net or put into a bucket and cannot bend its body to break the surface of the water to breathe, it drowns.

Plants Can Eat Animals

A Venus flytrap plant produces a sticky substance inside its pods that attract insects. When the plant detects that an insect is inside, the pod slowly closes, trapping the insect. The plant then digests the insect.

A Woman's Uterus Expands More Than Six Times Its Normal Size During Pregnancy

Although it feels like 60 times to a pregnant woman, the extremely elastic tissue of the uterus expands just enough to hold a fetus, placenta, and amniotic fluid.

Humans and Chimpanzees Have 99 Percent of Their Genetic Material in Common

The chimpanzee is the closest relative to a human. Chimps and humans were thought to have evolved together from a line of apes and then split apart about 6 million years ago. So, when you say you could be a monkey's uncle, you're not too far off!

For as Advanced as Humans Are, Human Offspring Are Helpless

To survive, most newborn animals quickly develop the ability to walk and fend for themselves. Foals, fawns, chicks, and baby elephants all exit their mother's womb (or egg) and get right into the swing of things. Even human's closest relatives, the chimpanzees, produce offspring that do not need years of nurturing before they can get food for themselves, walk, communicate, and defend themselves.

But, as humans evolved, the size of the skull and brain enlarged. However, the size of the female pelvis remained about the same. If a baby's head is too large to pass through the pelvis, there is a big problem for both the mother and baby. So, human babies have evolved to be born before their heads get too big to fit through the pelvic bones. Consequently, they are much more immature than the offspring of human ancestors. Being born earlier means that human infants are unable to walk, get food, seek shelter, or defend themselves. This neediness requires that the parents (both mother for food, warmth, and shelter and father for defense and food when the child is older) take care of the infant for an extended period of time — years!! But, that need has strengthened the bonds between humans, creating family units that are rarely seen in other animals and help to make us human, after all.

Earthworms Can Be Strangled to Death by Parasites That Live in the Soil

The parasites form loops in the soil, and when an unsuspecting earthworm wriggles through the loop, the parasite puts on the squeeze.

Plants Look Green Because They Reflect the Green Light Rays from the Sun

Chlorophyll is the predominant plant pigment that absorbs every color of the color spectrum except for green. Therefore, plants reflect green light, so that is what you see. When leaves change colors in autumn, they are really losing their chlorophyll. Then, other pigments are able to absorb all the colors of the spectrum except for what they reflect, such as yellow or orange. Finally, when all the pigments have been depleted, the leaf is unable to absorb many light rays. Instead, it reflects most of them at the same time, appearing brown.

Chopping a Starfish into Pieces Won't Kill It

But, you would have lots of new starfish around. Starfish have an incredible ability to regenerate. If a starfish becomes entrapped by a predator, it can leave a limb behind and grow a new one. Even if just one arm remains, a whole starfish can regenerate. Starfish, which are very adept at opening shells and eating the mollusks, love to dine on oysters. Men who fished for oysters used to chop up starfish that got tangled in their nets, until they realized that they were only adding more predators to the oyster beds. Maybe pearls are so expensive because starfish thinned the oyster population?

Appendix A

Classification of Living Things

Scientists like order. They want to organize the chaos of life. They like to name things, and they look for ways of classifying natural things all the time: chemicals, elements, drugs, diseases, hormones, and yes, every plant, animal, bacteria, slime mold, and fungus. Taxonomy, derived from the Greek words for "rank" and "law," is the orderly classification of living things. The categories are called taxa, and the biologists who determine them are called taxonomists (not to be confused with a taxidermist, who stuffs dead animals that proud hunters turn into home decor). The categories used in classifying living organisms are as follows: Kingdom, Phylum, Class, Order, Genus, and Species. Organisms are slotted into these taxa based on structural similarities and differences. The Greek philosopher Aristotle began classifying plants into major groups such as trees, shrubs, and herbs. But, the organization that is still followed today was devised by the 18th century botanist Carolus Linnaeus and modified by R. H. Whittaker of Cornell University. Do not feel obliged to memorize these. I am providing this information simply so that you can get a sense of what taxonomy is and how diverse the life forms are here on earth.

Kingdom Monera

Prokaryotes: Bacteria and cyanobacteria (blue-green algae)

Kingdom Protista

Eukaryotes, unicellular or colonies without tissue differentiation

[Protozoa—heterotrophic, unicellular protists]

Phylum Mastigophora. Flagellated protozoa

Trichonympha, Trypanosoma

Phylum Sarcodina. Pseudopodial protozoa

Amoeba, Entamoeba, Arcella, foraminiferans, rediolarians

Phylum Sporozoa. Spore-forming protozoa

Taxoplasma, Plasmodium

Phylum Ciliophora. Ciliated protozoa

Euplotes, Paramecium, Stentor, Vorticella, Didinium, Tokopyra

[Unicellular algae]

Division Euglenophyta. Euglenoids

Euglena

Division Phyrrophyta. Dinoflagellates

Ceratium, Gonyaulax, Peridinium, Gymnodinium

Division Chrysophyta. Yellow-green and golden-brown algae

Diatoms

[Fungus-like protists]

Division Chytridiomycota. Chytrids

Allomyces, Blastocladadiella

Division Oomycota. Water molds, late blights, downy mildens

Saprolegnia, Phytophthora, Plasmopara,

[Slime molds]

Division Gymnomycota. Plasmodial and cellular slime molds

Kingdom Fungi

Eukaryotic heterotrophs (absorptive nutrition), mycelial organization

Division Zygomycota. Zygospore-forming fungi

Rhizopus, many common fruit and bread molds

Division Ascomycota. Sac-fungi

Saccharomyces and other yeasts, morels, truffles, apple scabs, powdery mildews, Dutch elm disease, ergot disease of rye

Division Basidiomycota. Club fungi

Mushrooms, puffballs, shelf and bracket fungi, rusts, smuts

Division Deuteromycota. "Imperfect" fungi

Penicillium, Aspergillus, Candida

Kingdom Plantae

Multicellular eukaryotes with walled cells, photosynthetic

Division Chlorophyta. Green algae

Chlamydomonas, Volvox, Ulothrix, Spirogyra, Oedogonium, Ulva

Division Phaeophyta. Brown algae

Fucus, Macrocystis, Nereocystis, Laminaria, Sagassum

Division Rhodophyta. Red algae

Division Bryophyta. Mosses and liverworts

Marchantia

Division Tracheophyta. Vascular plants

Subdivision Psilopsida. "Whisk ferns"

Psilotum

Subdivision Lycopsida. Club mosses

Lycopodium

Subdivision Sphenopsida. Horsetails

Equisetum

Subdivision Pteropsida. Plants with complex conducting systems and large complex leaves

Class Filicineae. Ferns

Pteridium

Class Gymnospermae. “Naked seed” plants

Cycads, *Ginkgo, Welwitschia, Pinus, Sequoia*

Class Angiospermae. “Enclosed seed” plants

Subclass Monocotyldoneae. Monocots

Lilies, palms, orchids, grasses

Subclass Dicotyledoneae. Dicots.

Buttercups, maples, carnations, roses

Kingdom Animalia

Multicellular, eukaryotic heterotrophs

Phylum Porifera. Sponges

Phylum Coelenterata.

Class Hydrozoa. Hydrozoans.

Hydras, *Obelia, Physalia*

Class Scyphozoa. “True jellyfish”.

Hydras, *Obelia, Physalia*

Class Hydrozoa. Hydrozoans.

Aurelia

Phylum Platyhelminthes. Flatworms.

Class Turbellaria. Planarians.

Dugesia

Class Trematoda. Flukes.

Opisthorchis, Schistosoma

Class Cestoda. Tapeworms.

Taenia

[Protostomes]

Phylum Nematoda. Roundworms

Ascaris, pinworms, *Trichinella*, filarial worms, guinea worms

Phylum Rotifera. "Wheel animals"

Phylum Gastratricha. Gastratrichs

Phylum Mollusca. Molluscs

Class Amphineura. Chitons.

Class Pelecypoda. Bivalves.

Clams, oysters, scallops

Class Gastropoda. "Belly foot" mollusks

Helix, Nassarius, Urosalpinx, slugs, nudibranchs

Class Cephalopoda. Terrestrial and freshwater annelids without parapodia

Earthworms

Class Hirudinea. Leeches

Phylum Onychophora.

Peripatus

Phylum Arthropoda. "Joint-footed" animals

Subphylum Chelicerata. First mouthparts are chelicerae; antennae are absent

Class Xiphosura.

Limulus

Class Arachnida.

Spiders, scorpions, ticks, mites, daddy longlegs

Subphylum Mandipulata. Possess mandibles and not chelicerae, antennae are present

Class Crustacea.

Crabs, lobseters, crayfish, shrimp, copepods, barnacles

Class Chilpoda.

Centipedes

Class Diplopoda.

Millipedes

Class Insecta.

Silverfish, grasshoppers, termites, bugs, beetles, butterflies and moths, flies, bees, ants

[Deuterostomes]

Phylum Echinodermata.

Class Asteroidea.

Sea stars

Class Ophiurodea.

Brittle stars and basket stars

Class Echinoidea.

Sea urchins and sand dollars

Class Holothuroidea.

Sea cucumbers

Phylum Hemichordata. Acorn worms

Phylum Chordata. Chordates

Subphylum Urochordata.

Sea squirts

Subphylum Cephalochodata.

Amphioxus

Subphylum Vertebrata.

Class Agnatha. Cyclostomes

Lampreys, hagfishes, ostracoderms

Class Placodermi. Extinct, armored, jawed fish

Class Chondrichthyes. "Cartilage fishes"

Sharks, skates, rays

Class Osteichthyes. "Bony fishes"

Lobe-fin fishes, lungfishes, ray-fin fishes

Class Amphibia. Amphibians

Order Urodela. Tailed amphibians

Salamanders

Order Anura. Tail-less amphibians

Frogs, toads

Order Apoda. Limbless amphibians.

Class Reptilia. Reptiles

Order Chelonia.

Turtle

Order Crocodilia.

Crocodiles and alligators

Order Squamata.

Snakes and lizards

Order Rhynchocephalia.

Tuatara

Class Aves. Birds

Class Mammalia. Mammals

Subclass Prototheria. Egg-laying mammals

Platypus, echidna

Subclass Metatheria. Pouched mammals (marsupials)

Kangaroos, koala bears, opossums

Subclass Eutheria. Placental mammals

Shrews, bats, monkeys, apes, humans, rats, rabbits, elephants, whales, horses, antelopes, cats, dogs, bears, walruses

Appendix B

Units of Measure

This appendix exists simply to provide you with ways of figuring out information on your own. If you come across a measurement used in the text that you are unsure of, this appendix is the place to look to see where it fits in with other measurements. If you want to convert a value, you can find the way to do it here.

Length

1 km (kilometer) = 10^{3} meters (m)

1 m (meter) = 0.9144 yards = 32.92 inches

1 cm (centimeter) = 10^{-2} m

1 mm (millimeter) = 10^{-3} m

1 (m (micrometer) = 10^{-6} m

1 nm (nanometer) = 10^{-9} m

1 Å (Ångstrom) = 10^{-10} m

Mass

1 kg (kilogram) = 10^{3} grams

1 g (gram) = 28.353 ounces

1 mm (milligram) = 10^{-3} g

1 (m (microgram) = 10^{-6} g

1 nm (nanogram) = 10^{-9} g

Volume

1 l (liter) = $(10^{-1}$ m$)^3$= 33.8 fluid ounces

1 ml (milliliter) = 10^{-3} l = $(10^{-2}$ m$)^3$ = 1 cm^3

1(l (microliter) = 10^{-6} l = $(10^{-3}$ m$)^3$ = 1 mm^3

1 nl (nanoliter) = 10^{-9} l = $(10^{-4}$ m$)^3$

Concentration

1 M (molar) = 1 mole/l = 6.02 x 10^{23} molecules/l

1 mM (millimolar) = 10^{-3} M

1 (M (micromolar) = 10^{-6} M

1 nM (nanomolar) = 10^{-9} M

Useful Constants, Conversions, and Definitions

1 mole = 6.02 x 10^{23} molecules (called Avogadro's number)

1 c (calorie) = heat needed to raise the temperature of 1 g water by 1º C

1 kcal (kilocalorie) = 10^3 c = 4.18 kJ (kilojoules)

1 l water = 1 kg (at 4º C)

1 g carbohydrate = 4 calories

1 g protein = 4 calories

1 g fats = 9 calories

0 degrees Celsius = 32 degrees Fahrenheit

37 degrees Celsius = 98.6 degrees Fahrenheit (normal body temperature for humans)

100 degrees Celsius = 212 degrees Fahrenheit

1 d (dalton) = approximate mass of one hydrogen atom (1.7x10^{-24}g)

1 kd (kilodalton) = 10^3 d

Mass of earth = 10^{24} kg

Bacterial genome = 0.5-5 x 10^{24} nucleotide pairs

Human genome = 3 x 10^9 nucleotide pairs (haploid)

Index

• A •

acetylcholine, 185
acid, 46–47
acid rain, 303
acidosis, 48
action potential, 182, 183
activation energy, 172, 173
active site, 60
active transport, 30, 104
adaptive radiation, 279–280, 289
Addison, Thomas, 179
adenosine triphosphate (ATP)
  electron transport chain and, 117
  glucose and, 82–83
  Krebs cycle and, 327
  mitochondria and, 34
  muscle contraction and, 195–196
  phosphorus and, 301
  reduction and, 119
adhesion, 100, 113
adipose tissue, 66, 67
aerobic (cellular) respiration, 35, 122, 154, 155
aerobic exercise, 66–67, 67
aging, 274–276
AIDS (acquired immunodeficiency syndrome), 315–316
air pollution, 303
alcoholic fermentation, 125–126
algae, 218
alkalosis, 48
all-or-none phenomenon, 182
allantois, 240
allele, 213–214, 244
allosteric enzyme, 175
alpha-interferon, 260
alveoli, 30, 95
amino acids
  essential, 84–85
  genetic code and, 256, 257
  nitrogenous waste and, 164
  nonessential, 84–85
  structure of, 59, 83–84
aminopeptidases, 110
ammonification, 302–303
ammonium, 164
amniocentesis, 240–241
amniotic fluid, 240
amylases, 162
anaerobic respiration, 125–126, 154–155
anaphase of mitosis, 207
anemia, 141, 142
angina pectoris, 140
angiosperms, 96, 219–222
Anichkov, Nikolai, 17
animal cells compared to plant cells, 98
animal protein, 85
animals. *See also* mammals
  anaerobic respiration in, 125, 126
  excretion of nitrogenous wastes, 166–167
  heterothermic, 153–154
  homeothermic, 154
  Kingdom Animalia, 340–344
annelid, 131
annual plant, 145
antibiotics, 310–311
antibody, 68
antigen, 68
antioxidant, 276
antiviral therapy, 314
aphids, 148
arachnid, 317
archaeobacteria, 307
archaeology, 44
Ardipithecus ramidus, 289
Aristotle, 337
arthritis, 275
arthropod, 317–318
artificial selection, 283
-ase (suffix), 162
asexual reproduction, 217–218, 225–226
aspirated food, 108
atherosclerosis, 17, 140, 276
atomic mass, 43
atomic number, 42
atoms
  Bohr's model of, 41
  description of, 39–40
  electrons, 42–43
  functional group, 53

atoms *(continued)*
orientation in space, 56
oxidation-reduction equations, 115
protons and neutrons, 42
ATP. *See* adenosine triphosphate (ATP)
ATP synthases, 158
attraction, 232–233
Australopithecus afarensis, 287
Australopithecus anamensis, 287
autodigestion, 33
autoimmune disease, 275
autonomic nervous system, 186
autosomes, 216
autotrophic organism, 72, 299
axon, 180

• B •

B cell, 275
bacteria
bacteriophage and, 311–312
beneficial functions of, 307–309
categories of, 306–307
characteristics of, 307
discovery of, 323
Gram stain and, 327
in large intestine, 163
overview of, 305–306
pathogens, 309–310
Prokaryotae, 306
war against, 310–311
bacteriology, 11
bacteriophage, 311–312
basal body, 197
basal metabolic rate, 78–79
base, 47
basophils, 142
beaker, 21
bees, mating act of, 237–238
Berthold, A. A., 179
beta-interferon, 260
bicarbonate ion, 49
bile, 109
bio- (prefix), 10
biochemistry, 11
biogeochemical cycles
carbon, 301
hydrologic, 300
nitrogen, 301–303
overview of, 300
phosphorus, 301
biogeography, 280
biologist, 9
biology, definition of, 9
biosphere and humans, 303–304
biotic potential, 297
birds, 236–237, 267, 279–280, 326
bivalve, 106
bladder, 165
blastocyst, 238–239
blind spot, 191
blood
clotting of, 144–145
donating, 68
oxygenated, 132, 136
path of through body, 134–138
plasma, 143
platelets, 143, 144
purpose of, 140–141
red blood cells, 141
white blood cells, 142–143
blood group antigens, 68
blood pressure, 134, 135, 140
blood vessels, 95, 130, 131, 193
blue-green bacteria, 308–309
body mass index, 79
body parts, use and disuse of, 290
body temperature and muscle, 194
body weight of human, 163
Bohr, Neils, 41
bolus, 162
bonds, 46
botulism, 310–311
brain, 186–188, 286, 287, 288–289
breathing, 35, 90–91, 93–96, 154
Briggs, Robert, 268
bronchi, 94
bronchioles, 94
bubonic plague, 309, 311
buffer system, 48–49
bulk flow, 113
Bunsen burner, 21

• C •

calories, 77–81, 346
Calvin-Benson cycle, 118–120
cancer, 275–276
capillary, 30, 31, 95
capillary action, 100, 113
capillary exchange, 110, 137–138
carbohydrate synthesis, 120

carbohydrates. *See also* glucose
cellulose, 58
description of, 53
in diet, 82–83
disaccharide, 54, 55–56
monosaccharide, 54
plants, movement of through, 148–149
polysaccharide, 54, 56
types of, 54, 55
carbon, 51–52
carbon-14, 44
carbon cycle, 301
carbon-fixation reaction, 99
carbonic acid, 49
carboxylation, 119
cardiac cycle, 133–134
cardiac muscle, 195
cardio- (prefix), 10
cardiologist, 10
carnivore, 72
carrier, 30
carrots and night vision, 190
carrying capacity, 298
catalyst, 60, 86, 171
CD4 marker, 315
cell cleavage, 208
cell division
asexual reproduction, 217–218, 225–226
crossing-over, 211, 213
cytokinesis, 203, 208
determination of sex and, 216
independent assortment process, 211, 214, 246, 326
interphase, 203, 204–206
meiosis, 201, 208–212, 219
mitosis, 201, 203, 206–207, 210
mutation, 213, 225, 251–252, 284
nondisjunction, 214–216
nucleus and, 202–203
overview of, 26, 203–204
segregation, 211, 213–214, 246, 326
cell plate, 208
cell theory, 326
cell turnover, 248
cells, 25, 26–27, 32, 202, 292
cellular level, 3, 25
cellulose, 58, 74–75
central nervous system, 186
centrifuge, 19, 143
centriole, 197, 207
cerebellum, 187
cerebral palsy, 187
cerebrovascular fluid, 186
cerebrum, 187
Chain, Ernst, 324
channel in plasma membrane, 29
chemical digestion, 74
chemical evolution, 290–292
chemical properties, 53
chemical reactions, 60, 172–173
chemiosmotic theory, 157–158
chemistry, overview of, 9, 37–38
chemoreceptor, 188, 189
chimpanzees, 287, 335
chlorophyll, 114–115, 336
chloroplasts, 26, 155, 157, 292
cholesterol
atherosclerosis and, 140
in diet, 88–89
eggs and, 17
as steroid molecule, 66
chromatid, 205
chromatin, 31, 32, 244
chromosomal mutation, 213
chromosomes
chromatin and, 31, 32
description of, 244
determination of sex and, 216
DNA and, 202–203
gametes and, 208–209
interphase and, 205
mitosis and, 207
nondisjunction and, 214–216
chrysalis, 319
chylomicrons, 89, 110–111
chymotrypsin, 110
cilia, 197
circadian rhythm, 198
circulatory systems. *See also* blood
capillary exchange, 110, 137–138
cardiac cycle, 133–134
heart, 131–133
heart disease, 89–90, 139–140
heartbeat generation, 139
lymphatic system, 143–144
open vs. closed, 130
overview of, 129–130
path of blood, 134–138
pulmonary circulation, 136–137
systemic circulation, 137
two-circuit, 132, 133
citric acid cycle. *See* Krebs cycle

cloning, 259, 268
closed circulatory system, 130
Clostridium botulinum, 310
clotting of blood, 144–145
cocoon, 319
codon, 256, 257
coenzyme, 60, 174
coenzyme A, 156
coevolution, 289
cofactor, 173–174
cohesion-tension, 101, 113
collagen, 61
collenchyma cells, 97
collision theory, 172
colon, 162
color blindness, 191
community, 296
comparative anatomy, 281
complete digestive tract, 105–106
complete protein, 85
complex food chain, 74
compound
  elements and, 39
  intermediate, 99, 122
  molecules and, 45
  reactivity of, 52
concentration, units of, 346
conception, 239
conclusion, 17
condensation reaction, 84
connective tissue, 61
constants, 346
consumer, 73
continuous feeder, 106
contraction of muscles, 195–196
control, 17
conversions, 346
copulate, 232
coronary arteries, 140
corporate scientist, 11–12, 13
corpus luteum, 231
cotyledons, 96, 145, 146, 223
covalence, 46
covalent bond, 45, 46
creatine, 196
creatinine, 164
crenation, 31
Crick, Francis, 258, 324
cristae, 155
crop, 106
crossing-over, 211, 213
crustacean, 317
culturing sample, 20
Curtis, S., 269–270
cyanide, 174
cyanobacteria, 307
cyclic adenosine monophosphate, 178
cystic fibrosis, 245, 251, 325
cytochrome c, 284
cytochrome oxidase, 174
cytokines, 314
cytokinesis, 203, 208
cytoplasm, 147, 203, 269–270

• D •

dandelion, 218
dark reaction, 99, 118–119
Dart, Raymond, 285
Darwin, Charles, 278–283, 285, 326
dating with carbon-14, 44
daughter cells, 205
death, 274
decomposer, 300–303
dehydration synthesis, 55, 88
deletion type of mutation, 251
denatured enzyme, 61, 174–175
dendrites, 179–180
denitrification, 302–303
density (D), 38
density of population, 296
deoxygenated blood, 135, 136
deoxyribonucleic acid. *See* DNA (deoxyribonucleic acid)
dependent variable, 14–15, 16
depolarized neuron, 182
dermal tissue of plants, 97
dermis, 61
Descent of Man, The (Darwin), 285
designing experiments, 13–18
determination, 266
detritivore, 300
development. *See also* differentiation
  aging, 274–276
  totipotency, 268
dextrose, 54
diamonds, 51
diapause, 229
diaphragm, 94
diastole, 134
dicot, 145–146
dicotyledon, 96

differentiation
  cytoplasm and, 269–270
  definition of, 241
  during development, 265–267
  embryonic induction, 269
  gender in humans, 271–273
  homeotic genes and, 270
  hormones and, 270–273
  totipotency, 268
diffusion
  concentration and, 95
  description of, 166
  as intracellular digestion method, 104
  as passive transport method, 30
digestion. *See also* digestive systems
  chemical, 74
  extracellular, 104–105
  gizzard and, 75
  intracellular, 104
  mechanical, 74, 107
  ruminant and, 74–75
  teeth and, 75
  types of, 103–104
digestive systems
  bloodstream and, 163
  continuous vs. discontinuous feeders, 106
  feces and, 162–163
  incomplete vs. complete, 105–106
  large intestine and, 111–112
  liver and, 112
  mouth and, 107–108
  muscles and, 193
  nutrients and, 110–111
  overview of, 161–162
  pepsin and, 108–109
  of plants, 113
  small intestine and, 109–110
  taste and, 190
dihydrotestosterone (DHT), 273
dipeptide, 84
diploid, 226
directional selection, 282
disaccharidases, 109
disaccharide, 54, 55–56
discontinuous feeder, 106
disease
  autoimmune type, 275
  bacteria and, 309–310
  dominant and recessive, 245
  of heart, 89–90, 139–140
  homeostasis and, 160–161
  hospital-acquired infection, 311
dispersal, 218, 299
dispersion of population, 297
disruptive selection, 283
dissect, 20
divergent evolution, 289
DNA (deoxyribonucleic acid)
  cell nucleus and, 31, 32
  description of, 64–65, 244
  discovery of, 324–325
  error recognition and repair mechanisms, 250–251
  of eukaryotes, 32
  evolution and, 283–284
  genetic code, 31, 32, 256, 257, 258–259, 281
  genetic pioneering and, 258, 261
  genetically engineered products, 259–260
  Human Genome Project, 14, 258–259, 324
  mutation, 213, 225, 251–252, 284
  nitrogenous base and, 63
  nucleus of cell and, 202–203
  replication of, 248–250
  RNA compared to, 252
  transcription and, 253–255
dominant allele, 244
dominant disease, 245
dominant trait, 247, 326
donating blood, 68
dopamine, 185
double helix, 64
doves, 232
Down syndrome, 215–216
Dubois, Eugene, 285
dye, 20

• E •

E. coli, 111, 163, 260, 312
ear, 190–191
earthworms, 236, 335
ecdysone, 271
ecologist, 12
ecology, 296–299
ecosystem, 296, 299–300
ectopic pregnancy, 238
eggs and cholesterol, 17
ejaculation, 234, 235
electrical impedence, 67
electrolytes, 44, 111
electron, 42–43
electron microscope, 18–19
electron transport chain, 115–117

element, 39, 40, 42, 43
embryo, 239–241
embryologist, 12
embryology, 11
embryonic induction, 269
empid fly, 319
endergonic reactions, 172
endocrine system, 175–176
endometrium, 231
endoplasmic reticulum (ER), 32–33
endosperm, 223
endospore, 310
endosymbiotic theory, 292, 307
energy, law of, 80
entomologist, 12
entropy, 80
enzymes
  allosteric control and feedback inhibition, 174–175
  catalysts and activation energy, 172–173
  cofactors and coenzymes, 173–174
  description of, 60–61, 171
  digestive system, 162
  metabolic reactions and, 171
  types of, 86
  urease, 173
eosinophils, 142
epigenesis theory, 266
epinephrine, 185
equilibrium, 172
equipment, 18–21
ER (endoplasmic reticulum), 32–33
error, 17–18
Escherichia (E.) coli, 111, 163, 260, 312
essential amino acids, 84–85
eubacteria, 306
eukaryotes, 26–27, 32, 202, 292
eutrophication, 304
evolution
  of apes to humans, 287–288
  of brain, 286, 287, 288–289
  chemical type, 290–292
  coevolution, 289
  divergent type, 289
  DNA and, 283–284
  evidence for, 280–281
  of humans, 285–289
  theory of, 280, 326
excreted waste, 73, 166–168
exergonic reactions, 172
exhalation, 95
exocrine gland, 176
exons, 255
exoskeleton, 317
experiments, designing. *See also* laboratory equipment
  conclusion, 17
  control, 17
  error, 17–18
  graphing data, 16
  hypothesis, 13, 14
  statistical significance, 15
  variables, 14–15, 16
expressed gene, 244
extracellular digestion, 104–105
extracellular fluid, 27, 163
eye, 191–192, 269

• F •

fallopian tube, 235, 238
fat, 67, 86–90
fat-soluble vitamins, 82, 88
fatty acids, 87
feces, 162–163
feedback inhibition, 175
fertilization
  of human egg cell, 210–211, 214, 235, 239
  of plants, 222
fertilizer, 304
fetal period, 241
fever, 160–161
fiber, 58
fibrinogen, 143
filter feeder, 106
filtration, 31
fish, 91–92, 132, 166–167, 334
flagella, 197
flask, 21
flatworm, 153
Fleming, Alexander, 323–324
Florey, Howard, 324
flowering plant, 219–222
flu virus, 315
fluid-mosaic model, 28–29
follicle-stimulating hormone (FSH), 230
food, 34, 71–72. *See also* digestion; digestive systems; nutrition
food chain, 72–74, 299–300
Food Pyramid, 76–77
food vesicle, 104
forceps, 20

forensic science, 44
fossil records, 281
frameshift mutation, 252
fraternal twins, 240
free energy, 172
free radicals, 276
frog, 267, 271
fructose, 54
fruit fly, 270
functional group, 53
fungus, 105, 338–339

• G •

G_1 phase, 204–205
G_2 phase, 205
Galapagos Islands, 279, 326
gametes, 208–209, 219, 226–227
gametogenesis, 226–228
gametophyte, 221
gamma globulin, 143
gas, 38
gas exchange, 151–152, 153. *See also* respiration
gastrulation, 239
gender differentiation in humans, 271–273
gene, 64–65, 244
gene pool, 219
Genentech, 13
genetic code
  evolution and, 281
  Human Genome Project, 258–259
  interpretation of, 31, 32
  nucleotide bases, 256
  table of, 257
genetic crosses, 247–248
genetic material, 63
genetic pioneering, 258, 261
genetic tests, 325
genetically engineered products, 259–260
genome, 258
genotype, 247
geographical isolation, 279–280, 289
germination, 223
gestation period, 229
gills, 91–92, 132
gizzard, 75
globular proteins, 62–63
glomerulus, 164–165
glucogen, 57
glucose
  as carbohydrate, 54
  description of, 34
  in diet, 82–83
  formula for, 53
  glucogen and, 57
  photosynthesis and, 120
  starch and, 57–58
  storage forms of, 56–57
glycerol, 87
glycolysis, 82–83, 100, 121–122, 155
glycoproteins, 86
Golgi apparatus, 33
gonadotropin-releasing hormone (GnRH), 230
gradient, 158
Gram, Hans Christian Joachim, 327
gram (g), 38
graphing data, 16
gravitropism, 198
greenhouse effect, 303
ground tissue of plants, 97
growth hormone, 176
growth of population, 297
guard cells of plants, 167–169
Gurdon, J. B., 268
Gutenberg, Johann, 278
guttation, 147
gymnosperm, 96

• H •

habitat, 296
hand washing, 111
haploid number of chromosomes, 209, 226
hearing, 190–191
heart, 131–134, 139
heart attack, 90, 140
heart disease, 89–90, 139–140
heartwood, 146
Heimlich maneuver, 108
Helicobacter (H.) pylori, 109
hemato- (prefix), 10
hematocrit, 10
hematologist, 10
hemocoel, 130
hemoglobin
  description of, 62–63, 86, 131
  function of, 95
  red blood cells and, 141
hemolymph, 130

hemolysis, 31
herbaceous stems of plants, 145, 146
herbivore, 72, 299
d'Hérelle, Félix, 311
hermaphrodite, 236, 273
heterogeneous nuclear RNA (hnRNA), 255
heterothermic animal, 153–154
heterotroph theory, 290–292
heterotrophic organism, 72
hexose, 54
high-density lipoprotein (HDL), 89–90
histone, 284
HIV virus, 315–316
homeobox, 270
homeostasis
  blood and, 141
  carbonic acid reaction, 49
  description of, 154, 159
  disease and, 160–161
homeothermic animal, 154
homeotic gene, 270
Homo erectus, 285–286
Homo habilis, 285
Homo sapiens, 285
homology, 284
homunculus theory, 266
hormones
  differentiation and, 270–273
  discovery of, 179
  endocrine system and, 175
  fats and, 87
  functions of, 176–177
  gametogenesis and, 226
  in mammals, 177
  ovarian cycle and, 229–231
  in plants, 177, 198
  workings of, 177–178
hospital-acquired infection, 311
human chorionic gonadotropin (hCG), 230
Human Genome Project, 14, 258–259, 324
humans
  biosphere and, 303–304
  body weight of, 163
  evolution of, 285–289
  fertilization of egg cell, 210–211, 214, 235, 239
  gender differentiation, 271–273
  heart, 132–133
  mating act, 234–235
  offspring, helplessness of, 335
  organogenesis, 267
  population growth of, 298–299
  pregnancy, 230, 238, 334
  reproductive cycle, 229–231
humulin, 260
hunt for food, 71–72
Huntington's disease, 245, 252
hydrocarbon, 52
hydrochloric acid (HCl), 46–47
hydrogen bond, 45
hydrogen peroxide, 34
hydrolases, 86
hydrologic cycle, 300
hydrolysis, 11, 56, 57, 88
hyperpolarization of neuron, 183
hypertension, 140
hyperthermia, 61
hypertonic, 31, 166
hypocotyl, 223
hypothalamus, 179, 187, 230, 231
hypothermia, 61
hypothesis, 13, 14
hypotonic, 30–31, 166

• I •

identical twins, 240
Ignatowski, A. I., 17
immune response, 68
immune system
  aging and, 275
  stressed population and, 298–299
  white blood cells and, 142
immuno- (prefix), 10
immunoglobulins, 86
immunological hypothesis of aging, 274
immunologist, 10
impermeable, 29
incomplete digestive tract, 105
incomplete dominance, 248
incomplete protein, 85
incontinence, 168
independent assortment process, 211, 214, 246, 326
independent variable, 14–15, 16
indicator, 20
inoculate, 20
inorganic chemistry, 37

inositol triphosphate, 178
insect, 131, 167, 271, 316–319
insertion type of mutation, 252
insulin, 179
integumentary exchange, 91
intercourse, 234
interferon, 314
intermediate, 99, 122
intermediate filament, 197
Internet, 22
interneuron, 180
interphase, 203, 204–206
interstitial fluid, 163
intracellular digestion, 104
intracellular fluid, 27, 163
intron, 255
iodine, 20
ion, 45
ionic bond, 46
ischemia, 140
isomer, 54
isomerases, 86
isotonic, 30, 166
isotope, 43–44

• J •

Jenner, Edward, 314, 324
journals, 21–22
juvenile hormone, 271

• K •

kangaroo rat, 333
kidneys, 164–165
kilocalories, 77
King, T. J., 268
Kingdom Animalia, 340–344
Kingdom Fungi, 338–339
Kingdom Monera, 337
Kingdom Plantae, 339–340
Kingdom Protista, 337–338
Koch, Robert, 323–324
Krebs, Hans, 156, 327
Krebs cycle
  cyanide and, 174
  description of, 155–158, 327
  enzymes and, 173
  pyruvate and, 82–83
  respiration and, 35, 122–124

• L •

laboratory equipment, 18–21
lactic acid fermentation, 126
lactose, 55
Lamarck, Jean Baptiste de, 290
large intestine, 111–112, 162, 163
larvae, 319
Law of Independent Assortment, 246
Law of Segregation, 246
Leakey, Louis and Mary, 285
Leakey, Maeve, 287
Leakey, Richard, 285–286, 287
Leeuwenhoek, Antoni Van, 323
length, units of, 345
leprosy, 309
levulose, 54
life forms
  classification of, 337–344
  diversity of, 280
  oxygen and, 326
  similarity of, 281
ligases, 86
light microscope, 18, 19
light reaction, 99
Linnaeus, Carolus, 337
lip- (prefix), 162
lipase, 109
lipids, 65–67
lipoproteins, 86, 89–90
liquid, 38
litmus paper, 20
liver, 112
lock-and-key model, 60
losing weight, 67, 79–80
low-density lipoprotein (LDL), 89–90
lungfish, 334
lungs, 30, 93–96
luteinizing hormone (LH), 230
lyases, 86
Lyell, Charles, 278–279
lymphatic system, 143–144
lymphocyte, 142
lysosome, 33, 104

• M •

maceration, 75
macromolecule, 51
macular degeneration, 192

major minerals, 81
malphigian tubes, 316
maltose, 55
mammals. *See also* animals; humans
gender differentiation in, 271–273
hormones in, 177
kangaroo rat, 333
margarine, conflicting reports on, 10
mass, 38, 345
mass flow hypothesis, 148–149
mastication, 107
mating act
bees, 237–238
birds, 236–237
humans, 234–235
overview of, 233
sea urchins, 236
worms, 236
mating rituals, 228–229, 231–233
matrix, 27, 155
matter
atom, 39–40, 41–43
categories of, 39
characteristics of, 38–39
description of, 38
elements, 39–40
isotopes, 39–40, 43–44
maturation and hormones, 176
measurement, units of, 345–346
mechanical digestion, 74, 107
mechanoreceptor, 188, 190–191
meiosis
description of, 201, 208–211
meiosis I stage, 211–212
meiosis II stage, 212
mitosis compared to, 210
in plants, 219
Mendel, Gregor, 243, 245–246, 325–326
meninges, 186
menstrual cycle, 227–228, 229–231
messenger RNA (mRNA), 31, 32, 33, 253–256
metabolic rate, 153–154
metabolic reactions and enzymes, 171
metabolism, 159–160. *See also* Krebs cycle
metamorphosis, 176, 271, 318–319
metaphase of mitosis, 207
micro- (prefix), 10
microbiologist, 10
microbiology, 11
microfilament, 197
microscope, 18–19
microtubule, 197, 205
Miescher, Friedrich, 325
milliliters (ml), 38
minerals, 80–81, 101–102
mitochondria, 26, 34–35, 155, 284, 292
mitosis, 201, 203, 206–207, 210
mixture, 39
molecular biologist, 12
molecule, 39, 45, 53
molting, 176, 271, 317
monoclonal antibody, 260
monocot, 145–146
monocotyledon, 96
monocyte, 143
monohybrid cross, 247
monosaccharide, 54
Montezuma's revenge, 308
morphogenesis, 240
motility, 197–198
motion sickness, 191
motor neuron, 180, 181
mouth, 107–108
multicellular organism, 218
multiple sclerosis, 184
muscle, 67, 193–196
muscular dystrophy, 251
mutation, 213, 225, 251–252, 284
Mycobacterium leprae, 309
Mycobacterium tuberculosis, 309
myelin sheath, 184
myocardial infarction, 90, 140

• N •

naiad, 319
natural selection, 281–283
nephron, 164
nerve cell body, 179
nervous system
brain, 186–188
cells of, 179–180
divisions of, 186
endocrine system compared to, 175–176
impulses, creating and carrying, 181–185
overview of, 178–179
sense organs, 188–193
types of neurons, 180
neural tube, 241
neurofibromatosis, 252
neuroglial cells, 179
neurons, 179–180, 181–185

neurosecretion, 178–179
neurotransmitter, 184–185
neutron, 42
neutrophils, 143
niacin, 174
nicotinamide adenine dinucleotide (NAD), 156, 174
nicotinamide adenine dinucleotide phosphate (NADPH), 117, 119
night vision and carrots, 190
nitrification, 302
nitrogen cycle, 301–303
nitrogen fixation, 302
nitrogenous base, 63, 244, 250
nitrogenous wastes, 164–169
nodes of heart, 139
nondisjunction, 214–216
nonessential amino acids, 84–85
norepinephrine, 185
normal flora, 308
nosocomial infection, 311
nuclear envelope or nuclear membrane, 32
nucleic acid, 63. *See also* DNA (deoxyribonucleic acid); ribonucleic acid (RNA)
nucleolus, 32, 203
nucleoplasm, 203
nucleotide, 63
nucleotide bases, 256
nucleus, 26, 31–32
nutrients and digestive system, 110–111
nutrition. *See also* digestion; digestive systems; food
  calories, 77–81
  carbohydrates, 82–83
  fats, 87–90
  Food Pyramid, 76–77
  proteins, 83–86
  resources on, 78
  serving size, 76–77
  vitamins and minerals, 80–82
nymph, 319

• O •

observation, 13, 18
olfaction, 189
oligosaccharide, 54
-ologist (suffix), 11–12
omnivore, 72
On the Origin of Species (Darwin), 280, 326
oogenesis, 227–228
open circulatory system, 130
operculum, 92
organelles
  description of, 25, 27
  endosymbiotic theory and, 292
  interphase of cell division, 205
  plasma membrane, 27–31
organic chemistry, 37, 51–52
organizer cell, 269
organogenesis, 266–267
orgasm, 234–235
origin of species, beliefs about, 277–278
-ose (suffix), 54
osmoreceptor, 188, 193
osmosis, 30–31, 101, 113, 166
osmotic pressure, 147
ostia, 130
ovarian cycle, 229–231
ovulation, 228
oxidation-reduction equations, 115, 124
oxidative phosphorylation, 125, 155
oxidoreductases, 86
oxygen
  in air vs. in water, 152
  body size and shape and, 153
  living organism and, 326
  metabolic rate and, 153–154
oxygen debt, 126
oxygenated blood, 132, 136
ozone layer, 303–304

• P •

pain receptor, 193
paleontology, 44, 281
pancreatic amylase, 109
pancreatic juice, 109–110
parasite, 313, 335
parasympathetic nervous system, 186
parenchyma cells, 97, 146
parthenogenesis, 237–238, 319
passive transport, 30–31, 166
Pasteur, Louis, 326
pathogen, 308, 309–310
pathologist, 12
peer review, 22
penicillin, 323–324, 327
pepsin, 108–109
peptide, 84
peptide hormones, 178

perennial plant, 145
Periodic Table of Elements, 42, 43
peripheral nervous system, 186
peristalsis, 161, 193
permeability, 29–30
peroxisome, 34
pesticide, 304
petals, 220–221
petri dish, 20
pH scale, 47–49
phage therapy, 311–312
phagocytosis, 104
phenotype, 247
pheromones, 317
phloem, 97, 99–100, 145, 146, 147–149
phosphates, 304
phosphoglyceraldehyde (PGAL), 99, 119–120
phosphoglycerate (PGA), 119
phospholipid, 28, 66, 88
phosphoprotein, 86
phosphorus cycle, 301
photochemical reaction, 99
photolysis, 117–118
photoperiodism, 198
photophosphorylation, 115–117
photoreceptor, 188, 191
photosynthesis
  description of, 98–99, 113–115
  formula for, 100, 121
  storage of glucose and, 57
phototropism, 198
physical properties, 53
physics, definition of, 9
phytochrome, 198
pigment, 114–115
pinocytosis, 104
pistil, 221
pituitary gland, 187, 230
placenta, 230, 240
planarian worms, 236
plant cells compared to animal cells, 98
plant protein, 85
plant tissue, 97
plants
  anaerobic respiration in, 125–126
  Calvin-Benson cycle, 118–120
  digestive system, 113
  electron transport chain, 115–117
  excretion of waste by, 167–169
  flowering, 219–222
  green color of, 336
  Kingdom Plantae, 339–340
  life cycle of, 219
  minerals and, 101–102
  motility of, 198
  movement of fluids and minerals through, 147–148
  overwatering, 118
  photolysis, 117–118
  photophosphorylation, 115–117
  photosynthesis, 57, 98–99, 100, 113–115, 121
  pollination and fertilization, 222
  seed production, 223
  sexual reproduction, 218–219
  stems, 145–146
  structure of, 96–98
  totipotency, 268
  transpiration, 101, 113–115, 147
  transportation of water, 100–101
  Venus flytrap, 334
  xylem and phloem, 97, 99–100, 145, 146, 147–149
  zygote into embryo, developing, 222–223
plaques, 140
plasma, 131, 163
plasma membrane
  active transport, 30, 104
  description of, 27
  fluid-mosaic model of, 28–29
  passive transport, 30–31, 166
  transport through, 29–30
platelets, 143
poikilotherm, 154
point mutation, 251
polarity, 53
polarized neuron, 181–182
poliomyelitis, 315
pollen, 221, 222
pollination, 222
polygenic trait, 245
polymerize, 62
polypeptides, 83–84
polysaccharide, 54, 56
popular press, 22–23
population, 296
population ecology, 296–298
praying mantis, 319
prefixes, 10–11
preformation theory, 266
pregnancy, 230, 238, 334
primary consumer, 299
primary producer, 299

primary sex characteristic, 232
primordial germ cell, 270
Principles of Geology (Lyell), 278–279
printing press, 278
probe, 20
producer, 73
product, 172
progesterone, 231
prokaryotes, 26, 291
prophase of mitosis, 207
proprioceptor, 188, 193
prostate gland, enlargement of, 168
protein synthesis, 254
proteins
  amino acids and, 59, 83–84
  collagen, 61
  description of, 59, 83–84
  enzymes, 60–61
  functions of, 85–86
  genes and production of, 177–178
  genetic information and, 65
  hemoglobin, 62–63
  nucleus of cell and, 32
proton, 42
publishing research, 21–23
pulmonary capillary, 30
pulmonary circulation, 132, 136–137
Punnett squares, 247
pupa, 319
purine, 63
pyloric sphincter, 109
pyloric valve, 109
pyrimidine, 63
pyruvate, 82–83, 156

• R •

radioactive, 44
rain forest, 304
reactant, 172
reactivity, 52
receptor, 29, 188
recessive allele, 244
recessive disease, 245
recessive trait, 247
red blood cells, 141
reduction, 119
referred pain, 193
reflex arc, 175–176, 187–188
refractory period, 183
regeneration, 120
replication of DNA, 248–250
repolarization of neuron, 183
repressed gene, 244
reproduction. *See also* cell division; sexual reproduction
  asexual, 217–218, 225–226
  fertilization, 214
  hormones and, 177
  insects, 318–319
  overview of, 201–202
reproductive isolation, 289
research
  conflicting findings, 10
  publishing, 21–23
  scientific method, 13–18
respiration
  aerobic type, 35, 122, 154–155
  anaerobic type, 125–126, 154–155
  chemiosmotic theory, 157–158
  description of, 91, 154
  formula for, 121
  gas exchange, 151–152, 153
  gills, 91–92, 132
  glycolysis, 82–83, 100, 121–122, 155
  integumentary exchange, 91
  Krebs cycle, 122–124
  lungs, 93–96
  overview of, 151
  oxidative phosphorylation, 125
  respiratory chain, 124
  steps of, 121–125
  tracheal exchange system, 93
  types of, 154–155
resting potential, 182, 183
retrovirus, 313
ribonucleic acid (RNA)
  description of, 65, 252–253
  genetic code and, 31, 32
  nitrogenous base and, 63
  processing, 255
  transcription and, 253–254
  translation and, 256–257
ribosomal RNA (rRNA), 32, 33
ribosome, 31, 32, 33, 203
ribs, 94
rigor mortis, 196
rings in woody stems, 146
RNA. *See* ribonucleic acid (RNA)
root pressure, 147
ruminant, 74–75

• S •

S phase, 205
salivary amylase, 107, 108, 161
salivary lipase, 107
sap, 147
saprophytes, 105
sapwood, 146
scalpel, 20
scanning electron microscopy, 19
Schrodinger, Erwin, 325
Schwann, Theodor, 326
science, 9
scientific journals, 21–22
scientific method, designing experiments using, 13–18
scientist, types of, 11–13
sclerenchyma cells, 97
sea urchin, 236
secondary consumer, 300
secondary sex characteristic, 232–233
seed plant, 96
seed production, 223
segregation, 211, 213–214, 246, 326
selectively permeable, 29–30
semen, 234–235
sense organs, 188–193
sensory neuron, 180, 181
sepal, 220–221
sepsis, 111
serotonin, 185
sex-linked trait, 245
sexual reproduction
- act of mating, 233–234
- embryonic period, 239–241
- fetal period, 241
- flowering plants, 219–222
- gametogenesis, 226–228
- human reproductive cycles, 229–231
- mating rituals, 228–229, 231–233
- meiosis and, 208
- plants and, 218–219
- from single cells to blastocyst, 238–239

sexual selection, 283
shivering, 160
sickle cell anemia, 62, 245
side chain, 84
sieve-tube element, 100, 147–148
sight, 191–192
silent mutation, 251
simple food chain, 74
sink, movement to, 148–149
skeletal muscle, 195
skin, 61, 152
slides, 19–20
sliding filament theory, 195
small intestine, 109–110, 162
smallpox, 314–315, 324
smell, 189
smooth muscle, 195
sodium chloride (NaCl), 46
sodium hydroxide (NaOH), 47
solid, 38
somatic nervous system, 186
sound, 190–191
source and sink model, 148–149
space and matter, 38
species and mating, 233
spermatogenesis, 227
sphygmomanometer, 135
spina bifida, 241
spinal cord, 187
spindles, 205, 207
spiracles, 93
spleen, 144
splicing, 255
spontaneous abortion, 230
spores, 219, 221
stabilizing selection, 282
stamen, 221
standard metabolic rate, 153–154
starch, 57–58, 149
starfish, 336
statistical significance, 15
stele, 113
steroid hormones, 66, 178
sterols, 88–89
stigma, 222
stolon, 218
stoma, 167, 168
stomach, 106, 162
stomach ulcer, 109
stomata, 101, 167
strawberry plant, 218
Streptococcus pneumoniae, 309
stressed population, 298–299
stretch receptor, 188, 193
stroke, 90
stylet, 148
subatomic particle, 39, 42
substance, 39
substitution type of mutation, 251

substrate, 60, 175
succinate dehydrogenase, 173
sucrose, 55
suffixes, 10–11
sugar, 54
sun and food chain, 72–73, 74
surfactant, 240
survival, requirements for, 1
survival of the fittest, 281–283
survivorship, types of, 297
symbiotic relationship, 292
sympathetic nervous system, 186
synapse, 184, 185
synapsis process, 211
synthase, 158
systemic circulation, 132, 137
systole, 134

• T •

T cell, 275
T-helper cell, 315
taste, 189–190
taxonomy, 337–344
teeth, 75
telophase of mitosis, 207
terminology, 10–11
tertiary consumer, 300
tertiary source, 23
test tube, 19–20
testes, 210
tetrad, 211
textbooks, 22
thalamus, 187
thermodynamics, 80
threshold level of neuron, 182
thromboembolism, 140
thyroxine, 271
tissue
  adipose, 66, 67
  connective, 61
  muscle, 195
  of plants, 97, 145
tissue plasminogen activator, 260
tortoise, sea, 279, 326
totipotency, 268
touch, 192–193
trace element, 81
trachea, 94, 108
tracheal exchange system, 93
trait, inheritance and evolution of, 290
transcription, 253–255
transfatty acids, 10
transfer RNA (tRNA), 31, 32, 33, 256–257
transferases, 86
transition state theory, 173
translation, 256–257
translocation, 100, 147–149
transmission electron microscopy, 19
transpiration, 101, 113–115, 147
traveler's diarrhea, 308
tricarboxylic acid (TCA) cycle. *See* Krebs cycle
triglycerides, 66, 88
tripeptide, 84
trisomy, 215
trophic levels, 299–300
trophoblast, 238, 239
tropism, 198
tryp- (prefix), 162
trypsin, 110
tuberculosis, 309, 311
tumor, 276
Turner syndrome, 273
twins, 240
Tyndall, John, 323–324

• U •

ulcer of stomach, 109
umbilical cord, 230, 240
unicellular organism, 217
unit of mass (m), 38
universal donor, 68
universal recipient, 68
university scientist, 12
urea, 164
urease, 173
ureter, 164, 165
urethra, 165, 234
uric acid, 164
urinary incontinence, 168
urine, 164–165
use it or lose it, 290
uterus, 334

• V •

vaccines, 314, 324
valence, 46
valence number, 115
variables, 14–15, 16

vascular bundles of plants, 99
vascular cambium, 146
vascular cylinder, 113
vascular plant, 96
vascular tissue of plants, 97, 145
vegetative reproduction, 218
Venus flytrap, 334
very low-density lipoprotein (VLDL), 89–90
vesicle, 33, 208
vessel element, 100
virus
  aging and, 275
  attack against cell by, 313–315
  bacteriophage, 311–312
  description of, 26, 305–306, 312–313
  HIV, 315–316
vision, 190, 191–192
Vitamin A poisoning, 286
vitamin K, 308
vitamins, 80–82, 174
Volpe, E. P., 269–270
volume (v), 38, 346

• W •

Wallace, Alfred, 280
washing hands, 111
water, 45
water-soluble vitamins, 82
Watson, James, 258, 324
Web sites
  Arizona.edu, 330–331
  Astrobiology.com, 331
  Biochemlinks.com, 331
  Cellbio.com, 329
  Discover.com, 332
  Euronet.nl, 330
  Hoflink.com, 330
  Madsci.org, 331
  Meer.org, 332
  Periodic Table of Elements, 43
  subatomic particles, 40
  Tripod.com, 329–330
  USDA Nutrient Database, 78
weight
  losing, 67, 79–80
  mass and, 38
white blood cells, 142–143
Whittaker, R. H., 337
woody stems of plants, 145, 146
worm
  circulatory system, 131
  earthworm, 236, 335
  flatworm, 153
  mating act, 236
  planarian, 236
  urinary system, 166, 167

• X •

x-axis, 16
X chromosomes, 216, 272
xylem, 97, 99–100, 145, 146, 147

• Y •

y-axis, 16
Y chromosomes, 216, 272
yolk sac, 237

• Z •

zero population growth, 297
zona pellucida, 233
zoospore, 218
zygote, 208, 238